LONDON MATHEMATICAL SOCIETY LECTURE NOTE SERIES

Managing Editor: Professor M. Reid, Mathematics Institute,
University of Warwick, Coventry CV4 7AL, United Kingdom

The titles below are available from booksellers, or from Cambridge University Press at www.cambridge.org/mathematics

London Mathematical Society Lecture Note series: 381

Symmetries and Integrability of Difference Equations

Edited by

DECIO LEVI
Università degli Studi Roma Tre

PETER OLVER
University of Minnesota

ZORA THOMOVA
SUNY Institute of Technology

PAVEL WINTERNITZ
Université de Montréal

CAMBRIDGE
UNIVERSITY PRESS

CAMBRIDGE
UNIVERSITY PRESS

University Printing House, Cambridge CB2 8BS, United Kingdom

One Liberty Plaza, 20th Floor, New York, NY 10006, USA

477 Williamstown Road, Port Melbourne, VIC 3207, Australia

314-321, 3rd Floor, Plot 3, Splendor Forum, Jasola District Centre, New Delhi - 110025, India

103 Penang Road, #05-06/07, Visioncrest Commercial, Singapore 238467

Cambridge University Press is part of the University of Cambridge.

It furthers the University's mission by disseminating knowledge in the pursuit of education, learning and research at the highest international levels of excellence.

www.cambridge.org
Information on this title: www.cambridge.org/9780521136587

© Cambridge University Press 2011

First published 2011

A catalogue record for this publication is available from the British Library

Library of Congress Cataloging in Publication data
Symmetries and integrability of difference equations / edited by Decio Levi . . . [et al.].
p.   cm. – (London Mathematical Society lecture note series ; 381)
Includes bibliographical references.
ISBN 978-0-521-13658-7 (pbk.)
1. Difference equations.   2. Symmetry (Mathematics)   3. Integrals.   I. Levi, D. (Decio)
QA431.S952   2011
515´.625 – dc22   2011006852

ISBN 978-0-521-13658-7 Paperback

Contents

List of figures

Contributors

Vladimir Dorodnitsyn   *Keldysh Institute of Applied Mathematics, Russian Academy of Sciences, Miusskaya Pl. 4, Moscow, 125047, Russia;* `dorod@spp.keldysh.ru`

Basile Grammaticos   *IMNC, Université Paris VII & XI, CNRS, UMR 8165, Bât. 104, 91406 Orsay, France;* `grammati@paris7.jussieu.fr`

Jarmo Hietarinta   *Department of Physics and Astronomy, University of Turku, 20014 Turku, Finland;* `hietarin@utu.fi`

Mourad E. H. Ismail   *Department of Mathematics, University of Central Florida, Orlando, FL 32816, USA* and *Department of Mathematics, King Saud University, Riyadh, Saudi Arabia;* `ismail@math.ucf.edu`

Alexander Its   *Department of Mathematical Sciences, 402 N. Blackford Street, Indiana University–Purdue University Indianapolis, Indianapolis, IN 46202-3216, USA;* `itsa@math.iupui.edu`

Roman Kozlov   *Department of Finance and Management Science, Norwegian School of Economics and Business Administration, Helleveien 30, N-5045, Bergen, Norway;* `Roman.Kozlov@nhh.no`

Decio Levi   *Dipartimento di Ingegneria Elettronica, Università degli Studi Roma Tre and Sezione INFN Roma Tre, Via della Vasca Navale 84, 00146 Roma, Italy;* `levi@roma3.infn.it`

Sergey P. Novikov   *University of Maryland, College Park, USA, Landau Institute* and *Steklov Institute, Moscow, Russia;* `novikov@ipst.umd.edu`

Peter J. Olver   *School of Mathematics, University of Minnesota, Minneapolis, MN 55455, USA;* `olver@math.umn.edu`

Jiří Patera   *Département de mathématiques et statistique* and *Centre de recherches mathématiques, Université de Montréal, C.P. 6128, succ. Centre ville, Montréal, H3C 3J7, QC Canada;* `patera@crm.umontreal.ca`

Alfred Ramani   *Centre de Physique Théorique, École Polytechnique, CNRS, 91128 Palaiseau, France;* `ramani@cpht.polytechnique.fr`

Yuri B. Suris   *Institut für Mathematik, MA 7-2, Technische Universität Berlin, Str. des 17. Juni 136, 10623 Berlin, Germany;* suris@math.tu-berlin.de

Pavel Winternitz   *Département de mathématiques et statistique* and *Centre de recherches mathématiques, Université de Montréal, C.P. 6128, succ. Centre-ville, Montréal, QC H3C 3J7 Canada;* wintern@crm.umontreal.ca

Ravil I. Yamilov   *Ufa Institute of Mathematics, Russian Academy of Sciences, 112 Chernyshevsky Street, Ufa 450077, Russian Federation;* RvlYamilov@matem.anrb.ru

Preface

This book is based upon lectures delivered during the Summer School on Symmetries and Integrability of Difference Equations at the Université de Montréal, Canada, June 8, 2008–June 21, 2008. The lectures are devoted to methods that have been developed over the last 15–20 years for discrete equations. They are based on either the inverse spectral approach or on the application of geometric and group theoretical techniques. The topics covered in this volume can be summarized in the following categories:

- Integrability of difference equations
- Discrete differential geometry
- Special functions and their relation to continuous and discrete Painlevé functions
- Discretization of complex analysis
- General aspects of Lie group theory relevant for the study of difference equations. Specifically, two such subjects are treated: 1. Cartan's method of moving frames 2. Lattices in Euclidean space, symmetrical under the action of semisimple Lie groups
- Lie point symmetries and generalized symmetries of discrete equations

Twelve distinct lecture series were presented at the Summer School of which eleven are included in this volume. Close to 50 registered graduate students and researchers from twelve different countries participated.

The Summer School, Séminaire de mathématiques supérieures, is a yearly event at the Département de Mathématiques, Université de Montréal. The organizing committee for the year 2008 consisted of Pavel Winternitz (Université de Montréal, Canada), Vladimir Dorodnitsyn (Keldysh Institute of Applied Mathematics, Russian Academy of Sciences), Decio Levi (Universitá degli Studi Roma Tre, Italy) and Peter Olver (University of Minnesota, USA). The two scientific directors were Pavel Winternitz and Vladimir Dorodnitsyn. The financial support for the Summer School was provided by Université de Montréal, the National Foundation Grant No. 0737765, the Centre de Recherches Mathématiques, and the Institut de Sciences Mathématiques.

The editors would like to thank André Montpetit (CRM) for technical help in preparing this volume. We also acknowledge the support of David Tranah, Suzanne Parker and Clare Dennison from Cambridge University Press.

Introduction

The concept of integrability of Hamiltonian systems goes back at least to the 19th century. The idea of integrability in classical mechanics was formalised by J. Liouville. A finite-dimensional Hamiltonian system with n degrees of freedom is called "Liouville integrable" or "completely integrable" if it allows n functionally independent integrals of motion that are well defined functions on phase space and are in involution. In classical mechanics the equations of motion for a Liouville integrable system can be, at least in principle, reduced to quadratures. A completely integrable system in quantum mechanics is defined similarly. It should allow n commuting integrals of motion (including the Hamiltonian) that are well defined operators in the enveloping algebra of the Heisenberg algebra, or some generalization of this enveloping algebra. In quantum mechanics complete integrability does not guarantee that the spectral problem for the Schrödinger operator can be solved explicitly, or even that the energy levels can be calculated algebraically.

An n-dimensional integrable Hamiltonian system that admits more than n integrals of motion is called "superintegrable". Systems with $2n - 1$ integrals, with at least one subset of n of them in involution, are "maximally superintegrable". Such systems, namely the Kepler-Coulomb system and the harmonic oscillator, played a pivotal role in the development of physics and mathematics. Trajectories in classical maximally superintegrable systems can at least in principle be calculated algebraically (without using any calculus). Ironically, calculus was invented in order to calculate orbits in the Kepler system. In quantum mechanics these systems are exactly solvable: their energy spectra can be calculated algebraically, their wave functions expressed in terms of polynomials (in appropriate variables) multiplied by a known function. The $2n - 1$ integrals of motion generate (under Lie or Poisson commutation) a finite-dimensional nonabelian algebra that is usually an associative algebra rather than a Lie algebra. Superintegrability has also been called "nonabelian integrability".

The theory of nonlinear infinite dimensional integrable systems went through a rapid development since the creation of "soliton theory" in the famous 1967 paper by C.S. Gardner, J.M. Green, M.D. Kruskal, and R.M. Miura. This paper introduced the inverse scattering transform as a method of solving certain nonlinear partial differential equations by essentially linear techniques. It rapidly became clear that there exists an infinite number of non-linear partial differential equations that can be solved in this manner. These equations are usually called "integrable" nonlinear partial differential equations. They can actually be considered to be infinite-dimensional analogues of superintegrable finite dimensional systems. Indeed, they not only allow infinitely many integrals of motion but these integrals form nonabelian algebras. The usual infinite families of commuting flows actually form infinite dimensional abelian subalgebras of these larger non-abelian ones. Moreover, the corresponding soliton equations are exactly solvable in the sense that the inverse scattering transform provides exact solutions for large classes of initial data.

The use of group theory to solve ordinary and partial differential equations also has a long history going back to S. Lie. Lie group methods are applicable to a much wider class of equations than methods based on integrability. Whether a partial differential equation is integrable or not, Lie theory allows one to reduce the number of independent variables and to obtain special exact analytical solutions. For an ordinary differential equation admitting a symmetry group, Lie group methods enable one to decrease the order of the equation and, under appropriate solvability conditions, obtain the general solution. When both integrability and symmetry methods are applicable, they interact and complement each other fruitfully. In particular, group theory provides criteria of integrability.

A vigorous application of the ideas of integrability and of symmetry to discrete equations started much later, around 1990. Pioneering work on the integrability of difference and differential-difference equations was done 20 years earlier by M.J. Ablowitz and J.F. Ladik, R. Hirota, and others. The first applications of Lie group theory to difference equations are due to S. Maeda already in 1980.

This volume is devoted to recent developments in the theory of integrability and symmetries of discrete equations of all types: difference equations, q-difference equations, differential-difference equations, ultradiscrete equations and others. The contributions are ordered alphabetically by authors although by content they could be subdivided into several overlapping themes.

The first chapter, by V. Dorodnitsyn and R. Kozlov, is devoted to a specific aspect of the application of continuous Lie point symmetries to difference systems involving one discrete independent variable. An ordinary difference scheme consists of two difference equations, determining both the lattice and the actual difference equation. The authors develop discrete Lagrangian and Hamiltonian formalisms. They then use them to investigate the relation between continuous symmetries and conservation laws and first integrals for discrete Hamiltonian and Euler-Lagrange equations.

Chapter 2, by B. Grammaticos and A. Ramani, shifts the focus to the field of integrability and constitutes a comprehensive review of the Painlevé equations and the properties of their solutions. The authors give parallel derivations of continuous and discrete Painlevé equations and emphasize their shared integrability properties. The discrete equations considered are difference equations, q-difference equations and ultra-discrete ones. Two descriptions are presented. A "top-down" approach, starting from a Hamiltonian formulation and an isomonodromy deformation problem. The complementary "bottom-up" approach consists of applying certain integrability criteria to chosen classes of equations. For discrete systems the criteria selected in this chapter are singularity confinement and algebraic entropy.

By alphabetic coincidence, Chapter 3, written by J. Hietarinta, is closely related to Chapter 2 and presents different definitions of integrability and integrability criteria for difference equations. The author considers both ordinary and partial difference equations (with two discrete independent variables) and provides algorithmic tools for deciding whether a discrete equation is integrable, partially integrable, or chaotic. A section is devoted to conserved quantities, i.e. constants of motion. Singularity confinement and algebraic entropy are presented as algorithmic tools. When applying the algebraic entropy criterion, linear, polynomial, and exponential growth of complexity are associated with linearizable, integrable and chaotic equations, respectively. Finally, the author shows how the "consistency-around-a cube" criterion discussed by Yu.B. Suris in Chapter 10 can be applied to equations on square lattices to obtain Lax pairs and multisoliton solutions.

Chapters 4 and 5 of this book are related in that they both deal with orthogonal polynomials and their relation to discrete and continuous Painlevé functions. In both cases the orthogonal polynomials satisfy three term linear recurrence relations, i.e. second order linear difference equations. The coefficients in the recursion relations satisfy discrete or continuous Painlevé equations.

Chapter 4, written by M.E.H. Ismail, considers in particular the case when the polynomials are orthogonal with respect to an exponential measure and the recursion coefficients satisfy the discrete Painlevé I equation. The emphasis is on the spectral theory of orthogonal polynomials and on applications.

In Chapter 5, written by A. Its, the emphasis is on the connection between orthogonal polynomials, integrable systems and random matrices. The Riemann-Hilbert formalism for orthogonal polynomials is explained and used to introduce discrete Painlevé equations systematically. This setting is also used to perform a global asymptotic analysis of the solutions of discrete Painlevé equations.

Chapter 6, by D. Levi and R. Yamilov, is one of the three chapters in the book specifically devoted to Lie symmetries of difference equations. More specifically it deals with generalized symmetries. The authors consider generalized symmetries of partial difference equations with two independent variables on fixed non-transforming lattices. They make use of the formalism of evolutionary vector fields, acting on the dependent variables only. A method of constructing generalized symmetries for integrable multivariable difference equations or differential-difference equations is presented. It makes use of integrability properties of the equations, in particular recursion operators. A subclass of generalized symmetries is identified that in the continuous limit "contracts" to point symmetries. A section in Chapter 6 is devoted to how formal symmetries provide an integrability criterion for equations on lattices.

In Chapter 7, S.P. Novikov reviews an ambitious program that amounts to a discretization of complex analysis. After emphasizing the role of linear operators and their factorization properties for continuous nonlinear integrable systems, the author proceeds to their discretization. This is done on square lattices for hyperbolic equations and on equilateral triangular lattices for elliptic ones. The concept of triangle equations is introduced as well as that of GL_n connections. The discretization of complex analysis on square lattices was introduced by Ferrand in 1944 and has been used for discrete integrable systems by many authors. The approach of S.P. Novikov and his collaborators is instead based on the properties of an equilateral triangle lattice. The discretization is carried out both on flat and hyperbolic planes.

Chapters 8 and 9 are devoted to some rather general aspects of Lie group theory that are relevant, in particular, to the study of difference equations.

In Chapter 8, P.J. Olver gives an exposition of the method of moving frames. The modern development of this method goes back to Élie Cartan. The chapter starts with a definition of a moving frame as an equivariant map from a manifold M to a transformation group G. This definition turns the method into an algorithm applicable to very general group actions. The treatment is restricted to finite dimensional Lie groups (though the author refers to his work on moving frames for pseudogroups as well). Many aspects and applications of the method, and obtained differential invariants, joint invariants and joint differential invariants are discussed. The most relevant application from the point of view of this volume is Section 8.6 on invariant numerical approximations, more specifically on symmetry preserving approximations. An example is given in which Runge-Kutta schemes are compared. This use of moving frames provides an example of geometric numerical integration techniques for ordinary and partial differential equations.

Chapter 9, by J. Patera, provides an algorithm for constructing n-dimensional lattices in real Euclidean space E_n that are symmetrical with respect to the action of a compact semi-simple Lie group of rank n. A symmetric lattice is first constructed in a finite region of the space. The symmetry of the lattice is given by the symmetry of the weight lattice of the chosen Lie group and the density of points can be chosen a priori. The action of the affine Weyl group then extends the lattice to an infinite one on the entire space E_n. The motivation provided is the construction of functions that are orthogonal on the lattices. These in turn are needed in the treatment of digital data on lattices. The construction of symmetric lattices can also be related to the construction of the discrete integrable systems on other lattices than simple rectangular ones.

Chapter 10, by Yu.B. Suris, is on discrete differential geometry, a new subject emerging on the border between differential and discrete geometry. Discrete differential geometry is not only related to the topic of integrability of difference equations, but it actually provides new insights into the concept of integrability, both for discrete and continuous equations. It also leads to new integrability criteria. The author introduces basic notions like that of discrete nets, Q-nets and circular nets. The concept of integrability for discrete systems is introduced in terms of a multidimensional consistency principle. Namely, the discretization of surfaces, coordinate systems and all related concepts should be extendable to multidimensional consistent nets. The usual fundamental attributes of integrable systems like the existence of Lax pairs, Bäcklund transformations, permutability theorems, infinite families of commuting flows here appear as consequences of multidimensional consistency requirements.

The last chapter, Chapter 11, by P. Winternitz, concentrates on point symmetries of difference and differential-difference equations. It is thus related to Chapter 1 by V. Dorodnitsyn and R. Kozlov, Chapter 6 by D. Levi and R. Yamilov and partially to Chapter 8 by P.J. Olver. Sections 11.1-4 are on the symmetry preserving discretization of ordinary differential equations on symmetry adapted lattices. In particular, Section 11.3 discusses examples of geometric integration methods. It is shown that symmetry adapted numerical methods (so far for ordinary differential equations) provide qualitatively superior solutions, specially in the neighbourhood of singularities. Sections 11.5 and 11.6 are devoted to Lie point symmetries of differential-difference equations. The discrete independent variables in these sections are defined on uniform non-transforming lattices.

In summary, the papers in this volume provide a comprehensive overview of the current state of the art in integrability and symmetry for discrete equations. Our hope is that it will inspire the reader to further develop these fascinating and important theories and their applications.

The Editors

1

Lagrangian and Hamiltonian Formalism for Discrete Equations: Symmetries and First Integrals

Vladimir Dorodnitsyn and Roman Kozlov

Abstract

In this chapter the relation between symmetries and first integrals of discrete Euler–Lagrange and discrete Hamiltonian equations is considered. These results are built on those for continuous Euler–Lagrange and canonical Hamiltonian equations. First, the well-known Noether theorem which provides conservation laws for continuous Euler–Lagrange equations is reviewed. Then, its discrete analog is presented. Further, it is mentioned that continuous and discrete Hamiltonian equations can be obtained by the variational principle from action functionals. This is used to develop Noether-type theorems for canonical Hamiltonian equations and their discrete counterparts (discrete Hamiltonian equations). The approach based on symmetries of the discrete action functionals provides a simple and clear way to construct first integrals of discrete Euler–Lagrange and discrete Hamiltonian equations by means of differentiation of discrete Lagrangian (or Hamiltonian) and algebraic manipulations. It can be used to conserve structural properties of underlying differential equations under a discretization procedure that is useful for numerical implementation. The results are illustrated by a number of examples.

1.1 Introduction

It has been known since E. Noether's fundamental work that conservation laws of differential equations are connected with their symmetry properties [28]. For convenience we present here some well-known results (see also, for example, [1, 3, 18]) for the Lagrangian approach to conservation laws (first integrals). We restrict ourselves to the case with one independent variable.

Let us consider the functional

$$\mathbb{L}(\mathbf{u}) = \int_{t_1}^{t_2} L(t, \mathbf{u}, \dot{\mathbf{u}}) \, dt, \tag{1.1}$$

where t is the independent variable, $\mathbf{u} = (u^1, u^2, \ldots, u^n)$ are dependent variables, $\dot{\mathbf{u}} = (\dot{u}^1, \dot{u}^2, \ldots, \dot{u}^n)$ are first-order derivatives and $L(t, \mathbf{u}, \dot{\mathbf{u}})$ is a *first-order* Lagrangian. The functional (1.1) achieves its extremal values when $\mathbf{u}(t)$ satisfies the Euler–Lagrange equations

$$\frac{\delta L}{\delta u^i} = \frac{\partial L}{\partial u^i} - D\left(\frac{\partial L}{\partial \dot{u}^i}\right) = 0, \qquad i = 1, \ldots, n, \tag{1.2}$$

where

$$D = \frac{\partial}{\partial y} + \dot{u}^k \frac{\partial}{\partial u^k} + \ddot{u}^k \frac{\partial}{\partial \dot{u}^k} + \cdots$$

is the total differentiation operator. Here and below we assume summation over repeated indexes. Note that (1.2) are second-order ODEs.

We consider a Lie point transformation group G generated by the infinitesimal operator

$$X = \xi(t, \mathbf{u}) \frac{\partial}{\partial t} + \eta^i(t, \mathbf{u}) \frac{\partial}{\partial u^i} + \cdots, \tag{1.3}$$

where dots mean an appropriate prolongation of the operator to derivatives [5, 21, 29, 30]. The group G is called a variational symmetry of the functional $\mathbb{L}(\mathbf{u})$ if and only if the Lagrangian satisfies [28]

$$X(L) + LD(\xi) = 0, \tag{1.4}$$

where X is the first prolongation, i.e., the prolongation of the vector field X to the first derivatives $\dot{\mathbf{u}}$. We will actually need a weaker invariance condition than given by (1.4). The vector field X is a divergence symmetry of the functional $\mathbb{L}(\mathbf{u})$ if there exists a function $V(t, \mathbf{u}, \dot{\mathbf{u}})$ such that [4] (see also [5, 21, 29])

$$X(L) + LD(\xi) = D(V). \tag{1.5}$$

Generally, (1.5) should hold on the solutions of the Euler–Lagrange equations (1.2).

Noether's theorem [28] can be based on the following Noether-type identity [21], which holds for any vector field and any smooth function $L(t, \mathbf{u}, \dot{\mathbf{u}})$:

$$X(L) + LD(\xi) \equiv (\eta^i - \xi \dot{u}^i) \frac{\delta L}{\delta u^i} + D\left(\xi L + (\eta^i - \xi \dot{u}^i) \frac{\partial L}{\partial \dot{u}^i}\right). \tag{1.6}$$

The theorem states that for a Lagrangian satisfying the condition (1.4) there exists a first integral of the Euler–Lagrange equations (1.2):

$$I = \xi L + (\eta^i - \xi \dot{u}^i)\frac{\partial L}{\partial \dot{u}^i}. \tag{1.7}$$

This result can be generalized [4]: If X is a divergence symmetry of the functional $\mathbb{L}(\mathbf{u})$, i.e., (1.5) is satisfied, then there exists a conservation law

$$I = \xi L + (\eta^i - \xi \dot{u}^i)\frac{\partial L}{\partial \dot{u}^i} - V \tag{1.8}$$

of the corresponding Euler–Lagrange equations.

The strong version of Noether's theorem [21] states that there exists a conservation law of the Euler–Lagrange equations (1.2) in the form (1.7) if and only if the condition (1.4) is satisfied on the solutions of (1.2).

The goal of this chapter is to extend the results presented above to discrete equations in the Lagrangian and Hamiltonian frameworks. We will need to consider canonical Hamiltonian equations before we start to treat their discrete counterparts. It is known that the preservation of first integrals (conservation laws) in numerics is of great importance (see, for example, [19, 31]). Therefore, there is a strong motivation to establish discrete analogs of the conservation properties of the continuous Euler–Lagrange and Hamiltonian equations.

In the next section we will comment on invariance of the Euler–Lagrange equations. In Section 1.3 we will present the Lagrangian formalism for second-order difference equations, which are a discrete analog of the second-order ordinary differential equations. Canonical Hamiltonian equations are considered in Section 1.4. We will develop an analog of Noether's theorem which is based on invariance properties of the action functional, generating canonical Hamiltonian equations. The discrete Hamiltonian equations and their conservation properties are treated in Section 1.5. Section 1.6 presents applications of the theoretical results to a number of examples. Finally Section 1.7 contains concluding remarks.

1.2 Invariance of Euler–Lagrange equations

There exists a relation between the invariance of the Lagrangian function and invariance of the corresponding Euler–Lagrange equations:

Theorem 1.1 ([21, 29])   *If the Lagrangian L is invariant with respect to operator* (1.3), *i.e., condition* (1.4) *is satisfied, then the Euler–Lagrange equations* (1.2) *are also invariant.*

Remark 1.2   If the Lagrangian L is divergence invariant, i.e., satisfies the condition (1.5), then the Euler–Lagrange equations (1.2) are also invariant. This follows from the fact that full divergences belong to the kernel of variational operators.

Thus, if X is a variational or divergence symmetry of the functional $\mathbb{L}(\mathbf{u})$, it is also a symmetry of the corresponding Euler–Lagrange equations (1.2). The symmetry group of the Euler–Lagrange equations can of course be larger than the group generated by variational and divergence symmetries of the Lagrangian.

It is interesting to establish the necessary and sufficient condition for invariance of the Euler–Lagrange equations. We will need the following lemma:

Lemma 1.3   *For any smooth function $L(t, \mathbf{u}, \dot{\mathbf{u}})$ the following identity holds*

$$\frac{\delta}{\delta u^j}(X(L) + LD(\xi)) \equiv X\left(\frac{\delta L}{\delta u^j}\right) + \left(\frac{\partial \eta^i}{\partial u^j} + \delta_{ij}D(\xi) - \frac{\partial \xi}{\partial u^j}\dot{u}^i\right)\frac{\delta L}{\delta u^i}, \qquad j = 1,\dots,n,$$
(1.9)

where the notation δ_{ij} stands for the Kronecker symbol.

Proof   The result can be established by a direct computation.   □

Theorem 1.1 and Remark 1.2 follow from this lemma. The lemma also provides the *necessary and sufficient* condition for the invariance of the Euler–Lagrange equations:

Theorem 1.4   *The Euler–Lagrange equations (1.2) are invariant with respect to a symmetry (1.3) if and only if the following conditions are true (on the solutions of the equations):*

$$\frac{\delta}{\delta u^j}(X(L) + LD(\xi))\bigg|_{\delta L/\delta u^1 = \cdots = \delta L/\delta u^n = 0} = 0, \qquad j = 1,\dots,n.$$
(1.10)

Proof   The statement follows from the identities of Lemma 1.3.   □

Example 1.5   Equation

$$\ddot{u} = \frac{1}{u^2}$$
(1.11)

is the Euler–Lagrange equation for the Lagrangian function

$$L(t, u, \dot{u}) = \frac{\dot{u}^2}{2} - \frac{1}{u}.$$

The equation admits symmetries

$$X_1 = \frac{\partial}{\partial t}, \qquad X_2 = 3t\frac{\partial}{\partial t} + 2u\frac{\partial}{\partial u}.$$

The operator X_1 is a symmetry of Lagrangian L and, consequently, a symmetry of (1.11). The symmetry X_2 is not a symmetry of the Lagrangian:

$$X_2(L) + LD(\xi_2) = L.$$

However, it is a symmetry of the equation as follows from Theorem 1.4:

$$\frac{\delta}{\delta u}(X_2(L) + LD(\xi_2))\Big|_{\delta L/\delta u=0} = \frac{\delta L}{\delta u}\Big|_{\delta L/\delta u=0} = 0.$$

In the next section we will develop discrete analogs of these results.

1.3 Lagrangian formalism for second-order difference equations

Let us present the results concerning the variational formulation of discrete Euler–Lagrange equations [9–11, 13, 14]. The notations are clear from the following picture:

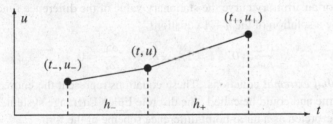

We consider a finite-difference functional

$$\mathbb{L}_h = \sum_{\Omega} \mathcal{L}(t, t_+, \mathbf{u}, \mathbf{u}_+)h_+, \tag{1.12}$$

defined on some one-dimensional lattice Ω with steplength $h_+ = t_+ - t$. Generally, the lattice can depend on the solution, for example, as

$$\Omega(t, t_-, t_+, \mathbf{u}, \mathbf{u}_-, \mathbf{u}_+) = 0. \tag{1.13}$$

Functional (1.12) must be considered together with lattice (1.13). On different lattices it can have different continuous limits.

Let us take a variation of the difference functional (1.12) along some curve $u^i = \phi_i(t)$, $i = 1, \ldots, n$ at some point $(t, \mathbf{u})$. The variation will affect only two terms in the sum (1.12):

$$\mathbb{L}_h = \cdots + \mathcal{L}(t_-, t, \mathbf{u}_-, \mathbf{u})h_- + \mathcal{L}(t, t_+, \mathbf{u}, \mathbf{u}_+)h_+ + \cdots. \tag{1.14}$$

Thus, we get the following expression for the variation of the difference functional

$$\delta \mathbb{L}_h = \frac{\delta \mathcal{L}}{\delta u^i} \delta u^i + \frac{\delta \mathcal{L}}{\delta t} \delta t, \tag{1.15}$$

where $\delta u^i = \phi'_i \delta t, i = 1, \ldots, n$ and

$$\begin{aligned}
\frac{\delta \mathcal{L}}{\delta u^i} &= h_+ \frac{\partial \mathcal{L}}{\partial u^i} + h_- \frac{\partial \mathcal{L}^-}{\partial u^i}, &\quad i = 1, \ldots, n, \\
\frac{\delta \mathcal{L}}{\delta t} &= h_+ \frac{\partial \mathcal{L}}{\partial t} + h_- \frac{\partial \mathcal{L}^-}{\partial t} + \mathcal{L}^- - \mathcal{L},
\end{aligned} \tag{1.16}$$

with $\mathcal{L} = \mathcal{L}(t, t_+, \mathbf{u}, \mathbf{u}_+)$ and $\mathcal{L}^- = \underset{-h}{S}(\mathcal{L}) = \mathcal{L}(t_-, t, \mathbf{u}_-, \mathbf{u})$. For convenience we will use the following total left and right shift operators

$$\underset{-h}{S} f(t, \mathbf{u}) = f(t_-, \mathbf{u}_-), \qquad \underset{+h}{S} f(t, \mathbf{u}) = f(t_+, \mathbf{u}_+)$$

and left and right total difference operators

$$\underset{+h}{D} = \frac{\underset{+h}{S} - 1}{h_+}, \qquad \underset{-h}{D} = \frac{1 - \underset{-h}{S}}{h_-}.$$

Thus, for an arbitrary curve the stationary value of the difference functional is given by a solution of the $n + 1$ equations

$$\frac{\delta \mathcal{L}}{\delta u^i} = 0, \quad i = 1, \ldots, n, \qquad \frac{\delta \mathcal{L}}{\delta t} = 0, \tag{1.17}$$

called *global extremal* equations. These equations represent the entire difference scheme and could be called "the discrete Euler–Lagrange system." They can be interpreted as a three-point difference scheme of the form

$$F_i(t, t_-, t_+, \mathbf{u}, \mathbf{u}_-, \mathbf{u}_+) = 0, \qquad i = 1, \ldots, n,$$
$$\Omega(t, t_-, t_+, \mathbf{u}, \mathbf{u}_-, \mathbf{u}_+) = 0.$$

Here the first n equations are approximations of differential equations (1.2) and the last equation provides a lattice, on which these approximations are considered. In the continuous limit the lattice equation vanishes (turns into an identity like $0 = 0$). Given two points, for instance $(t, \mathbf{u})$ and $(t_-, \mathbf{u}_-)$, we can calculate $(t_+, \mathbf{u}_+)$.

Note that the variational equations (1.17) can be obtained by the action of discrete variational operators

$$\frac{\delta}{\delta u^i} = \frac{\partial}{\partial u^i} + \underset{-h}{S} \frac{\partial}{\partial u^i_+}, \qquad i = 1, \ldots, n, \tag{1.18}$$

$$\frac{\delta}{\delta t} = \frac{\partial}{\partial t} + \underset{-h}{S} \frac{\partial}{\partial t_+} \tag{1.19}$$

on the discrete elementary action $\mathcal{L}(t, t_+, \mathbf{u}, \mathbf{u}_+) h_+$.

Now let us consider a variation of the functional (1.12) along the orbit of a group generated by the operator (1.3). Then, we have $\delta t = \xi \delta a$, $\delta u^i = \eta^i \delta a$, $i = 1, \ldots, n$, where δa is the variation of the group parameter. A stationary value of the difference functional (1.12) along the flow generated by this vector field is given by the equation

$$\eta^i \frac{\delta \mathcal{L}}{\delta u^i} + \xi \frac{\delta \mathcal{L}}{\delta t} = 0, \tag{1.20}$$

which depends explicitly on the coefficients of the generator. This equation is called a *quasiextremal* equation. If we have a Lie algebra of vector fields of dimension $n + 1$ or more, then the stationary value of difference functional (1.12) along the entire flow will be achieved on the intersection of the solutions of all quasiextremal equations of the type (1.20), i.e., the system of equations (1.17).

Remark 1.6   Sometimes it is convenient to consider the variational equations (1.17) in a modified form

$$\frac{\partial \mathcal{L}}{\partial u^i} + \frac{h_-}{h_+} \frac{\partial \mathcal{L}^-}{\partial u^i} = 0, \qquad i = 1, \ldots, n,$$

$$\frac{\partial \mathcal{L}}{\partial t} + \frac{h_-}{h_+} \frac{\partial \mathcal{L}^-}{\partial t} - \underset{+h}{D}(\mathcal{L}^-) = 0, \tag{1.21}$$

obtained on dividing by h_+.

Let us consider a Lie group of point transformations, generated by a vector field (1.3). When acting on discrete equations and functionals, a vector field must be prolonged to variables at other points of the lattice. The prolongation is obtained by shifting the coefficients to the corresponding points. For three-point schemes we have

$$X = \xi \frac{\partial}{\partial t} + \xi_- \frac{\partial}{\partial t_-} + \xi_+ \frac{\partial}{\partial t_+} + \eta^i \frac{\partial}{\partial u^i} + \eta^i_- \frac{\partial}{\partial u^i_-} + \eta^i_+ \frac{\partial}{\partial u^i_+}$$

$$+ (\xi_+ - \xi) \frac{\partial}{\partial h_+} + (\xi - \xi_-) \frac{\partial}{\partial h_-}, \tag{1.22}$$

where coefficients are given as follows

$$\xi_- = \xi(t_-, \mathbf{u}_-), \qquad \eta^i_- = \eta^i(t_-, \mathbf{u}_-), \qquad \xi_+ = \xi(t_+, \mathbf{u}_+), \qquad \eta^i_+ = \eta^i(t_+, \mathbf{u}_+).$$

The infinitesimal invariance condition for the functional (1.12) on the lattice (1.13) is given by two equations [9–11, 14]:

$$X(\mathcal{L}) + \mathcal{L}\underset{+h}{D}(\xi) \Big|_{\Omega=0} = 0, \qquad X(\Omega)\big|_{\Omega=0} = 0, \tag{1.23}$$

which are valid on the lattice (1.13). Generally, the lattice is provided by the global extremal equations (1.17). Therefore, we need to require their invariance to consider the invariance of the functional.

A useful operator identity, valid for any Lagrangian $\mathcal{L}(t, t_+, \mathbf{u}, \mathbf{u}_+)$ and any vector field X is [9, 11]

$$X(\mathcal{L}) + \mathcal{L}\,\underset{+h}{D}(\xi) \equiv \xi\left(\frac{\partial \mathcal{L}}{\partial t} + \frac{h_-}{h_+}\frac{\partial \mathcal{L}^-}{\partial t} - \underset{+h}{D}(\mathcal{L}^-)\right)$$
$$+ \eta^i\left(\frac{\partial \mathcal{L}}{\partial u^i} + \frac{h_-}{h_+}\frac{\partial \mathcal{L}^-}{\partial u^i}\right) + \underset{+h}{D}\left(h_-\eta^i\frac{\partial \mathcal{L}^-}{\partial u^i} + h_-\xi\frac{\partial \mathcal{L}^-}{\partial t} + \xi\mathcal{L}^-\right). \quad (1.24)$$

The identity is a discrete analog of Noether identity (1.6) and can be called *the discrete Noether identity*. From this relation we obtain the following discrete analog of Noether's theorem.

Theorem 1.7 ([9, 11, 14]) *The global extremal equations* (1.17), *invariant under the Lie group G of local point transformations generated by vector fields X of the form* (1.3), *possess a first integral*

$$\mathcal{I} = h_-\eta^i\frac{\partial \mathcal{L}^-}{\partial u^i} + h_-\xi\frac{\partial \mathcal{L}^-}{\partial t} + \xi\mathcal{L}^- \quad (1.25)$$

if and only if the Lagrangian density $\mathcal{L}$ is invariant with respect to the same group on the solutions of (1.17).

Remark 1.8　If the Lagrangian density $\mathcal{L}$ is divergence invariant under Lie group G of local point transformations, i.e.,

$$X(\mathcal{L}) + \mathcal{L}\,\underset{+h}{D}(\xi) = \underset{+h}{D}(V) \quad (1.26)$$

for some function $V(t, \mathbf{u})$, then each element X of the Lie algebra corresponding to group G provides us with a first integral of the global extremal equations (1.17), namely

$$\mathcal{I} = h_-\eta^i\frac{\partial \mathcal{L}^-}{\partial u^i} + h_-\xi\frac{\partial \mathcal{L}^-}{\partial t} + \xi\mathcal{L}^- - V. \quad (1.27)$$

Remark 1.9　In a particular case when the discrete Lagrangian is invariant with respect to time translations, i.e., $\mathcal{L} = \mathcal{L}(h_+, \mathbf{q}, \mathbf{q}_+)$, where $h_+ = t_+ - t$ is the step size, there is a conservation of energy

$$\mathcal{E} = -\mathcal{L}^- - h_-\frac{\partial \mathcal{L}^-}{\partial h_-} = -\mathcal{L} - h_+\frac{\partial \mathcal{L}}{\partial h_+}.$$

In this case we get symplectic-momentum-energy preserving variational integrators [22].

It has been shown elsewhere [9–11], that if the functional (1.12) is invariant or divergence invariant under some group G, then the global extremal equations (1.17) are also invariant with respect to G:

Theorem 1.10   *If the Lagrangian $\mathcal{L}$ is invariant with respect to the operator* (1.3), *then the global extremal equations* (1.17) *are also invariant.*

Remark 1.11   If the Lagrangian $\mathcal{L}$ is divergence invariant, then the global extremal equations (1.17) are also invariant. This follows from the fact that total finite differences belong to the kernel of discrete variational operators.

As in the continuous case, the global extremal equations can be invariant with respect to a larger group than the corresponding Lagrangian.

Now we are in a position to establish the necessary and sufficient condition for the invariance of global extremal equations. We will obtain new identities and a new theorem.

Lemma 1.12   *The following identities hold for any smooth function* $\mathcal{L}(t, t_+, \mathbf{u}, \mathbf{u}_+)$:

$$\frac{\delta}{\delta u^j}\left(\left(X(\mathcal{L}) + \mathcal{L}\underset{+h}{D}(\xi)\right)h_+\right) \equiv X\left(\frac{\delta\mathcal{L}}{\delta u^j}\right) + \frac{\partial\eta^i}{\partial u^j}\frac{\delta\mathcal{L}}{\delta u^i} + \frac{\partial\xi}{\partial u^j}\frac{\delta\mathcal{L}}{\delta t}, \qquad j = 1,\ldots,n,$$
(1.28)

$$\frac{\delta}{\delta t}\left(\left(X(\mathcal{L}) + \mathcal{L}\underset{+h}{D}(\xi)\right)h_+\right) \equiv X\left(\frac{\delta\mathcal{L}}{\delta t}\right) + \frac{\partial\eta^i}{\partial t}\frac{\delta\mathcal{L}}{\delta u^i} + \frac{\partial\xi}{\partial t}\frac{\delta\mathcal{L}}{\delta t}.$$
(1.29)

Proof   The identities can be verified directly.   □

The lemma allows us to obtain not only the sufficient (Theorem 1.10) but also the necessary and sufficient condition for the invariance of the global extremal equations.

Theorem 1.13   *The global extremal equations* (1.17) *are invariant with respect to a symmetry* (1.3) *if and only if the following conditions are true (on the solutions of the equations):*

$$\frac{\delta}{\delta u^j}\left(\left(X(\mathcal{L}) + \mathcal{L}\underset{+h}{D}(\xi)\right)h_+\right)\bigg|_{(1.17)} = 0, \qquad j = 1,\ldots,n,$$
(1.30)

$$\frac{\delta}{\delta t}\left(\left(X(\mathcal{L}) + \mathcal{L}\underset{+h}{D}(\xi)\right)h_+\right)\bigg|_{(1.17)} = 0.$$
(1.31)

Proof   The statement follows from identities of Lemma 1.12.   □

Many examples of applications of the discrete version of Noether's theorem in Lagrangian framework can be found in [14]. It should be noted that the discrete Lagrangian formalism and the corresponding Noether's theorem are not restricted to ordinary equations. They can also be used for discretizations of partial differential equations [6].

We note that there exists an alternative approach to conservation laws of discrete equations on fixed meshes based on direct methods [20].

1.4 Hamiltonian formalism for differential equations

In this chapter we will also present the Hamiltonian formalism for discrete Hamiltonian equations. Before that we consider the canonical Hamiltonian equations

$$\dot{q}^i = \frac{\partial H}{\partial p_i}, \qquad \dot{p}_i = -\frac{\partial H}{\partial q^i}, \qquad i = 1, \ldots, n \qquad (1.32)$$

and rewrite results concerning their invariance and conservation properties in a nonstandard way, that provides us with a simple "translation" of the Lagrangian formalism into the Hamiltonian one. We also present a new criterion (Theorem 1.23) for the invariance of the Hamiltonian equations.

1.4.1 Canonical Hamiltonian equations

It is well known that canonical Hamiltonian equations (1.32) can be obtained by the variational principle from the action functional

$$\delta \int_{t_1}^{t_2} (p_i \dot{q}^i - H(t, \mathbf{q}, \mathbf{p})) \, dt = 0 \qquad (1.33)$$

in the phase space $(\mathbf{q}, \mathbf{p})$, where $\mathbf{q} = (q^1, q^2, \ldots, q^n)$ and $\mathbf{p} = (p_1, p_2, \ldots, p_n)$ [17, 27]. Let us note that the canonical Hamiltonian equations (1.32) can be derived by action of the variational operators

$$\frac{\delta}{\delta p_i} = \frac{\partial}{\partial p_i} - D \frac{\partial}{\partial \dot{p}_i}, \qquad i = 1, \ldots, n, \qquad (1.34)$$

$$\frac{\delta}{\delta q^i} = \frac{\partial}{\partial q^i} - D \frac{\partial}{\partial \dot{q}^i}, \qquad i = 1, \ldots, n, \qquad (1.35)$$

where D is the operator of total differentiation with respect to time

$$D = \frac{\partial}{\partial t} + \dot{q}^k \frac{\partial}{\partial q^k} + \dot{p}_k \frac{\partial}{\partial p_k} + \cdots,$$

on the function

$$p_i \dot{q}^i - H(t, \mathbf{q}, \mathbf{p}).$$

As an analog of Lagrangian elementary action $L \, dt$ [21, 29] we consider Hamiltonian elementary action [12], namely

$$p_i \, dq^i - H(t, \mathbf{q}, \mathbf{p}) \, dt, \qquad (1.36)$$

and investigate its invariance with respect to a point transformation group generated by an operator

$$X = \xi(t, \mathbf{q}, \mathbf{p}) \frac{\partial}{\partial t} + \eta^i(t, \mathbf{q}, \mathbf{p}) \frac{\partial}{\partial q^i} + \zeta_i(t, \mathbf{q}, \mathbf{p}) \frac{\partial}{\partial p_i}. \qquad (1.37)$$

It should be noted that such point symmetry operators might correspond to nonpoint symmetries in the Lagrangian framework.

Definition 1.14 A Hamiltonian function is called invariant with respect to a symmetry operator (1.37) if the elementary action (1.36) is an invariant of the group generated by this operator.

This definition makes it possible to develop the following proposition.

Theorem 1.15 ([12]) *A Hamiltonian is invariant with respect to a group generated by the operator (1.37) if and only if the following condition holds*

$$\zeta_i \dot{q}^i + p_i D(\eta^i) - X(H) - HD(\xi) = 0. \tag{1.38}$$

The basic identity, stated in [12], relates conservation properties of the canonical Hamiltonian equations to the invariance of the Hamiltonian function:

$$\zeta_i \dot{q}^i + p_i D(\eta^i) - X(H) - HD(\xi) \equiv \xi\left(D(H) - \frac{\partial H}{\partial t}\right)$$
$$- \eta^i\left(\dot{p}_i + \frac{\partial H}{\partial q^i}\right) + \zeta_i\left(\dot{q}^i - \frac{\partial H}{\partial p_i}\right) + D[p_i \eta^i - \xi H]. \tag{1.39}$$

This identity, called *the Hamiltonian identity*, is the well-known Noether identity rewritten for the Hamiltonian function. It allows us to state the following result.

Theorem 1.16 ([12]) *The canonical Hamiltonian equations (1.32) possess a first integral of the form*

$$J = p_i \eta^i - \xi H \tag{1.40}$$

if and only if the Hamiltonian function is invariant with respect to operator (1.37) on the solutions of the equations.

Theorem 1.16 corresponds to the strong version of the Noether theorem (i.e., necessary and sufficient condition) for invariant Lagrangians and Euler–Lagrange equations [21].

Remark 1.17 Theorem 1.16 can be generalized to the case of divergence invariance of the Hamiltonian action

$$\zeta_i \dot{q}^i + p_i D(\eta^i) - X(H) - HD(\xi) = D(V), \tag{1.41}$$

where $V = V(t, \mathbf{q}, \mathbf{p})$ is some function. If this condition holds on the solutions of the canonical Hamiltonian equations (1.32), then there is a first integral

$$J = p_i \eta^i - \xi H - V. \tag{1.42}$$

Remark 1.18   Let us note that according to Definition 1.14 any Hamiltonian is invariant with respect to the family of operators

$$X_* = \zeta_i(t, \mathbf{q}, \mathbf{p})\frac{\partial}{\partial p_i} \tag{1.43}$$

on the solutions of the corresponding Hamiltonian equations. These operators do not provide nontrivial first integrals (they give $J = 0$). Therefore, it makes sense to consider symmetry operators up to the set of operators (1.43). It should be mentioned that in general operators X_* are not symmetries of the Hamiltonian equations (1.32).

Remark 1.19   Alternatively, one can exploit Hamiltonian symmetries (evolutionary symmetries of a special form) to find first integrals of the canonical Hamiltonian equations. We refer to [29] for details of this approach.

1.4.2 The Legendre transformation

The Legendre transformation [3] of a Lagrangian density $L(t, \mathbf{q}, \dot{\mathbf{q}})$ for $\dot{\mathbf{q}}$ is the function

$$H(t, \mathbf{q}, \mathbf{p}) = p_i \dot{q}^i - L(t, \mathbf{q}, \dot{\mathbf{q}}), \tag{1.44}$$

in which $\dot{\mathbf{q}}$ is expressed from

$$\mathbf{p} = \frac{\partial L}{\partial \dot{\mathbf{q}}}. \tag{1.45}$$

This transformation provides us with the following relations for the derivatives:

$$\dot{\mathbf{q}} = \frac{\partial H}{\partial \mathbf{p}}, \qquad \frac{\partial H}{\partial \mathbf{q}} = -\frac{\partial L}{\partial \mathbf{q}}, \qquad \frac{\partial H}{\partial t} = -\frac{\partial L}{\partial t}.$$

Thus, the Euler–Lagrange equations (1.2) are transformed into the canonical Hamiltonian equations (1.32). One can establish equivalence of the Euler–Lagrange and Hamiltonian equations under some regularity conditions [3].

1.4.3 Invariance of canonical Hamiltonian equations

In the Lagrangian framework, the variational principle provides us with Euler–Lagrange equations. The invariance of the Euler–Lagrange equations follows from the invariance of the action functional. In Section 1.2 we also discussed conditions for the invariance of the Euler–Lagrange equations. Here we provide analogous results for the Hamiltonian framework.

Application of variational operators (1.34), (1.35) to the left-hand side of the invariance condition (1.38) yields the statement:

Lemma 1.20 *The following identities are true for any smooth function* $H = H(t, \mathbf{q}, \mathbf{p})$:

$$\frac{\delta}{\delta p_j}(\zeta_i \dot{q}^i + p_i D(\eta^i) - X(H) - H D(\xi))$$

$$\equiv D(\eta^j) - \dot{q}^j D(\xi) - X\left(\frac{\partial H}{\partial p_j}\right) + \frac{\partial \xi}{\partial p_j}\left(D(H) - \frac{\partial H}{\partial t}\right)$$

$$- \frac{\partial \eta^i}{\partial p_j}\left(\dot{p}_i + \frac{\partial H}{\partial q^i}\right) + \left(\frac{\partial \zeta_i}{\partial p_j} + \delta_{ij} D(\xi)\right)\left(\dot{q}^i - \frac{\partial H}{\partial p_i}\right),$$

$$j = 1, \ldots, n, \quad (1.46)$$

$$\frac{\delta}{\delta q^j}(\zeta_i \dot{q}^i + p_i D(\eta^i) - X(H) - H D(\xi))$$

$$\equiv -D(\zeta_j) + \dot{p}_j D(\xi) - X\left(\frac{\partial H}{\partial q^j}\right) + \frac{\partial \xi}{\partial q^j}\left(D(H) - \frac{\partial H}{\partial t}\right)$$

$$- \left(\frac{\partial \eta^i}{\partial q^j} + \delta_{ij} D(\xi)\right)\left(\dot{p}_i + \frac{\partial H}{\partial q^i}\right) + \frac{\partial \zeta_i}{\partial q^j}\left(\dot{q}^i - \frac{\partial H}{\partial p_i}\right),$$

$$j = 1, \ldots, n, \quad (1.47)$$

where the notation δ_{ij} *stands for the Kronecker symbol.*

Proof The result can be shown by a direct computation. $\square$

The lemma allows us to establish the following results concerning the invariance of canonical Hamiltonian equations.

Theorem 1.21 *If a Hamiltonian* H *is invariant with respect to a symmetry* (1.37), *then the canonical Hamiltonian equations* (1.32) *are also invariant.*

Proof For invariance of the canonical Hamiltonian equations (1.32) we need the equations

$$D(\eta^j) - \dot{q}^j D(\xi) = X\left(\frac{\partial H}{\partial p_j}\right), \qquad j = 1, \ldots, n,$$

$$D(\zeta_j) - \dot{p}_j D(\xi) = -X\left(\frac{\partial H}{\partial q^j}\right), \qquad j = 1, \ldots, n$$

to hold on the solutions of the Hamiltonian equations. These conditions follow from the identities (1.46) and (1.47). $\square$

Remark 1.22 The statement of Theorem 1.21 remains valid if we consider divergence symmetries of the Hamiltonian, i.e., condition (1.41) instead of (1.38), because the term $D(V)$ belongs to the kernel of variational operators (1.34), (1.35).

Invariance or divergence invariance of the Hamiltonian is a *sufficient* condition for the canonical Hamiltonian equations to be invariant. The symmetry group of the canonical Hamiltonian equations can include transformations which are not admitted by Hamiltonian action (see also Remark 1.18). The following Theorem 1.23 establishes the *necessary and sufficient* condition for canonical Hamiltonian equations to be invariant.

Theorem 1.23 *Canonical Hamiltonian equations* (1.32) *are invariant with respect to a symmetry* (1.37) *if and only if the following conditions are true (on the solutions of the canonical Hamiltonian equations):*

$$\frac{\delta}{\delta p_j}(\zeta_i \dot{q}^i + p_i D(\eta^i) - X(H) - HD(\xi))\Big|_{\substack{\dot{q}=H_p \\ \dot{p}=-H_q}} = 0, \qquad j = 1, \ldots, n, \quad (1.48)$$

$$\frac{\delta}{\delta q^j}(\zeta_i \dot{q}^i + p_i D(\eta^i) - X(H) - HD(\xi))\Big|_{\substack{\dot{q}=H_p \\ \dot{p}=-H_q}} = 0, \qquad j = 1, \ldots, n. \quad (1.49)$$

Proof   The statement follows from identities (1.46) and (1.47).   □

It should be noticed that conditions (1.48) and (1.49) are true for all symmetries of canonical Hamiltonian equations. But not all of those symmetries yield the "variational integral" of these conditions, i.e.,

$$(\zeta_i \dot{q}^i + p_i D(\eta^i) - X(H) - HD(\xi))\Big|_{\substack{\dot{q}=H_p \\ \dot{p}=-H_q}} = 0,$$

which gives first integrals in accordance with Theorem 1.16. *That is why not all symmetries of the canonical Hamiltonian equations provide first integrals.*

Example 1.24   Let us consider the canonical Hamiltonian equations

$$\dot{q} = p, \qquad \dot{p} = \frac{1}{q^2}, \qquad (1.50)$$

which correspond to the Hamiltonian function

$$H(t, q, p) = \frac{p^2}{2} + \frac{1}{q}.$$

The equations admit the symmetries

$$X_1 = \frac{\partial}{\partial t}, \qquad X_2 = 3t\frac{\partial}{\partial t} + 2q\frac{\partial}{\partial q} - p\frac{\partial}{\partial p}.$$

The invariance of Hamiltonian condition (1.38) is satisfied for operator X_1 only. Application of operator X_2 to the Hamiltonian action gives

$$\zeta_2 \dot{q} + p D(\eta_2) - X_2(H) - HD(\xi_2) = p\dot{q} - \left(\frac{p^2}{2} + \frac{1}{q}\right) \neq 0.$$

Meanwhile, in accordance with Theorem 1.23 we have

$$\frac{\delta}{\delta p}(\zeta_2 \dot{q} + pD(\eta_2) - X_2(H) - HD(\xi_2))\Big|_{\substack{\dot{q}=p \\ \dot{p}=1/q^2}} = (\dot{q} - p)\Big|_{\substack{\dot{q}=p \\ \dot{p}=1/q^2}} = 0,$$

$$\frac{\delta}{\delta q}(\zeta_2 \dot{q} + pD(\eta_2) - X_2(H) - HD(\xi_2))\Big|_{\substack{\dot{q}=p \\ \dot{p}=1/q^2}} = \left(-\dot{p} + \frac{1}{q^2}\right)\Big|_{\substack{\dot{q}=p \\ \dot{p}=1/q^2}} = 0.$$

1.5 Discrete Hamiltonian formalism

Now we are ready to consider discrete Hamiltonian equations at some point $(t, \mathbf{q}, \mathbf{p})$ of a lattice. Generally, the lattice is not regular. The notations are given in the picture:

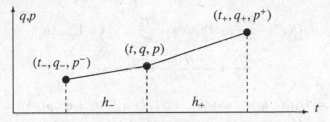

1.5.1 Discrete Legendre transform

Using the analogy with the continuous case, we can construct discrete Hamiltonian equations on the basis of the discrete equations in the Lagrangian framework. The discrete *Legendre transform* was introduced in [24]. Here we use a slightly modified version of that transformation. The discrete Legendre transform of $\mathcal{L}(t, t_+, \mathbf{q}, \mathbf{q}_+)$ for $\mathbf{q}_+$ is the function

$$\mathcal{H}(t, t_+, \mathbf{q}, \mathbf{p}^+) = p_i^+ \underset{+h}{D}(q^i) - \mathcal{L}(t, t_+, \mathbf{q}, \mathbf{q}_+), \tag{1.51}$$

where $\underset{+h}{D}(q^i) = (q_+^i - q^i)/h_+$ and $\mathbf{q}_+$ is defined implicitly by

$$\mathbf{p}^+ = h_+ \frac{\partial \mathcal{L}}{\partial \mathbf{q}_+}. \tag{1.52}$$

Remark 1.25 Alternatively, it is possible to consider discrete Legendre transformation for $\mathbf{q}$. Then,

$$\mathcal{H}(t, t_+, \mathbf{q}_+, \mathbf{p}) = p_i \underset{+h}{D}(q^i) - \mathcal{L}(t, t_+, \mathbf{q}, \mathbf{q}_+), \tag{1.53}$$

where $\mathbf{q}$ is found from

$$\mathbf{p} = -h_+ \frac{\partial \mathcal{L}}{\partial \mathbf{q}}. \tag{1.54}$$

In the terminology of [24] transformations (1.51), (1.52) and (1.53), (1.54) provide us with right and left Hamiltonians, respectively.

For discrete Legendre transform (1.51), (1.52) we get the following relations for the derivatives of the Hamiltonian function:

$$\frac{\partial \mathcal{H}}{\partial \mathbf{p}^+} = \underset{+h}{D}(\mathbf{q}), \qquad\qquad \frac{\partial \mathcal{H}}{\partial \mathbf{q}} = -\frac{\mathbf{p}^+}{h_+} - \frac{\partial \mathcal{L}}{\partial \mathbf{q}},$$

$$\frac{\partial \mathcal{H}}{\partial t} = \frac{p_i^+}{h_+}\underset{+h}{D}(q^i) - \frac{\partial \mathcal{L}}{\partial t}, \qquad \frac{\partial \mathcal{H}}{\partial t_+} = -\frac{p_i^+}{h_+}\underset{+h}{D}(q^i) - \frac{\partial \mathcal{L}}{\partial t_+}.$$

Using these relations as well as relations (1.51), (1.52), we can transform $n + 1$ global extremal equations (1.17) for the Lagrangian $\mathcal{L}(t, t_+, \mathbf{q}, \mathbf{q}_+)$ into the Hamiltonian framework. We arrive at the system of $2n + 1$ equations

$$\underset{+h}{D}(q^i) = \frac{\partial \mathcal{H}}{\partial p_i^+}, \quad \underset{+h}{D}(p_i) = -\frac{\partial \mathcal{H}}{\partial q^i}, \qquad i = 1, \ldots, n,$$

$$\frac{\partial \mathcal{H}}{\partial t} + \frac{h_-}{h_+}\frac{\partial \mathcal{H}^-}{\partial t} - \underset{+h}{D}(\mathcal{H}^-) = 0, \tag{1.55}$$

where $\mathcal{H} = \mathcal{H}(t, t_+, \mathbf{q}, \mathbf{p}^+)$ and $\mathcal{H}^- = \underset{-h}{S}(\mathcal{H}) = \mathcal{H}(t_-, t, \mathbf{q}_-, \mathbf{p})$. We will call them *discrete Hamiltonian* equations. Although introduced via the discrete Legendre transform, these equations can be considered without a relationship to the Lagrangian framework.

Let us note that the first $2n$ equations (1.55) are first-order discrete equations, which correspond to the canonical Hamiltonian equations (1.32) in the continuous limit. These (and equivalent) equations were considered in a number of papers [2, 15, 16, 33]. The last equation is of second order. It defines the lattice on which the canonical Hamiltonian equations are discretized. In the continuous limit the lattice equation itself vanishes. Being a second-order difference equation, it needs one more initial value (first spacing of the lattice) to state the initial-value problem.

Remark 1.26   The second version of the discrete Legendre transform (1.53), (1.54) provides us with discrete Hamiltonian equations

$$\underset{+h}{D}(q^i) = \frac{\partial \mathcal{H}}{\partial p_i}, \quad \underset{+h}{D}(p_i) = -\frac{\partial \mathcal{H}}{\partial q_+^i}, \qquad i = 1, \ldots, n,$$

$$\frac{\partial \mathcal{H}}{\partial t} + \frac{h_-}{h_+}\frac{\partial \mathcal{H}^-}{\partial t} - \underset{+h}{D}(\mathcal{H}^-) = 0, \tag{1.56}$$

where $\mathcal{H} = \mathcal{H}(t, t_+, \mathbf{q}^+, \mathbf{p})$ and $\mathcal{H}^- = \underset{-h}{S}(\mathcal{H}) = \mathcal{H}(t_-, t, \mathbf{q}, \mathbf{p}^-)$.

1.5.2 Variational formulation of the discrete Hamiltonian equations

The discrete Hamiltonian equations (1.55), which were obtained by application of the discrete Legendre transformation to the discrete Euler–Lagrange equations, follow from a variational principle. Indeed, let us consider the finite-difference functional

$$
\mathbb{H}_h = \sum_\Omega \left(p_i^+ \underset{+h}{D}(q^i) - \mathcal{H}(t, t_+, \mathbf{q}, \mathbf{p}^+) \right) h_+
$$

$$
= \sum_\Omega (p_i^+(q_1^i - q^i) - \mathcal{H}(t, t_+, \mathbf{q}, \mathbf{p}^+) h_+). \tag{1.57}
$$

A variation of this functional along a curve $q^i = \phi_i(t)$, $p_i = \psi_i(t)$, $i = 1, \ldots, n$ at some point $(t, \mathbf{q}, \mathbf{p})$ will affect only two terms of the sum (1.57):

$$
\mathbb{H}_h = \cdots + p_i(q^i - q_-^i) - \mathcal{H}(t_-, t, \mathbf{q}_-, \mathbf{p})h_-
$$
$$
+ p_i^+(q_+^i - q^i) - \mathcal{H}(t, t_+, \mathbf{q}, \mathbf{p}^+)h_+ + \cdots . \tag{1.58}
$$

Therefore, we get the following expression for the functional variation

$$
\delta \mathbb{H}_h = \frac{\delta \mathcal{H}}{\delta p_i} \delta p_i + \frac{\delta \mathcal{H}}{\delta q^i} \delta q^i + \frac{\delta \mathcal{H}}{\delta t} \delta t, \tag{1.59}
$$

where $\delta q^i = \phi_i' \delta t$, $\delta p_i = \psi_i' \delta t$, $i = 1, \ldots, n$ and

$$
\frac{\delta \mathcal{H}}{\delta p_i} = q^i - q_-^i - h_- \frac{\partial \mathcal{H}^-}{\partial p_i}, \qquad\qquad i = 1, \ldots, n,
$$
$$
\frac{\delta \mathcal{H}}{\delta q^i} = -\left(p_i^+ - p_i + h_+ \frac{\partial \mathcal{H}}{\partial q^i} \right), \qquad\qquad i = 1, \ldots, n, \tag{1.60}
$$
$$
\frac{\delta \mathcal{H}}{\delta t} = -\left(h_+ \frac{\partial \mathcal{H}}{\partial t} - \mathcal{H} + h_- \frac{\partial \mathcal{H}^-}{\partial t} + \mathcal{H}^- \right).
$$

For the stationary value of the finite-difference functional (1.57) we obtain the system of $2n + 1$ equations

$$
\frac{\delta \mathcal{H}}{\delta p_i} = 0, \qquad \frac{\delta \mathcal{H}}{\delta q^i} = 0, \qquad i = 1, \ldots, n, \qquad \frac{\delta \mathcal{H}}{\delta t} = 0, \tag{1.61}
$$

which are equivalent to the discrete Hamiltonian equations (1.55).

Let us note that the variational equations (1.60) can be derived by the action of the variational operators

$$\frac{\delta}{\delta p_i} = \frac{\partial}{\partial p_i} + S_{-h}\frac{\partial}{\partial p_i^+}, \qquad i = 1,\ldots,n, \tag{1.62}$$

$$\frac{\delta}{\delta q^i} = \frac{\partial}{\partial q^i} + S_{-h}\frac{\partial}{\partial q_+^i}, \qquad i = 1,\ldots,n, \tag{1.63}$$

$$\frac{\delta}{\delta t} = \frac{\partial}{\partial t} + S_{-h}\frac{\partial}{\partial t_+} \tag{1.64}$$

on the discrete Hamiltonian elementary action

$$p_i^+(q_+^i - q^i) - \mathcal{H}(t, t_+, \mathbf{q}, \mathbf{p}^+)h_+.$$

For a variation of the functional (1.57) along the orbit of a group generated by the operator (1.37) we have $\delta t = \xi \delta a$, $\delta q^i = \eta^i \delta a$, $\delta p_i = \zeta_i \delta a$, $i = 1,\ldots,n$, where δa is the variation of a group parameter. A stationary value of the finite-difference functional (1.57) along the flow generated by this vector field is given by the equation

$$\zeta_i\frac{\delta\mathcal{H}}{\delta p_i} + \eta^i\frac{\delta\mathcal{H}}{\delta q^i} + \xi\frac{\delta\mathcal{H}}{\delta t} = 0, \tag{1.65}$$

which depends explicitly on the coefficients of the generator. It corresponds to the quasiextremal equation (1.20) in the Lagrangian framework.

If we have a Lie algebra of vector fields of dimension $2n + 1$ or more, then the stationary value of the functional (1.57) along the entire flow is achieved on the solutions of the system (1.61).

Remark 1.27   Similarly, one can show that the discrete Hamiltonian equations (1.56) can be obtained by variations of the finite-difference functional

$$\mathbb{H}_h = \sum_{\Omega}(p_i\underset{+h}{D}(q^i) - \mathcal{H}(t, t_+, \mathbf{q}_+, \mathbf{p}))h_+$$

$$= \sum_{\Omega}(p_i(q_+^i - q^i) - \mathcal{H}(t, t_+, \mathbf{q}_+, \mathbf{p})h_+). \tag{1.66}$$

In the following sections we will consider invariance and conservation properties of the discrete Hamiltonian equations. For simplicity, we restrict ourselves to the version (1.55) of such equations. All results can be equivalently formulated for the other version, i.e., the discrete Hamiltonian equations (1.56).

1.5.3 Symplecticity of the discrete Hamiltonian equations

The canonical Hamiltonian equations generate symplectic transformations in the phase space $(\mathbf{q}, \mathbf{p})$. For the solution $(\mathbf{q}(t), \mathbf{p}(t))$ of the system (1.32) with initial data $\mathbf{q}(t_0) = \mathbf{q}_0$, $\mathbf{p}(t_0) = \mathbf{p}^0$ this property can be expressed as a conservation of the two-form

$$dp_i \wedge dq^i = dp_i^0 \wedge dq_0^i. \tag{1.67}$$

This property is used to select symplectic numerical integrators [25, 32] as numerical schemes with the property

$$dp_i^{n+1} \wedge dq_{n+1}^i = dp_i^n \wedge dq_n^i, \qquad n = 0, 1, \ldots . \tag{1.68}$$

The definition (1.68) for conservation of symplecticity can not be used for discretizations on solution-dependent meshes such as discrete Euler–Lagrange equations (1.17) and discrete Hamiltonian equations (1.55). Generally, variations of the dependent variables involve variations of the lattice points. It is clearly seen from the variational equations for the system (1.55):

$$dq_+^i - dq^i = \frac{\partial^2(\mathcal{H}h_+)}{\partial t \partial p_i^+} \, dt + \frac{\partial^2(\mathcal{H}h_+)}{\partial t_+ \partial p_i^+} \, dt_+ + \frac{\partial^2(\mathcal{H}h_+)}{\partial q^j \partial p_i^+} \, dq^j + \frac{\partial^2(\mathcal{H}h_+)}{\partial p_j^+ \partial p_i^+} \, dp_j^+,$$

$$i = 1, \ldots, n,$$

$$dp_i^+ - dp_i = -\frac{\partial^2(\mathcal{H}h_+)}{\partial t \partial q^i} \, dt - \frac{\partial^2(\mathcal{H}h_+)}{\partial t_+ \partial q^i} \, dt_+ - \frac{\partial^2(\mathcal{H}h_+)}{\partial q^j \partial q^i} \, dq^j - \frac{\partial^2(\mathcal{H}h_+)}{\partial p_j^+ \partial q^i} \, dp_j^+,$$

$$\iota = 1, \ldots, n,$$

$$\frac{\partial^2(\mathcal{H}h_+)}{\partial t^2} \, dt + \frac{\partial^2(\mathcal{H}h_+)}{\partial t_+ \partial t} \, dt_+ + \frac{\partial^2(\mathcal{H}h_+)}{\partial q^j \partial t} \, dq^j + \frac{\partial^2(\mathcal{H}h_+)}{\partial p_j^+ \partial t} \, dp_j^+$$

$$+ \frac{\partial^2(\mathcal{H}^- h_-)}{\partial t_- \partial t} \, dt_- + \frac{\partial^2(\mathcal{H}^- h_-)}{\partial t^2} \, dt + \frac{\partial^2(\mathcal{H}^- h_-)}{\partial q_-^j \partial t} \, dq_-^j + \frac{\partial^2(\mathcal{H}^- h_-)}{\partial p_j \partial t} \, dp_j = 0.$$

For dt_+, dq_+, dp^+, i.e., the variations in the next point of the lattice, these equations are a system of $2n + 1$ linear algebraic equations. Thus, the variational equations considered in the phase space (without variations of the independent variable) form an overdetermined system of $2n + 1$ equations for $2n$ variables, which in the general case has only trivial solutions.

Therefore, we are forced to look for symplecticity in the extended phase space $(t, \mathbf{q}, \mathbf{p})$ (see also general considerations for the continuous case in [7]).

Theorem 1.28 *The discrete Hamiltonian equations (1.55) possess the conservation of sympecticity*

$$dp_i^+ \wedge dq_+^i - d\mathcal{E}_+ \wedge dt_+ = dp_i \wedge dq^i - d\mathcal{E} \wedge dt, \tag{1.69}$$

where

$$\mathcal{E}_+ = \mathcal{H} + h_+ \frac{\partial \mathcal{H}}{\partial t_+}, \qquad \mathcal{E} = \mathcal{H}^- + h_- \frac{\partial \mathcal{H}^-}{\partial t}$$

are discrete energies for lattice points t_+ and t.

Proof From the first $2n$ variational equations we obtain

$$dp_i^+ \wedge dq_+^i - dp_i \wedge dq^i = \frac{\partial^2(\mathcal{H}h_+)}{\partial t \partial p_i^+} dp_i^+ \wedge dt + \frac{\partial^2(\mathcal{H}h_+)}{\partial t_+ \partial p_i^+} dp_i^+ \wedge dt_+$$

$$+ \frac{\partial^2(\mathcal{H}h_+)}{\partial t \partial q^i} dq^i \wedge dt + \frac{\partial^2(\mathcal{H}h_+)}{\partial t_+ \partial q^i} dq^i \wedge dt_+. \quad (1.70)$$

With the help of the relations for variations

$$d\mathcal{E}_+ = \frac{\partial^2(\mathcal{H}h_+)}{\partial t \partial t_+} dt + \frac{\partial^2(\mathcal{H}h_+)}{\partial t_+^2} dt_+ + \frac{\partial^2(\mathcal{H}h_+)}{\partial q^j \partial t_+} dq^j + \frac{\partial^2(\mathcal{H}h_+)}{\partial p_j^+ \partial t_+} dp_j^+$$

and

$$d\mathcal{E} = \frac{\partial^2(\mathcal{H}^- h_-)}{\partial t_- \partial t} dt_- + \frac{\partial^2(\mathcal{H}^- h_-)}{\partial t^2} dt + \frac{\partial^2(\mathcal{H}^- h_-)}{\partial q_-^j \partial t} dq_-^j + \frac{\partial^2(\mathcal{H}^- h_-)}{\partial p_j \partial t} dp_j$$

$$= -\frac{\partial^2(\mathcal{H}h_+)}{\partial t^2} dt - \frac{\partial^2(\mathcal{H}h_+)}{\partial t_+ \partial t} dt_+ - \frac{\partial^2(\mathcal{H}h_+)}{\partial q^j \partial t} dq^j - \frac{\partial^2(\mathcal{H}h_+)}{\partial p_j^+ \partial t} dp_j^+,$$

where the last variational equation was used, we get

$$d\mathcal{E}_+ \wedge dt_+ - d\mathcal{E} \wedge dt = \frac{\partial^2(\mathcal{H}h_+)}{\partial p_i^+ \partial t} dp_i^+ \wedge dt + \frac{\partial^2(\mathcal{H}h_+)}{\partial p_i^+ \partial t_+} dp_i^+ \wedge dt_+$$

$$+ \frac{\partial^2(\mathcal{H}h_+)}{\partial q^i \partial t} dq^i \wedge dt + \frac{\partial^2(\mathcal{H}h_+)}{\partial q^i \partial t_+} dq^i \wedge dt_+. \quad (1.71)$$

Comparing right-hand sides of (1.70) and (1.71), we obtain the statement of the theorem. $\qquad\square$

Remark 1.29 The conservation of symplecticity can be transferred to the Lagrangian framework. The discrete Euler–Lagrange equations (1.17) possess the conservation of symplecticity (1.69), where the conjugate momenta

$$\mathbf{p}^+ = h_+ \frac{\partial \mathcal{L}}{\partial \mathbf{u}_+}, \qquad \mathbf{p} = h_- \frac{\partial \mathcal{L}^-}{\partial \mathbf{u}}$$

and discrete energies

$$\mathcal{E}_+ = -\mathcal{L} - h_+ \frac{\partial \mathcal{L}}{\partial t_+}, \qquad \mathcal{E} = -\mathcal{L}^- - h_- \frac{\partial \mathcal{L}^-}{\partial t}$$

are expressed in terms of the Lagrangian function. This relation holds on the solutions of the variational equations for the discrete Euler–Lagrange equations (1.17):

$$\frac{\partial^2(\mathcal{L}h_+)}{\partial t \partial u^i}\,dt + \frac{\partial^2(\mathcal{L}h_+)}{\partial t_+ \partial u^i}\,dt_+ + \frac{\partial^2(\mathcal{L}h_+)}{\partial u^j \partial u^i}\,du^j + \frac{\partial^2(\mathcal{L}h_+)}{\partial u^j_+ \partial u^i}\,du^j_+$$

$$+ \frac{\partial^2(\mathcal{L}^-h_-)}{\partial t_- \partial u^i}\,dt_- + \frac{\partial^2(\mathcal{L}^-h_-)}{\partial t \partial u^i}\,dt + \frac{\partial^2(\mathcal{L}^-h_-)}{\partial u^j_- \partial u^i}\,du^j_- + \frac{\partial^2(\mathcal{L}^-h_-)}{\partial u^j \partial u^i}\,du^j = 0,$$

$$i = 1,\ldots,n,$$

$$\frac{\partial^2(\mathcal{L}h_+)}{\partial t^2}\,dt + \frac{\partial^2(\mathcal{L}h_+)}{\partial t_+ \partial t}\,dt_+ + \frac{\partial^2(\mathcal{L}h_+)}{\partial u^j \partial t}\,du^j + \frac{\partial^2(\mathcal{L}h_+)}{\partial u^j_+ \partial t}\,du^j_+$$

$$+ \frac{\partial^2(\mathcal{L}^-h_-)}{\partial t_- \partial t}\,dt_- + \frac{\partial^2(\mathcal{L}^-h_-)}{\partial t^2}\,dt + \frac{\partial^2(\mathcal{L}^-h_-)}{\partial u^j_- \partial t}\,du^j_- + \frac{\partial^2(\mathcal{L}^-h_-)}{\partial u^j \partial t}\,du^j = 0.$$

Substituting the expressions for the momenta and energies into (1.69), we obtain the noncanonical form of the symplecticity

$$\frac{\partial^2(\mathcal{L}h_+)}{\partial u^j \partial u^i_+}\,du^j \wedge du^i_+ + \frac{\partial^2(\mathcal{L}h_+)}{\partial t \partial u^i_+}\,dt \wedge du^i_+ + \frac{\partial^2(\mathcal{L}h_+)}{\partial u^i \partial t_+}\,du^i \wedge dt_+ + \frac{\partial^2(\mathcal{L}h_+)}{\partial t \partial t_+}\,dt \wedge dt_+$$

$$= \frac{\partial^2(\mathcal{L}^-h_-)}{\partial u^j_- \partial u^i}\,du^j_- \wedge du^i + \frac{\partial^2(\mathcal{L}^-h_-)}{\partial t_- \partial u^i}\,dt_- \wedge du^i$$

$$+ \frac{\partial^2(\mathcal{L}^-h_-)}{\partial u^i_- \partial t}\,du^i_- \wedge dt + \frac{\partial^2(\mathcal{L}^-h_-)}{\partial t_- \partial t}\,dt_- \wedge dt,$$

which holds on the variational equations for the discrete Euler–Lagrange equations.

1.5.4 Invariance of the Hamiltonian action

To consider discrete Hamiltonian equations we will need three points of a lattice. The prolongation of Lie group operator (1.37) for neighboring points $(t_-, \mathbf{q}_-, \mathbf{p}^-)$ and $(t_+, \mathbf{q}_+, \mathbf{p}^+)$ is the following:

$$X = \xi \frac{\partial}{\partial t} + \xi_- \frac{\partial}{\partial t_-} + \xi_+ \frac{\partial}{\partial t_+} + \eta^i \frac{\partial}{\partial q^i} + \eta^i_- \frac{\partial}{\partial q^i_-} + \eta^i_+ \frac{\partial}{\partial q^i_+}$$

$$+ \zeta_i \frac{\partial}{\partial p_i} + \zeta^-_i \frac{\partial}{\partial p^-_i} + \zeta^+_i \frac{\partial}{\partial p^+_i} + (\xi_+ - \xi)\frac{\partial}{\partial h_+} + (\xi - \xi_-)\frac{\partial}{\partial h_-}, \quad (1.72)$$

where

$$\xi_- = \xi(t_-, \mathbf{q}_-, \mathbf{p}^-), \qquad \eta^i_- = \eta^i(t_-, \mathbf{q}_-, \mathbf{p}^-), \qquad \zeta^-_i = \zeta_i(t_-, \mathbf{q}_-, \mathbf{p}^-),$$

$$\xi_+ = \xi(t_+, \mathbf{q}_+, \mathbf{p}^+), \qquad \eta^i_+ = \eta^i(t_+, \mathbf{q}_+, \mathbf{p}^+), \qquad \zeta^+_i = \zeta_i(t_+, \mathbf{q}_+, \mathbf{p}^+).$$

Let us consider the functional (1.57) on some lattice, given by the equation

$$\Omega(t, t_-, t_+, \mathbf{q}, \mathbf{q}_-, \mathbf{q}_+, \mathbf{p}, \mathbf{p}^-, \mathbf{p}^+) = 0. \tag{1.73}$$

Definition 1.30   We call a discrete Hamiltonian function $\mathcal{H}$ considered on the mesh (1.73) *invariant* with respect to a group generated by the operator (1.37), if the action functional (1.57) considered on the mesh (1.73) is an invariant manifold of the group.

Theorem 1.31   *A Hamiltonian function considered together with the mesh (1.73) is invariant with respect to a group generated by the operator (1.37) if and only if the following conditions hold*

$$\zeta_i^+ \underset{+h}{D}(q^i) + p_i^+ \underset{+h}{D}(\eta^i) - X(\mathcal{H}) - \mathcal{H} \underset{+h}{D}(\xi)\Big|_{\Omega=0} = 0, \qquad X(\Omega)\big|_{\Omega=0} = 0. \tag{1.74}$$

Proof   The invariance condition follows directly from the action of X on the functional:

$$X(\mathbb{H}_h) = X\left(\sum_\Omega (p_i^+(q_+^i - q^i) - \mathcal{H}h_+)\right)$$

$$= \sum_\Omega (\zeta_i^+(q_+^i - q^i) + p_i^+(\eta_+^i - \eta^i) - X(\mathcal{H})h_+ - \mathcal{H}(\xi_+ - \xi))$$

$$= \sum_\Omega \left(\zeta_i^+ \underset{+h}{D}(q^i) + p_i^+ \underset{+h}{D}(\eta^i) - X(\mathcal{H}) - \mathcal{H} \underset{+h}{D}(\xi)\right)h_+ = 0.$$

It should be provided with the invariance of a mesh, which is obtained by the action of the symmetry operator on mesh equation (1.73).   □

In the general case, the lattice is provided by the discrete Hamiltonian equations (1.55). Therefore, we need to require their invariance to consider the invariance of the Hamiltonian function.

1.5.5 Discrete Hamiltonian identity and discrete Noether theorem

As in the continuous case, the invariance of a discrete Hamiltonian on a specified mesh yields first integrals of discrete Hamiltonian equations.

Lemma 1.32   *The following identity is true for any smooth function* $\mathcal{H} = \mathcal{H}(t, t_+, \mathbf{q}, \mathbf{p}^+)$:

$$\zeta_i^+ \underset{+h}{D}(q^i) + p_i^+ \underset{+h}{D}(\eta^i) - X(\mathcal{H}) - \mathcal{H} \underset{+h}{D}(\xi)$$

$$\equiv -\xi\left(\frac{\partial \mathcal{H}}{\partial t} + \frac{h_-}{h_+}\frac{\partial \mathcal{H}^-}{\partial t} - \underset{+h}{D}(\mathcal{H}^-)\right) - \eta^i\left(\underset{+h}{D}(p_i) + \frac{\partial \mathcal{H}}{\partial q^i}\right)$$

$$+ \zeta_i^+\left(\underset{+h}{D}(q^i) - \frac{\partial \mathcal{H}}{\partial p_i^+}\right) + \underset{+h}{D}\left[\eta^i p_i - \xi\left(\mathcal{H}^- + h_-\frac{\partial \mathcal{H}^-}{\partial t}\right)\right]. \tag{1.75}$$

Proof The identity can be established by a direct calculation. □

We call this identity *the discrete Hamiltonian identity*. It makes it possible to state the following result.

Theorem 1.33 *The discrete Hamiltonian equations* (1.55) *which are invariant with respect to the symmetry operator* (1.37) *possess a first integral*

$$\mathcal{J} = \eta^i p_i - \xi \left(\mathcal{H}^- + h_- \frac{\partial \mathcal{H}^-}{\partial t} \right) \tag{1.76}$$

if and only if the Hamiltonian function is invariant with respect to the same symmetry on the solutions of (1.55).

Proof This result is a consequence of the identity (1.75). The invariance of the discrete Hamiltonian equations is needed to guarantee the invariance of the mesh, which is defined by these equations. □

Remark 1.34 Theorem 1.33 can be generalized to the case of divergence invariance of the Hamiltonian action, i.e., for the case

$$\zeta_i^+ \underset{+h}{D}(q^i) + p_i^+ \underset{+h}{D}(\eta^i) - X(\mathcal{H}) - \mathcal{H} \underset{+h}{D}(\xi) = \underset{+h}{D}(V), \tag{1.77}$$

where $V = V(t, \mathbf{q}, \mathbf{p})$. If this condition holds on the solutions of the discrete Hamiltonian equations (1.55), then there is a first integral

$$\mathcal{J} = \eta^i p_i - \xi \left(\mathcal{H}^- + h_- \frac{\partial \mathcal{H}^-}{\partial t} \right) - V. \tag{1.78}$$

Remark 1.35 For discrete Hamiltonian equations with Hamiltonian functions invariant with respect to time translations, i.e., $\mathcal{H} = \mathcal{H}(h_+, \mathbf{q}, \mathbf{p}^+)$, there is a conservation of energy

$$\mathcal{E} = \mathcal{H}^- + h_- \frac{\partial \mathcal{H}^-}{\partial h_-} = \mathcal{H} + h_+ \frac{\partial \mathcal{H}}{\partial h_+}.$$

In this case the discrete Hamiltonian equations (1.55) are related to symplectic-momentum-energy preserving variational integrators introduced for the discrete Lagrangian framework in [22]. Note that $\mathcal{H}$ is not the discrete energy, it has a meaning of a generating function for the discrete Hamiltonian flow.

Remark 1.36 Similarly, we can consider the following identity

$$\zeta_i \underset{+h}{D}(q^i) + p_i \underset{+h}{D}(\eta^i) - X(\mathcal{H}) - \mathcal{H} \underset{+h}{D}(\xi)$$

$$\equiv -\xi \left(\frac{\partial \mathcal{H}}{\partial t} + \frac{h_-}{h_+} \frac{\partial \mathcal{H}^-}{\partial t} - \underset{+h}{D}(\mathcal{H}^-) \right) - \eta_+^i \left(\underset{+h}{D}(p_i) + \frac{\partial \mathcal{H}}{\partial q_+^i} \right)$$

$$+ \zeta_i \left(\underset{+h}{D}(q^i) - \frac{\partial \mathcal{H}}{\partial p_i} \right) + \underset{+h}{D} \left[\eta^i p_i - \xi \left(\mathcal{H}^- + h_- \frac{\partial \mathcal{H}^-}{\partial t} \right) \right], \tag{1.79}$$

which allows us to formulate Noether's theorem for the second version of the discrete Hamiltonian equations (1.56).

1.5.6 Invariance of the discrete Hamiltonian equations

Application of discrete variational operators (1.62), (1.63), (1.64) to the expression obtained by the action of the symmetry operator X on the elementary action

$$\zeta_i^+(q_+^i - q^i) + p_i^+(\eta_+^i - \eta^i) - X(\mathcal{H})h_+ - \mathcal{H}(\xi_+ - \xi)$$

$$\equiv \left(\zeta_i^+ \underset{+h}{D}(q^i) + p_i^+ \underset{+h}{D}(\eta^i) - X(\mathcal{H}) - \mathcal{H}\underset{+h}{D}(\xi)\right)h_+$$

provides us with the result:

Lemma 1.37   *The following identities hold for any smooth function* $\mathcal{H} = \mathcal{H}(t, t_+, \mathbf{q}, \mathbf{p}_+)$:

$$\frac{\delta}{\delta p_j}(\zeta_i^+(q_+^i - q^i) + p_i^+(\eta_+^i - \eta^i) - X(\mathcal{H})h_+ - \mathcal{H}(\xi_+ - \xi))$$

$$\equiv X\left(\frac{\delta\mathcal{H}}{\delta p_j}\right) + \frac{\partial\zeta_i}{\partial p_j}\frac{\delta\mathcal{H}}{\delta p_i} + \frac{\partial\eta^i}{\partial p_j}\frac{\delta\mathcal{H}}{\delta q^i} + \frac{\partial\xi}{\partial p_j}\frac{\delta\mathcal{H}}{\delta t}, \qquad j = 1, \ldots, n, \quad (1.80)$$

$$\frac{\delta}{\delta q^j}(\zeta_i^+(q_+^i - q^i) + p_i^+(\eta_+^i - \eta^i) - X(\mathcal{H})h_+ - \mathcal{H}(\xi_+ - \xi))$$

$$\equiv X\left(\frac{\delta\mathcal{H}}{\delta q^j}\right) + \frac{\partial\zeta_i}{\partial q^j}\frac{\delta\mathcal{H}}{\delta p_i} + \frac{\partial\eta^i}{\partial q^j}\frac{\delta\mathcal{H}}{\delta q^i} + \frac{\partial\xi}{\partial q^j}\frac{\delta\mathcal{H}}{\delta t}, \qquad j = 1, \ldots, n, \quad (1.81)$$

$$\frac{\delta}{\delta t}(\zeta_i^+(q_+^i - q^i) + p_i^+(\eta_+^i - \eta^i) - X(\mathcal{H})h_+ - \mathcal{H}(\xi_+ - \xi))$$

$$\equiv X\left(\frac{\delta\mathcal{H}}{\delta t}\right) + \frac{\partial\zeta_i}{\partial t}\frac{\delta\mathcal{H}}{\delta p_i} + \frac{\partial\eta^i}{\partial t}\frac{\delta\mathcal{H}}{\delta q^i} + \frac{\partial\xi}{\partial t}\frac{\delta\mathcal{H}}{\delta t}. \qquad (1.82)$$

Using the lemma, we can relate invariance of the discrete Hamiltonian equations to that of the Hamiltonian.

Theorem 1.38   *If the discrete Hamiltonian* $\mathcal{H}$ *is invariant with respect to the operator* (1.37), *then the discrete Hamiltonian equations* (1.55) *are also invariant.*

Proof   If the discrete Hamiltonian $\mathcal{H}$ is invariant, the left-hand sides of the identities of Lemma 1.37 are zeros. It follows that the variational equations (1.61) are invariant. Consequently, the discrete Hamiltonian equations, which are equivalent to these variational equations, are also invariant. $\square$

Remark 1.39   If the discrete Hamiltonian $\mathcal{H}$ is divergence invariant, then the discrete Hamiltonian equations (1.55) are also invariant. This follows from the fact that total finite differences belong to the kernel of discrete variational operators.

With the help of the identities of Lemma 1.37 we can refine the result of Theorem 1.38 and formulate the necessary and sufficient condition for discrete Hamiltonian equations to be invariant. This explicitly shows the difference between invariance of Hamiltonians and invariance of Hamiltonian equations.

Theorem 1.40   *The discrete Hamiltonian equations* (1.55) *are invariant with respect to a symmetry* (1.37) *if and only if the following conditions are true (on the solutions of the discrete Hamiltonian equations):*

$$\frac{\delta}{\delta p_j}(\zeta_i^+(q_+^i - q^i) + p_i^+(\eta_+^i - \eta^i) - X(\mathcal{H})h_+ - \mathcal{H}(\xi_+ - \xi))\Big|_{(1.55)} = 0,$$

$$j = 1,\ldots,n, \quad (1.83)$$

$$\frac{\delta}{\delta q^j}(\zeta_i^+(q_+^i - q^i) + p_i^+(\eta_+^i - \eta^i) - X(\mathcal{H})h_+ - \mathcal{H}(\xi_+ - \xi))\Big|_{(1.55)} = 0,$$

$$j = 1,\ldots,n, \quad (1.84)$$

$$\frac{\delta}{\delta t}(\zeta_i^+(q_+^i - q^i) + p_i^+(\eta_+^i - \eta^i) - X(\mathcal{H})h_+ - \mathcal{H}(\xi_+ - \xi))\Big|_{(1.55)} = 0. \quad (1.85)$$

Proof   We use the fact that the discrete Hamiltonian equations (1.55) are equivalent to the variational equations (1.61). Then, the statement follows from the identities of Lemma 1.37.   □

1.6 Examples

In this section we present examples of the theoretical results presented in the previous sections applied to a number of differential equations and their discrete counterparts. We note, that discrete models in Lagrangian and Hamiltonian frameworks are related by the discrete Legendre transform, described in Section 1.5.1.

1.6.1 Nonlinear motion

As the first example we consider the second-order ODE

$$\ddot{u} = \frac{1}{u^3}, \quad (1.86)$$

which admits a Lie algebra with basis operators

$$X_1 = \frac{\partial}{\partial t}, \qquad X_2 = 2t\frac{\partial}{\partial t} + u\frac{\partial}{\partial u}, \qquad X_3 = t^2\frac{\partial}{\partial t} + tu\frac{\partial}{\partial u}.$$

Lagrangian framework

The Lagrangian function

$$L(t, u, \dot{u}) = \frac{1}{2}\left(\dot{u}^2 - \frac{1}{u^2}\right),$$

which provides (1.86) as its Euler–Lagrange equation, is invariant with respect to X_1 and X_2. Therefore, by means of Noether's theorem there exist first integrals

$$I_1 = -\frac{1}{2}\left(\dot{u}^2 + \frac{1}{u^2}\right), \qquad I_2 = u\dot{u} - t\left(\dot{u}^2 + \frac{1}{u^2}\right).$$

The action of the third operator X_3 yields a divergence invariance condition

$$X_3 L + LD(\xi_3) = u\dot{u} = D\left(\frac{u^2}{2}\right).$$

Due to the divergence invariance of the Lagrangian we can find the following first integral

$$I_3 = -\frac{1}{2}\left(\frac{t^2}{u^2} + (u - t\dot{u})^2\right).$$

It should be mentioned that independence of first integrals obtained with the help of the Noether theorem is guaranteed only in the case when there is one Lagrangian which is invariant with respect to all symmetries. This condition is broken in the considered example. Therefore, the integrals obtained are not independent, they are connected by the relation

$$4I_1 I_3 - I_2^2 = 1. \tag{1.87}$$

Thus, any two of the integrals I_1, I_2 and I_3 are independent. Putting $I_1 = A/2$, $I_2 = B$ and excluding $\dot{u}$, we find the general solution of (1.86) as

$$Au^2 + (At - B)^2 + 1 = 0. \tag{1.88}$$

Now let us consider the discrete case. We choose the discrete Lagrangian function

$$\mathcal{L}(t, t_+, u, u_+) = \frac{1}{2}\left(\left(\frac{u_+ - u}{h_+}\right)^2 - \frac{1}{uu_+}\right),$$

which admits the same symmetries as Lagrangian L in the continuous case:

$$X_1 \mathcal{L} + \mathcal{L} \underset{+h}{D}(\xi_1) = 0, \qquad X_2 \mathcal{L} + \mathcal{L} \underset{+h}{D}(\xi_2) = 0,$$

$$X_3 \mathcal{L} + \mathcal{L} \underset{+h}{D}(\xi_3) = \underset{+h}{D}\left(\frac{u^2}{2}\right).$$

The Lagrangian generates the invariant global extremal equations

$$
\begin{aligned}
\frac{\delta \mathcal{L}}{\delta u} : &\qquad 2\left(\frac{u_+ - u}{h_+} - \frac{u - u_-}{h_-}\right) = \frac{h_+}{u^2 u_+} + \frac{h_-}{u^2 u_-}, \\
\frac{\delta \mathcal{L}}{\delta t} : &\qquad \left(\frac{u_+ - u}{h_+}\right)^2 + \frac{1}{uu_+} - \left(\frac{u - u_-}{h_-}\right)^2 - \frac{1}{uu_-} = 0.
\end{aligned}
\tag{1.89}
$$

The global extremal equations have three first integrals

$$
\begin{aligned}
\mathcal{I}_1 &= -\frac{1}{2}\left(\left(\frac{u_+ - u}{h_+}\right)^2 + \frac{1}{uu_+}\right), \\
\mathcal{I}_2 &= \frac{u_+^2 - u^2}{2h_+} - \frac{t + t_+}{2}\left(\left(\frac{u_+ - u}{h_+}\right)^2 + \frac{1}{uu_+}\right), \\
\mathcal{I}_3 &= -\frac{1}{2}\left(\frac{tt_+}{uu_+} + \left(\frac{t + t_+}{2}\frac{u_+ - u}{h_+} - \frac{u + u_+}{2}\right)^2\right).
\end{aligned}
$$

In the discrete case the integrals $\mathcal{I}_1$, $\mathcal{I}_2$ and $\mathcal{I}_3$ are functionally independent. Relation (1.87) no longer holds and instead we have

$$4\mathcal{I}_1 \mathcal{I}_3 - \mathcal{I}_2^2 = 1 - \frac{1}{4}\left(\frac{h_+}{uu_+}\right)^2. \tag{1.90}$$

In order to integrate the discrete system (1.89) we use all three first integrals. Setting

$$\mathcal{I}_1 = A/2, \qquad \mathcal{I}_2 = B, \qquad \frac{h_+}{uu_+} = \varepsilon,$$

we can eliminate t_+ and u_+ and obtain the solution

$$Au^2 + (At - B)^2 + 1 = \frac{\varepsilon^2}{4}. \tag{1.91}$$

This solution agrees with the solution of the underlying differential equation, given by (1.88), up to order ε^2. Complete integration of the discrete equations (1.89), where the positions of the lattice points are also obtained, can be found in [14].

Hamiltonian framework

Let us transfer the differential equation (1.86) into the Hamiltonian form. We change variables

$$q = u, \qquad p = \frac{\partial L}{\partial \dot{u}} = \dot{u}.$$

The corresponding Hamiltonian is

$$H(t, q, p) = \dot{u}\frac{\partial L}{\partial \dot{u}} - L = \frac{1}{2}\left(p^2 + \frac{1}{q^2}\right).$$

Thus, we get the Hamiltonian equations

$$\dot{q} = p, \qquad \dot{p} = \frac{1}{q^3}, \qquad\qquad (1.92)$$

which admit symmetries

$$X_1 = \frac{\partial}{\partial t}, \qquad X_2 = 2t\frac{\partial}{\partial t} + q\frac{\partial}{\partial q} - p\frac{\partial}{\partial p}, \qquad X_3 = t^2\frac{\partial}{\partial t} + tq\frac{\partial}{\partial q} + (q - tp)\frac{\partial}{\partial p}.$$

We check invariance of H in accordance with Theorem 1.15 and find that condition (1.38) is satisfied for the operators X_1 and X_2. Using Theorem 1.16, we calculate the corresponding first integrals

$$J_1 = -H = -\frac{1}{2}\left(p^2 + \frac{1}{q^2}\right), \qquad J_2 = pq - t\left(p^2 + \frac{1}{q^2}\right).$$

For the third symmetry operator the Hamiltonian is divergence invariant with $V_3 = q^2/2$. In accordance with Remark 1.17, it yields the following conserved quantity

$$J_3 = -\frac{1}{2}\left(\frac{t^2}{q^2} + (q - tp)^2\right).$$

Note that no integration is needed to provide solutions of (1.92). As we noticed before in the Lagrangian case, only two first integrals are functionally independent. They are connected by the relation

$$4J_1J_3 - J_2^2 = 1. \qquad\qquad (1.93)$$

Putting $J_1 = A/2$ and $J_2 = B$, we find the solution as

$$Aq^2 + (At - B)^2 + 1 = 0, \qquad p = \frac{B - At}{q}. \qquad\qquad (1.94)$$

Let us consider a discrete model for (1.92). We choose a discretization

$$\frac{q_+ - q}{h_+} = \frac{qp + q_+p_+}{q + q_+}, \qquad \frac{p_+ - p}{h_+} = \frac{1}{q^2q_+^2}\frac{q + q_+}{2}, \qquad\qquad (1.95)$$

which is invariant with respect to the three symmetry operators. The discretization will be considered on the invariant lattice

$$\frac{h_+}{qq_+} = \frac{h_-}{qq_-}. \tag{1.96}$$

This discrete system can be found with the help of the method of finite-difference invariants [8].

The difference equations (1.95) can be rewritten as

$$\frac{q_+ - q}{h_+} = p_+ - \frac{h_+}{2q\tilde{q}^2}, \qquad \frac{p_+ - p}{h_+} = \frac{1}{q^2\tilde{q}^2}\frac{q+\tilde{q}}{2}, \tag{1.97}$$

where $\tilde{q} - \tilde{q}(h_+, q, p_+)$ is the solution of the cubic equation

$$\tilde{q}^3 - (q + h_+ p_+)\tilde{q}^2 + \frac{h_+^2}{2q} = 0 \tag{1.98}$$

expressed in terms of the equation parameters h_+, q and p_+ (the expression for $\tilde{q}$ can be written down explicitly; we do not provide it here because of the size of the expression). These equations are generated by the Hamiltonian function

$$\mathcal{H}(t, t_+, q, p_+) = \frac{1}{2}\left(\left(p_+ - \frac{h_+}{2q\tilde{q}^2}\right)^2 + \frac{2\tilde{q} - q}{q\tilde{q}^2}\right), \qquad \tilde{q} = \tilde{q}(h_+, q, p_+).$$

The last difference Hamiltonian equation of (1.55) takes the form

$$-\frac{1}{2}\left(\left(p_+ - \frac{h_+}{2q\tilde{q}^2}\right)^2 + \frac{1}{q\tilde{q}}\right) + \frac{1}{2}\left(\left(p - \frac{h_-}{2\tilde{q}_-^2 q_-}\right)^2 + \frac{1}{\tilde{q}_- q_-}\right) = 0, \tag{1.99}$$

where $\tilde{q}_- = \tilde{q}(h_-, q_-, p)$ solves the equation

$$\tilde{q}_-^3 - (q_- + h_- p)\tilde{q}_-^2 + \frac{h_-^2}{2q_-} = 0, \tag{1.100}$$

which is (1.98) shifted to the left. It can be shown that lattice equation (1.99), generated by the Hamiltonian, is equivalent to lattice equation (1.96) on the solutions of (1.97).

The Hamiltonian function is invariant with respect to symmetry operators X_1 and X_2. For symmetry X_3 we have divergence invariance with $V_3 = q^2/2$. Thus, these symmetries yield three first integrals

$$\mathcal{J}_1 = -\frac{1}{2}\left(\left(p - \frac{h_-}{2q^2 q_-}\right)^2 + \frac{1}{qq_-}\right),$$

$$\mathcal{J}_2 = qp - t\left(\left(p - \frac{h_-}{2q^2 q_-}\right)^2 + \frac{1}{qq_-}\right),$$

$$\mathcal{J}_3 = tqp - \frac{t^2}{2}\left(\left(p - \frac{h_-}{2q^2 q_-}\right)^2 + \frac{1}{qq_-}\right) - \frac{q^2}{2}.$$

Note that on the solutions of (1.95) we have the relation

$$4\mathcal{J}_1\mathcal{J}_3 - \mathcal{J}_2^2 = 1 - \frac{1}{4}\left(\frac{h_-}{qq_-}\right)^2 \tag{1.101}$$

that explains the choice of the lattice (1.96). In contrast to the differential case (see relation (1.93)) the first integrals $\mathcal{J}_1$, $\mathcal{J}_2$ and $\mathcal{J}_3$ are independent.

In order to integrate the discrete system (1.95), (1.96) we use all three first integrals. Fixing values of first integrals

$$\mathcal{J}_1 = A/2, \qquad \mathcal{J}_2 = B, \qquad \frac{h_-}{qq_-} = \varepsilon,$$

we obtain the solution

$$Aq^2 + (At - B)^2 + 1 = \frac{\varepsilon^2}{4}, \qquad p = \frac{B - At}{q}. \tag{1.102}$$

The solution agrees with the solution of the underlying differential equation, given by (1.94), up to order ε^2.

1.6.2 A nonlinear ODE

The equation

$$\ddot{u} = \dot{u}^{3/2} \tag{1.103}$$

is invariant with respect to symmetry operators

$$X_1 = \frac{\partial}{\partial t}, \qquad X_2 = \frac{\partial}{\partial u}, \qquad X_3 = t\frac{\partial}{\partial t} - u\frac{\partial}{\partial u}.$$

Lagrangian framework

This equation can be obtained by the usual variational procedure from the Lagrangian

$$L(t, u, \dot{u}) = -4\sqrt{\dot{u}} + u,$$

which is invariant for operators X_1 and X_3 and divergence invariant with $V_2 = t$ for operator X_2. Thus, we obtain three first integrals

$$I_1 = -2\sqrt{\dot{u}} + u, \qquad I_2 = -\frac{2}{\sqrt{\dot{u}}} - t, \qquad I_3 = \frac{2}{\sqrt{\dot{u}}}(u - t\dot{u}) + tu.$$

There is the following relation:

$$4 - I_1 I_2 - I_3 = 0. \tag{1.104}$$

Thus, the integral I_3 is not independent and is of no use in the present context. Setting $I_1 = A$, $I_2 = B$ and eliminating $\dot{u}$, we find the general solution as

$$u = A - \frac{4}{t + B}. \tag{1.105}$$

Let us turn to the discrete case. We choose the Lagrangian

$$\mathcal{L}(t, t_+, u, u_+) = -4\sqrt{\frac{u_+ - u}{h_+}} + \frac{u + u_+}{2}.$$

It is invariant for operators X_1 and X_3 and divergence invariant for X_2:

$$X_1\mathcal{L} + \mathcal{L}\underset{+h}{D}(\xi_1) = 0, \qquad X_3\mathcal{L} + \mathcal{L}\underset{+h}{D}(\xi_3) = 0,$$

$$X_2\mathcal{L} + \mathcal{L}\underset{+h}{D}(\xi_2) = 1 = \underset{+h}{D}(t).$$

From the Lagrangian we obtain the following global extremal equations

$$\frac{\delta\mathcal{L}}{\delta u}: \qquad -\frac{4}{h_- + h_+}\left(\sqrt{\frac{h_+}{u_+ - u}} - \sqrt{\frac{h_-}{u - u_-}}\right) = 1,$$

$$\frac{\delta\mathcal{L}}{\delta t}: \qquad 4\left(\sqrt{\frac{u_+ - u}{h_1}} - \sqrt{\frac{u - u_-}{h}}\right) - \frac{u + u_+}{2} + \frac{u_- + u}{2} = 0. \tag{1.106}$$

This system of equations is invariant with respect to all three symmetries.

The application of the difference analog of the Noether theorem gives us three first integrals:

$$I_1 = -2\sqrt{\frac{u - u_-}{h_-}} + \frac{u_- + u}{2}, \qquad I_2 = -2\sqrt{\frac{h_-}{u - u_-}} - \frac{t_- + t}{2},$$

$$I_3 = \sqrt{\frac{h_-}{u - u_-}}\left(u_- + u - (t_- + t)\frac{u - u_-}{h_-}\right) + \frac{tu_- + ut_-}{2}.$$

In contrast to the continuous case the three difference first integrals I_1, I_2 and I_3 are functionally independent and instead of relation (1.104) we have

$$4 - I_1 I_2 - I_3 = \frac{1}{4}h_-^2\frac{u - u_-}{h_-}. \tag{1.107}$$

This coincides with relation (1.104) in the continuous limit $h_- \to 0$. We see that the expression $h_- \sqrt{(u - u_-)/h_-}$ is also a first integral of (1.106). Fixing values of first integrals

$$I_1 = A, \qquad I_2 = B, \qquad \frac{1}{2}h_-\sqrt{\frac{u - u_-}{h_-}} = \varepsilon,$$

where the last equation specifies a lattice, we find the solution of the discrete scheme as

$$u = A - \frac{4}{t + B}\left(1 - \frac{\varepsilon^2}{4}\right).$$ (1.108)

It agrees with the solution of the underlying differential equation up to $O(\varepsilon^2)$. Integration of the lattice equation can be found in [14].

Hamiltonian framework

The equations

$$\dot{q} = \frac{4}{p^2}, \qquad \dot{p} = 1$$ (1.109)

are generated by the Hamiltonian

$$H(t, q, p) = -\frac{4}{p} - q.$$

These equations admit, in particular, symmetries

$$X_1 = \frac{\partial}{\partial t}, \qquad X_2 = \frac{\partial}{\partial q}, \qquad X_3 = t\frac{\partial}{\partial t} - q\frac{\partial}{\partial q} + p\frac{\partial}{\partial p}.$$

Using the invariance of the Hamiltonian with respect to symmetries X_1 and X_3 as well as the divergence invariance with $V_2 = t$ for X_2, we find three first integrals

$$J_1 = -H = \frac{4}{p} + q, \qquad J_2 = p - t, \qquad J_3 = -qp + t\left(\frac{4}{p} + q\right).$$

As in the Lagrangian framework only two first integrals are independent:

$$4 - J_1 J_2 - J_3 = 0.$$ (1.110)

Putting $J_1 = A$ and $J_2 = B$, we obtain the solution

$$q = A - \frac{4}{t + B}, \qquad p = t + B.$$ (1.111)

Now we discretize (1.109) as

$$\frac{q_+ - q}{h_+} = \frac{4}{(p_+ - h_+/2)(p + h_+/2)}, \qquad \frac{p_+ - p}{h_+} = 1$$ (1.112)

on the lattice

$$\frac{h_+}{p_+ - h_+/2} = \frac{h_-}{p - h_-/2}.$$ (1.113)

This scheme is invariant with respect to all three symmetries.

The difference equations (1.112) can be rewritten as

$$\frac{q_+ - q}{h_+} = \frac{4}{(p_+ - h_+/2)^2}, \qquad \frac{p_+ - p}{h_+} = 1.$$ (1.114)

These equations are generated by the discrete Hamiltonian function

$$\mathcal{H}(t, t_+, q, p_+) = -\frac{4}{p_+ - h_+/2} - q.$$

The last discrete Hamiltonian equation (1.55) is

$$\frac{4p_+}{(p_+ - h_+/2)^2} + q - \frac{4p}{(p - h_-/2)^2} - q_- = 0. \tag{1.115}$$

On the solutions of (1.112) this equation leads to the lattice equation (1.113).

The Hamiltonian function is invariant with respect to symmetry operators X_1 and X_3. For symmetry X_2 we have divergence invariance with $V_2 = t$. Therefore, these symmetries provide us with three first integrals

$$\mathcal{J}_1 = \frac{4p}{(p - h_-/2)^2} + q_-, \qquad \mathcal{J}_2 = p - t, \qquad \mathcal{J}_3 = -qp + t\left(\frac{4p}{(p - h_-/2)^2} + q_-\right).$$

The first integrals satisfy the relation

$$4 - \mathcal{J}_1\mathcal{J}_2 - \mathcal{J}_3 = \left(\frac{h_-}{p - h_-/2}\right)^2 \tag{1.116}$$

on the solutions of the difference equations (1.112) that justifies the lattice (1.113). Setting

$$\mathcal{J}_1 = A, \qquad \mathcal{J}_2 = B, \qquad \frac{h_-}{p - h_-/2} = \varepsilon,$$

we find the solution of the discrete model as

$$q = A - \frac{4}{t + B}\left(1 - \frac{\varepsilon^2}{4}\right), \qquad p = t + B. \tag{1.117}$$

1.6.3 Discrete harmonic oscillator

Let us consider the one-dimensional harmonic oscillator

$$\ddot{u} + u = 0. \tag{1.118}$$

The symmetry group admitted by this equation and the corresponding first integrals can be found, for example, in [26].

Lagrangian framework

As a discretization we consider the scheme

$$\frac{u_+ - 2u + u_-}{h^2} + \frac{u_+ + 2u + u_-}{4} = 0, \qquad h_+ = h_- = h, \tag{1.119}$$

where the second equation states that the discretization of (1.118) is considered on a uniform mesh. It is not difficult to verify that the discrete system (1.119) admits the symmetries generated by the operators

$$X_1 = \sin(\omega t)\frac{\partial}{\partial u}, \qquad X_2 = \cos(\omega t)\frac{\partial}{\partial u}, \qquad X_3 = \frac{\partial}{\partial t}, \qquad X_4 = u\frac{\partial}{\partial u},$$

where

$$\omega = \frac{\arctan(h/2)}{h/2}.$$

The discrete model (1.119) has the discrete Lagrangian

$$\mathcal{L}(t, t_+, u, u_+) = \frac{1}{2}\left(\left(\frac{u_+ - u}{h_+}\right)^2 - \left(\frac{u_+ + u}{2}\right)^2\right).$$

Although the global extremal equations

$$\frac{\delta \mathcal{L}}{\delta u}: \qquad \frac{u_+ - u}{h_+} - \frac{u - u_-}{h_-} + \frac{h_+}{2}\frac{u_+ + u}{2} + \frac{h_-}{2}\frac{u + u_-}{2} = 0,$$

$$\frac{\delta \mathcal{L}}{\delta t}: \qquad \left(\frac{u_+ - u}{h_+}\right)^2 + \left(\frac{u_+ + u}{2}\right)^2 - \left(\frac{u - u_-}{h_-}\right)^2 - \left(\frac{u + u_-}{2}\right)^2 = 0 \tag{1.120}$$

look different from (1.119), the global extremal equations provide $h_+ = \pm h_-$. We take $h_+ = h_-$, then the system of global extremal equations is equivalent to discrete model (1.119).

For symmetries X_1 and X_2 we get the divergence invariance conditions:

$$X_1(\mathcal{L}) + \mathcal{L}\underset{+h}{D}(\xi_1) = \underset{+h}{D}(u\cos(\omega t)),$$

$$X_2(\mathcal{L}) + \mathcal{L}\underset{+h}{D}(\xi_2) = \underset{+h}{D}(-u\sin(\omega t)).$$

Thus, we obtain two corresponding first integrals

$$\mathcal{I}_1 = \frac{u_+ \sin(\omega t) - u\sin(\omega t_+)}{\sin(\omega h)}, \qquad \mathcal{I}_2 = \frac{u_+ \cos(\omega t) - u\cos(\omega t_+)}{\sin(\omega h)}.$$

Symmetry X_3 satisfies the invariance condition

$$X_3(\mathcal{L}) + \mathcal{L}\underset{+h}{D}(\xi_3) = 0$$

and provides us with the first integral

$$\mathcal{I}_3 = -\frac{1}{2}\left(\frac{u_+ - u}{h_+}\right)^2 - \frac{1}{2}\left(\frac{u_+ + u}{2}\right)^2.$$

The operator X_4 is neither a variational nor a divergence symmetry of the Lagrangian and does not generate a first integral.

Using

$$I_1^2 + I_2^2 = \frac{u_+^2 + u^2 - 2u_+u\cos(\omega h)}{\sin^2(\omega h)} = \left(1 + \frac{h_+^2}{4}\right)\left(\left(\frac{u_+ - u}{h_+}\right)^2 + \left(\frac{u_+ + u}{2}\right)^2\right),$$

we see that the first integral I_3 can be taken in an equivalent form

$$\tilde{I}_3 = h_+.$$

The three first integrals I_1, I_2, $\tilde{I}_3$ are sufficient for integration of system (1.119). We obtain the solution

$$u = I_2 \sin(\omega t) - I_1 \cos(\omega t) \tag{1.121}$$

on the regular lattice

$$t_i = t_0 + ih, \qquad i = 0, \pm 1, \pm 2, \ldots, \quad h = \tilde{I}_3. \tag{1.122}$$

Hamiltonian framework

Now we consider the one-dimensional harmonic oscillator in Hamiltonian form:

$$\dot{q} = p, \qquad \dot{p} = -q. \tag{1.123}$$

These equations are generated by the Hamiltonian function

$$H(t, q, p) = \frac{1}{2}(q^2 + p^2).$$

As a discretization of (1.123) we consider the application of the midpoint rule

$$\frac{q_+ - q}{h_+} = \frac{p + p_+}{2}, \qquad \frac{p_+ - p}{h_+} = -\frac{q + q_+}{2} \tag{1.124}$$

on a uniform mesh $h_+ = h_-$. The presented discretization can be rewritten as the following system of equations

$$\underset{+h}{D}(q) = \frac{4}{4 - h_+^2}\left(p_+ + \frac{h_+}{2}q\right), \qquad \underset{+h}{D}(p) = -\frac{4}{4 - h_+^2}\left(q + \frac{h_+}{2}p_+\right), \tag{1.125}$$

$$h_+ = h_- = h.$$

It can be shown that this system is generated by the discrete Hamiltonian function

$$\mathcal{H}(t, t_+, q, p_+) = \frac{2}{4 - h_+^2}(q^2 + p_+^2 + h_+qp_+).$$

Indeed, the first and second equations of (1.55) are exactly the same as those of (1.125). The last equation of (1.55) takes the form

$$-\frac{2(4 + h_+^2)}{(4 - h_+^2)^2}(q^2 + p_+^2) - \frac{16h_+}{(4 - h_+^2)^2}qp_+ + \frac{2(4 + h_-^2)}{(4 - h_-^2)^2}(q_-^2 + p^2) + \frac{16h_-}{(4 - h_-^2)^2}q_-p = 0.$$

Using the first and second equations, we can rewrite it as

$$\left(-\frac{2}{4 + h_+^2} + \frac{2}{4 + h_-^2}\right)(q^2 + p^2) = 0.$$

Therefore, for the case $q^2 + p^2 \neq 0$ this equation can be taken in an equivalent form

$$h_+ = h_-.$$

The system of difference equations (1.125) admits, in particular, the following symmetries

$$X_1 = \sin(\omega t)\frac{\partial}{\partial q} + \cos(\omega t)\frac{\partial}{\partial p}, \qquad X_2 = \cos(\omega t)\frac{\partial}{\partial q} - \sin(\omega t)\frac{\partial}{\partial p},$$

$$X_3 = \frac{\partial}{\partial t}, \qquad X_4 = q\frac{\partial}{\partial q} + p\frac{\partial}{\partial p}, \qquad X_5 = p\frac{\partial}{\partial q} - q\frac{\partial}{\partial p},$$

where

$$\omega = \frac{\arctan(h/2)}{h/2}.$$

For symmetry operators X_1 and X_2 we have the divergence invariance conditions

$$\zeta_{+\atop +h}D(q) + p_{+\atop +h}D(\eta) - X(\mathcal{H}) - \mathcal{H}\underset{+h}{D}(\xi) = \underset{+h}{D}(V)$$

fulfilled on the solutions of (1.125) with functions $V_1 = q\cos(\omega t)$ and $V_2 = -q\sin(\omega t)$, respectively. Therefore, we obtain two corresponding first integrals

$$\mathcal{J}_1 = p\sin(\omega t) - q\cos(\omega t), \qquad \mathcal{J}_2 = p\cos(\omega t) + q\sin(\omega t).$$

The symmetry operator X_3 satisfies the invariance condition

$$\zeta_{+\atop +h}D(q) + p_{+\atop +h}D(\eta) - X(\mathcal{H}) - \mathcal{H}\underset{+h}{D}(\xi) = 0.$$

Thus, we get the first integral

$$\mathcal{J}_3 = -\frac{4}{4 - h_-^2}\left(\frac{4 + h_-^2}{4 - h_-^2}\frac{q_-^2 + p^2}{2} + \frac{4h_-}{4 - h_-^2}q_-p\right).$$

Using the first and second equations of (1.125), we can simplify it as

$$\mathcal{J}_3 = -\frac{4}{4 + h_-^2}\frac{q^2 + p^2}{2}.$$

Since from the first integrals $\mathcal{J}_1$ and $\mathcal{J}_2$ we have the conservation law

$$\mathcal{J}_1^2 + \mathcal{J}_2^2 = q^2 + p^2 = \text{const},$$

it follows that we can take the third first integral equivalently as

$$\tilde{\mathcal{J}}_3 = h_-$$

that allows us to use a regular lattice.

The three first integrals $\mathcal{J}_1, \mathcal{J}_2, \tilde{\mathcal{J}}_3$ are sufficient for integration of system (1.124). We obtain the solution

$$q = \mathcal{J}_2 \sin(\omega t) - \mathcal{J}_1 \cos(\omega t), \qquad p = \mathcal{J}_1 \sin(\omega t) + \mathcal{J}_2 \cos(\omega t) \qquad (1.126)$$

on the lattice

$$t_i = t_0 + ih, \qquad i = 0, \pm 1, \pm 2, \ldots, \qquad h = \tilde{\mathcal{J}}_3. \qquad (1.127)$$

1.6.4 Modified discrete harmonic oscillator (exact scheme)

The solutions of the discrete harmonic oscillator in the Lagrangian case (1.121), (1.122) and in the Hamiltonian case (1.126), (1.127) follow the same trajectory as the solution of the continuous harmonic oscillator, but with a different velocity. These discretization errors can be corrected by time reparametrization. In this case we will get *the exact discretization of the harmonic oscillator*, i.e., a discretization which gives the solution of the underlying ODE.

Lagrangian framework

We consider discrete equations

$$\frac{u_+ - 2u + u_-}{h^2} + \Omega^2 \frac{u_+ + 2u + u_-}{4} = 0, \qquad h_+ = h_- = h, \qquad (1.128)$$

where

$$\Omega = \frac{\tan(h/2)}{h/2} \qquad (1.129)$$

represents scaling of t

$$t \to \frac{\arctan(h/2)}{h/2} t,$$

applied to the previous example (h is the step from (1.119)).

It is easy to check that the discrete model (1.128) admits symmetries

$$X_1 = \sin t \frac{\partial}{\partial u}, \qquad X_2 = \cos t \frac{\partial}{\partial u}, \qquad X_3 = \frac{\partial}{\partial t}, \qquad X_4 = u \frac{\partial}{\partial u},$$

which are exactly the same as for the underlying ODE.

Similarly to the previous example, one can show that (1.128) are generated by the discrete Lagrangian

$$\mathcal{L}(t, t_+, u, u_+) = \frac{1}{2}\left(\frac{1}{\Omega}\left(\frac{u_+ - u}{h_+}\right)^2 - \Omega\left(\frac{u_+ + u}{2}\right)^2\right).$$

For symmetries X_1 and X_2, which satisfy the divergence invariance conditions

$$X_1(\mathcal{L}) + \mathcal{L}\underset{+h}{D}(\xi_1) = \underset{+h}{D}(u\cos t),$$

$$X_2(\mathcal{L}) + \mathcal{L}\underset{+h}{D}(\xi_2) = \underset{+h}{D}(-u\sin t),$$

we obtain two first integrals

$$\mathcal{I}_1 = \frac{u_+ \sin t - u \sin t_+}{\sin h}, \qquad \mathcal{I}_2 = \frac{u_+ \cos t - u \cos t_+}{\sin h}.$$

The symmetry X_3 satisfies the invariance condition

$$X_3(\mathcal{L}) + \mathcal{L}\underset{+h}{D}(\xi_3) = 0$$

and provides the first integral

$$\mathcal{I}_3 = -\frac{1}{2\Omega}\left(\frac{u_+ - u}{h_+}\right)^2 - \frac{\Omega}{2}\left(\frac{u_+ + u}{2}\right)^2.$$

Since

$$\mathcal{I}_1^2 + \mathcal{I}_2^2 = \frac{u_+^2 + u^2 - 2u_+ u \cos h}{\sin^2 h} = \frac{1 + (\Omega h/2)^2}{\Omega}\left(\frac{1}{\Omega}\left(\frac{u_+ - u}{h_+}\right)^2 + \Omega\left(\frac{u_+ + u}{2}\right)^2\right),$$

the first integral $\mathcal{I}_3$ can be chosen equivalently as

$$\tilde{\mathcal{I}}_3 = h_+.$$

There is no conservation property for the symmetry X_4.

The three first integrals $\mathcal{I}_1, \mathcal{I}_2, \tilde{\mathcal{I}}_3$ allow us to integrate the discrete equations. We get the solution

$$u = \mathcal{I}_2 \sin t - \mathcal{I}_1 \cos t \tag{1.130}$$

on the regular lattice

$$t_i = t_0 + ih, \qquad i = 0, \pm 1, \pm 2, \ldots, \qquad h = \tilde{\mathcal{I}}_3. \tag{1.131}$$

This solution is the solution of the harmonic oscillator (1.118).

Alternatively, one can verify that the discrete model (1.128), (1.129) provides an infinite-order approximation for the underlying differential equation by substitution of the Taylor series expressions into the discrete equations. Indeed, to estimate the approximation error we substitute expansions

$$u_+ = u + u'h + \frac{u''}{2!}h^2 + \cdots = \sum_{n=0}^{\infty} \frac{u^{(n)}}{n!}h^n,$$

$$u_- = u - u'h + \frac{u''}{2!}h^2 + \cdots = \sum_{n=0}^{\infty} \frac{u^{(n)}}{n!}(-h)^n$$

into (1.128)

$$\frac{u_+ - 2u + u_-}{h^2} + \Omega^2 \frac{u_+ + 2u + u_-}{4}$$

$$= \left(\frac{1}{h^2} + \frac{\Omega^2}{4}\right)(u_+ + u_-) + \left(-\frac{2}{h^2} + \frac{\Omega^2}{2}\right)u$$

$$= \left(\frac{1}{h^2} + \frac{\tan^2(h/2)}{4(h/2)^2}\right)2\sum_{n=0}^{\infty} \frac{u^{(2n)}}{(2n)!}h^{2n} + \left(-\frac{2}{h^2} + \frac{\tan^2(h/2)}{2(h/2)^2}\right)u$$

$$= \frac{2}{h^2\cos^2(h/2)}\left(\sum_{n=0}^{\infty} \frac{u^{(2n)}}{(2n)!}h^{2n} - (\cos h)u\right).$$

Substituting the relations

$$u'' = -u, \quad u^{(4)} = u, \quad \ldots, \quad u^{(2n)} = (-1)^n u, \quad \ldots$$

from the continuous oscillator equation, we complete the computation

$$\frac{2}{h^2\cos^2(h/2)}\left(\sum_{n=0}^{\infty} \frac{u^{(2n)}}{(2n)!}h^{2n} - (\cos h)u\right)$$

$$= \frac{2}{h^2\cos^2(h/2)}\left(\sum_{n=0}^{\infty} \frac{(-1)^n}{(2n)!}h^{2n} - \cos h\right)u = 0.$$

Hamiltonian framework

Similarly to the Lagrangian case it is possible to reparametrize the discrete harmonic oscillator. We consider the harmonic oscillator (1.123) discretized as

$$\frac{q_+ - q}{h_+} = \Omega\frac{p + p_+}{2}, \qquad \frac{p_+ - p}{h_+} = -\Omega\frac{q + q_+}{2}, \tag{1.132}$$

$$h_+ = h_- = h,$$

where Ω is given by (1.129). Similarly to Section 1.6.3 it can be shown that this discrete model of the harmonic oscillator is generated by the discrete Hamiltonian

$$\mathcal{H}(t, t_+, q, p_+) = \frac{2\Omega}{4 - \Omega^2 h_+^2}(q^2 + p_+^2 + \Omega h_+ q p_+).$$

The system of difference equations (1.132) admits the symmetries

$$X_1 = \sin t \frac{\partial}{\partial q} + \cos t \frac{\partial}{\partial p}, \qquad X_2 = \cos t \frac{\partial}{\partial q} - \sin t \frac{\partial}{\partial p},$$

$$X_3 = \frac{\partial}{\partial t}, \qquad X_4 = q \frac{\partial}{\partial q} + p \frac{\partial}{\partial p}, \qquad X_5 = p \frac{\partial}{\partial q} - q \frac{\partial}{\partial p}.$$

For symmetries X_1 and X_2, which satisfy the divergence invariance condition (1.77) on the solutions of the discrete equations (1.132) with functions $V_1 = q \cos t$ and $V_2 = -q \sin t$, we obtain two first integrals

$$\mathcal{J}_1 = p \sin t - q \cos t, \qquad \mathcal{J}_2 = p \cos t + q \sin t.$$

The operator X_3 satisfies the invariance condition (1.74) and provides us with the first integral $\mathcal{J}_3$, which (similarly to Section 1.6.3) can be taken in an equivalent form

$$\tilde{\mathcal{J}}_3 = h_-.$$

The scheme (1.132) gives the exact solution of the harmonic oscillator, which can be found with the help of first integrals $\mathcal{J}_1$, $\mathcal{J}_2$ and $\tilde{\mathcal{J}}_3$ as

$$q = \mathcal{J}_2 \sin t - \mathcal{J}_1 \cos t, \qquad p = \mathcal{J}_1 \sin t + \mathcal{J}_2 \cos t. \qquad (1.133)$$

This discrete solution is given on the lattice

$$t_i = t_0 + ih, \qquad i = 0, \pm 1, \pm 2, \ldots, \qquad h = \tilde{\mathcal{J}}_3. \qquad (1.134)$$

It is also possible to use Taylor series to check that the discrete model (1.132), (1.129) is an infinite-order approximation of the continuous oscillator equations.

The exact schemes for two- and four-dimensional harmonic oscillators were used in [23] to construct exact schemes for two- and three-dimensional Kepler motion, respectively.

1.7 Conclusion

In this chapter we present methods to find first integrals of discrete Euler–Lagrange (called the global extremal) and discrete Hamiltonian equations. We rewrite the Hamiltonian formalism for continuous equations similarly to the classical Lagrangian approach, i.e., using the variational formulation. In both cases we use the invariance of the action functional to obtain first integrals. The conservation properties of the canonical Hamiltonian equations are based on the newly written identity (called the Hamiltonian identity). This identity can be viewed as a "translation" of the Noether identity into the Hamiltonian framework. The identity makes it possible to establish one-to-one correspondence between invariance of the Hamiltonian and first integrals of the Hamiltonian equations (the strong version of Noether's theorem).

The approach presented in the continuous case was applied to discrete Euler–Lagrange and discrete Hamiltonian equations, which can be obtained by the variational principle from finite-difference functionals. Similarly to the continuous case we related invariance of discrete action functionals to first integrals of the discrete equations. In particular, energy conserving numerical schemes can be obtained as discrete equations generated by a Lagrangian or Hamiltonian function invariant with respect to time translations. The discrete systems also include one additional second-order equation which corresponds to the variation of the action functional with respect to the independent variable. This equation generates the lattice, which is, generally, nonuniform and can depend on the solution.

With the help of new operator identities (Lemmas 1.3, 1.12, 1.20, 1.37) we refine the invariance conditions for Euler–Lagrange and Hamiltonian equations in both the continuous and discrete settings by formulating the necessary and sufficient condition for the invariance of these equations. This shows explicitly the difference between the invariance of Lagrangians or Hamiltonians and the invariance of the corresponding Euler–Lagrange or Hamiltonian equations in both continuous and discrete cases.

We show that variations for discrete Hamiltonian systems, which can change the independent variable and hence a mesh, require a generalization of symplecticity. We establish conservation of symplecticity in the extended phase space, which holds on the solutions of the variational equations for the discrete Hamiltonian equations. We also rewrite this conservation of symplecticity for the discrete Euler–Lagrange equations.

The results presented here can be used to find first integrals of Euler–Lagrange and Hamiltonian equations in both continuous and discrete settings. They also provide guidelines on how to construct conservative finite-difference schemes that are important in numerical implementations.

Acknowledgments

V.D.'s research was sponsored in part by the Russian Fund for Basic Research under the research project no. 09-01-00610-a. The research of R.K. was partly supported by the Norwegian Research Council under SpadeAce contract no. 176891/V30.

References

[1] Abraham, R., and Marsden, J. E. 1978. *Foundations of Mechanics.* 2nd edn. Reading, MA: Benjamin/Cummings.

[2] Ahlbrandt, C. D. 1993. Equivalence of discrete Euler equations and discrete Hamiltonian systems. *J. Math. Anal. Appl.*, **180**(2), 498–517.

[3] Arnol'd, V. I. 1989. *Mathematical Methods of Classical Mechanics.* 2nd edn. Grad. Texts in Math., vol. 60. New York: Springer.

[4] Bessel-Hagen, E. 1921. Über die Erhaltungssätze der Elektrodynamik. *Math. Ann.*, **84**(3-4), 258–276.

[5] Bluman, G. W., and Kumei, S. 1989. *Symmetries and Differential Equations.* Appl. Math. Sci., vol. 81. New York: Springer.

[6] Budd, C., and Dorodnitsyn, V. 2001. Symmetry-adapted moving mesh schemes for the nonlinear Schrödinger equation. *J. Phys. A*, **34**(48), 10387–10400.

[7] Cartan, É. 1922. *Leçons sur les invariants intégraux.* Paris: Hermann.

[8] Dorodnitsyn, V. 1991. Transformation groups in difference spaces. *J. Soviet Math.*, **55**(1), 1490–1517.

[9] Dorodnitsyn, V. 1993. A finite-difference analogue of Noether's theorem. *Dokl. Akad. Nauk*, **328**(6), 678–682. Russian.

[10] Dorodnitsyn, V. 2001a. *The Group Properties of Difference Equations.* Moscow: Fizmatlit. Russian.

[11] Dorodnitsyn, V. 2001b. Noether-type theorems for difference equations. *Appl. Numer. Math.*, **39**(3-4), 307–321.

[12] Dorodnitsyn, V., and Kozlov, R. 2010. Invariance and first integrals of continuous and discrete Hamiltonian equations. *J. Engrg. Math.*, **66**(1-3), 253–270.

[13] Dorodnitsyn, V., Kozlov, R., and Winternitz, P. 2003. Symmetries, Lagrangian formalism and integration of second order ordinary difference equations. *J. Nonlinear Math. Phys.*, **10**(suppl. 2), 41–56.

[14] Dorodnitsyn, V., Kozlov, R., and Winternitz, P. 2004. Continuous symmetries of Lagrangians and exact solutions of discrete equations. *J. Math. Phys.*, **45**(1), 336–359.

[15] Elnatanov, N. A., and Schiff, J. 1996. The Hamilton-Jacobi difference equation. *Funct. Differ. Equ.*, **3**(3-4), 279–286.

[16] Erbe, L. H., and Yan, P. X. 1992. Disconjugacy for linear Hamiltonian difference systems. *J. Math. Anal. Appl.*, **167**(2), 355–367.

[17] Gelfand, I. M., and Fomin, S. V. 1963. *Calculus of Variations.* Englewood Cliffs, NJ: Prentice-Hall Inc.

[18] Goldstein, H. 1980. *Classical Mechanics.* 2nd edn. Addison-Wesley Series in Physics. Reading, MA: Addison-Wesley.

[19] Hairer, E., Lubich, C., and Wanner, G. 2006. *Geometric Numerical Integration.* 2nd edn. Springer Ser. Comput. Math., vol. 31. Berlin: Springer.

[20] Hydon, P. E. 2001. Conservation laws of partial difference equations with two independent variables. *J. Phys. A*, **34**(48), 10347–10355.

[21] Ibragimov, N. H. 1985. *Transformation Groups Applied to Mathematical Physics.* Math. Appl. (Soviet Ser.). Dordrecht: Reidel.

[22] Kane, C., Marsden, J. E., and Ortiz, M. 1999. Symplectic-energy-momentum preserving variational integrators. *J. Math. Phys.*, **40**(7), 3353–3371.

[23] Kozlov, R. 2007. Conservative discretizations of the Kepler motion. *J. Phys. A*, **40**(17), 4529–4539.

[24] Lall, S., and West, M. 2006. Discrete variational Hamiltonian mechanics. *J. Phys. A*, **39**(19), 5509–5519.

[25] Leimkuhler, B., and Reich, S. 2004. *Simulating Hamiltonian Dynamics*. Cambridge Monogr. Appl. Comput. Math., vol. 14. Cambridge: Cambridge Univ. Press.

[26] Lutzky, M. 1978. Symmetry groups and conserved quantities for the harmonic oscillator. *J. Phys. A*, **11**(2), 249–258.

[27] Marsden, J. E., and Ratiu, T. S. 1999. *Introduction to Mechanics and Symmetry*. 2nd edn. Texts Appl. Math., vol. 17. New York: Springer.

[28] Noether, E. 1918. Invariante Variationsprobleme. *Nachr. v. d. Ges. d. Wiss. zu Göttingen*, 235–257.

[29] Olver, P. J. 1993. *Applications of Lie Groups to Differential Equations*. 2nd edn. Grad. Texts in Math., vol. 107. New York: Springer.

[30] Ovsiannikov, L. V. 1982. *Group Analysis of Differential Equations*. New York: Academic Press.

[31] Samarskii, A. A. 2001. *The Theory of Difference Schemes*. Monogr. Textbooks Pure Appl. Math., vol. 240. New York: Marcel Dekker.

[32] Sanz-Serna, J. M., and Calvo, M. P. 1994. *Numerical Hamiltonian Problems*. Appl. Math. Math. Comput., vol. 7. London: Chapman & Hall.

[33] Shi, Y. 2002. Symplectic structure of discrete Hamiltonian systems. *J. Math. Anal. Appl.*, **266**(2), 472–478.

2

Painlevé Equations:
Continuous, Discrete and Ultradiscrete

Basil Grammaticos and Alfred Ramani

Abstract

We present a derivation of the continuous and discrete Painlevé equations and then proceed to establish a parallel between the special properties these equations possess, and which are related to their integrable character. The ultradiscrete forms of Painlevé equations are then derived and we show that their properties follow closely the ones of their continuous and discrete counterparts.

2.1 Introduction

Deriving integrable systems is a (very) delicate business. In the absence of a general, constructive theory the usual approach to discovering new integrable equations is to try to construct specific examples. Sometimes they are suggested by physical models, the KdV equation being the prototype of such a system. Once a sufficient number of examples are obtained one can formulate conjectures and proceed to propose integrability criteria. Painlevé equations are a minor exception to this approach. Their discovery is due to the inspired intuition of Painlevé [23]. He was faced with the problem of defining new functions from the solutions of differential equations, a challenge set by Picard [25], who thought that this would have been impossible for second-order equations. This pessimistic attitude was due to the fact that nonlinear differential equations possess multivaluedness-inducing singularities, the position of which depends on the initial conditions, thus making impossible any uniformisation treatment. The masterful solution of Painlevé was to look only for equations free of these "bad" singularities. This could have led to a trivial set, but as it turned out the opposite was true. Not only did he (with the help of his student, Gambier [6]) find dozens of integrable equations but he discovered new transcendents, the ones known today under Painlevé's name.

Once an integrability domain is well established, one can try to understand the underlying mathematical structures. One can then show that the systems previously obtained, almost by trial and error, can be most elegantly produced by some general theory. This was of course the case for the Painlevé equations. The same occured in an even more spectacular way for the discrete Painlevé equations. While scores of discrete Painlevé equations were obtained and steps towards their understanding in geometrical terms were undertaken, Sakai [33] solved the problem of discrete Painlevé equations classification in a most elegant way. As a bonus he discovered a new type of discrete Painlevé equations, the elliptical ones, which complemented the already known difference and q (multiplicative) ones [29].

In this review we shall present both approaches to Painlevé equations. We shall start with the *top-down* approach, roughly what is sketched in the previous paragraph. Then we shall move to the *bottom-up* approach, namely deriving integrable equations through the use of an integrability detector. Once the workings of the derivation of Painlevé equations, both continuous and discrete, are explained, we shall devote a section to the description of the properties of these equations, insisting on the parallels existing between continuous and discrete systems. A final section will present extensions to the Painlevé equations, in particular to the ultradiscrete domain.

2.2 A rough sketch of the top-down description of the Painlevé equations

In this section we shall sketch what we believe is a most elegant formulation of the Painlevé equations. We shall start from the continuous ones, and our presentation will be largely inspired from the works of Okamoto [21]. Then we shall present Sakai's approach to discrete Painlevé equations, an approach which solved in a definitive way the problem of classification of these discrete integrable systems.

The Hamiltonian formulation of Painlevé equations

In what follows we shall present the Hamiltonian approach to Painlevé equations, as introduced by Malmquist [17] and Bureau [2] and refined by

Okamoto. According to this approach a Painlevé equation for the variable x can be obtained from the Hamiltonian equations of motion:

$$f(t)\frac{dx}{dt} = \frac{\partial H}{\partial p}$$
$$f(t)\frac{dp}{dt} = -\frac{\partial H}{\partial x}$$

(2.1)

where $f(t)$ is 1, t and $t(t-1)$ for the Painlevé equations (I, II, IV), (III, V) and VI respectively. If we start for instance from

$$H = x(x-1)(x-t)p^2 - (\kappa_0(x-1)(x-t) + \kappa_1 x(x-t) + (\theta-1)x(x-1))p + \kappa(x-t)$$ (2.2)

and use (2.1) to eliminate the variable p, we obtain P_{VI} for x. One can, naturally, wonder as to what is obtained if one eliminates x. The answer is: another equation, for p, satisfying the Painlevé property. In the case of (2.2) this equation would be the Fokas–Ablowitz–Bureau equation [3, 4]. Thus system (2.1) can be viewed as defining a Miura transformation connecting these two Painlevé equations. (Sometimes the two equations turn out to be the same, in which case this Miura is just an auto-Bäcklund.)

A most interesting property of the Painlevé equations is that they can be associated to an isomonodromic deformation problem. One starts from a linear equation

$$\frac{d^2 y}{d\zeta^2} + p_1(\zeta; t)\frac{dy}{d\zeta} + p_2(\zeta; t)y = 0$$ (2.3)

where p_1 and p_2 are rational functions of ζ. The precise form of the p_i must of course be obtained for each Painlevé equation. Next one introduces the deformation of (2.3) in the form

$$\frac{dy}{dt} = a_1(\zeta; t)\frac{dy}{d\zeta} + a_2(\zeta; t)y$$ (2.4)

and finds a_1 and a_2. This isomonodromic deformation defines what is more customarily called the Lax pair of the Painlevé equation.

The continuous Painlevé equations, through coalescence of singularities, form a degeneration cascade [12]. Starting from the highest we can, through appropriate limiting processes, obtain the lower ones (after some rescalings and changes of variables):

$$
\begin{array}{ccccc}
P_{VI} & \longrightarrow & P_V & \longrightarrow & P_{IV} \\
\downarrow & & \downarrow & & \\
P_{III} & \longrightarrow & P_{II} & \longrightarrow & P_I
\end{array}
$$

(2.5)

Note that P_{IV} and P_{III} are at the same level since they can both be obtained from P_V. Moreover both P_{IV} and P_{III} degenerate to P_{II} with the appropriate coalescence.

In order to introduce the geometrical description of Painlevé equations we consider their birational transformations. This is better illustrated through a specific example. We start from the Hamiltonian for P_{II}

$$H(\alpha) = \frac{1}{2}p^2 - (x^2 + t/2)p - (\alpha + 1/2)x \tag{2.6}$$

It is straightfoward to show that $\underline{x}$ and $\underline{p}$ defined by the birational transformation

$$\underline{x} = -x + \frac{\alpha - 1/2}{p}, \qquad \underline{p} = -p \tag{2.7}$$

form a solution of P_{II} corresponding to Hamiltonian $H(\alpha - 1)$. Thus (2.7) defines an auto-Bäcklund for P_{II}. It can easily be shown that the group associated with the hamiltonian of P_{II} is generated the transformation $\alpha \to 1 - \alpha$, $\alpha \to -\alpha$. This is a realisation of the affine Weyl group of root system of type A_1. As Okamoto has shown the auto-Bäcklund (Schlesinger) transformations of the remaining Painlevé equations also generate extended affine Weyl groups. He has provided the following correspondence between equations and symmetries: $P_{II} - A_1^{(1)}$, $P_{III} - (2A_1^{(1)})$, $P_{IV} - A_2^{(1)}$, $P_V - A_3^{(1)}$, $P_{VI} - D_4^{(1)}$. (Equation P_I has no parameters and thus no auto-Bäcklund transformation).

Another fundamental notion, introduced by Okamoto [22], was that of the τ-function. The latter was defined through its relation to the Hamiltonian as follows:

$$f(t)\frac{d}{dt}\log\tau = H \tag{2.8}$$

The crucial observation is that, for Painlevé equations, the τ-function thus defined is an entire function on the complex plane of the independent variable. It is interesting to point out that the birational transformations (2.7) introduced for P_{II} can be expressed in terms of the τ-function in a very simple way

$$x = \frac{d}{dt}\log\frac{\tau(\alpha - 1)}{\tau(\alpha)} \tag{2.9}$$

Analogous results hold for the remaining Painlevé equations. As a matter of fact, through the successive application of auto-Bäcklund transformations one can introduce a sequence τ_m of τ-functions, corresponding to a translation in the space of parameters, a τ-sequence in Okamoto's terminology. One fundamental property of these sequences is that they obey the Toda equation:

$$\left(f(t)\frac{d}{dt}\right)^2 \log\tau_m = \frac{\tau_{m-1}\tau_{m+1}}{\tau_m^2} \tag{2.10}$$

which in bilinear formalism can be rewritten

$$(D_z^2 - 2e^{D_m})\tau_m \cdot \tau_m = 0 \qquad (2.11)$$

absorbing the $f(t)$ factor through a new variable z.

The last notion, also introduced by Okamoto, is that of the "space of initial conditions." Here is what is meant by this. The continuous Painlevé equations are second-order differential equations. Thus one would expect the space of their initial conditions to be $\mathbb{C}^2$ since for a given value t_0 of the independent variable the solution is specified by the data of the function and its derivative at this point (with some precautions concerning the points at which the coefficients of the equation become singular). However there exist solutions which diverge at t_0. Thus we must compactify $\mathbb{C}^2$. Once this is done, it may happen that several solutions pass through the point at infinity. We must then separate them. The procedure is through a blowing-up of the space (i.e., through the introduction of local coordinates which make the divergence disappear).

The Sakai theory of the discrete Painlevé equations

Sakai, has tackled the problem of the geometrical description of discrete Painlevé equations and presented a global answer [33]. His approach consisted in studying rational surfaces in connection to extended Weyl groups. Surfaces obtained by successive blow-ups of $\mathbb{P}^2$ or $\mathbb{P}^1 \times \mathbb{P}^1$ have been studied through the connections between Weyl groups and the groups of Cremona isometries on the Picard group of the surfaces. (The Picard group of a rational surface X is the group of isomorphism classes of invertible sheaves on X and it is isomorphic to the group of linear equivalent classes of divisors on X. A Cremona isometry is an isomorphism of the Picard group such that (a) it preserves the intersection number of any pair of divisors, (b) it preserves the canonical divisor K_X and (c) it leaves the set of effective classes of divisors invariant.) In the case where 9 points (for $\mathbb{P}^2$, or 8 points for $\mathbb{P}^1 \times \mathbb{P}^1$) are blown up, if the points are in a generic position, the group of Cremona isometries becomes isomorphic to an extension of the Weyl group of type $E_8^{(1)}$. When the 9 points are not in a generic position, the classification of connections between the group of Cremona isometries and the extended affine Weyl groups was studied in full generality by Sakai. Birational (bi-meromorphic) mappings on $\mathbb{P}^2$ (or $\mathbb{P}^1 \times \mathbb{P}^1$) are obtained by interchanging the procedure of blow-downs. Discrete Painlevé equations are recovered as birational mappings corresponding to translations of affine Weyl groups. We shall not go into the presentation of the work of Sakai (lest we sink into plagiarism). We urge the interested reader to seek out this excellent piece of work and study it carefully.

The net result of the Sakai approach is a complete classification of the d-$\mathbb{P}$'s in terms of affine Weyl groups. Starting from the exceptional Weyl group $E_8^{(1)}$ he obtained the systems corresponding to the degeneracy pattern below:

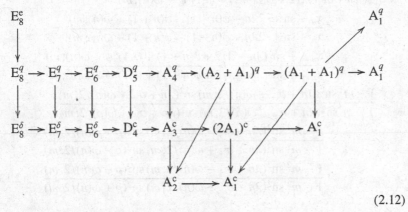

$$(2.12)$$

In this diagram, we assign to a Weyl group an upper index "e" if its supports a discrete equation involving elliptic functions, an upper index q if the equation is of q-type, an upper index δ if it is a difference equation not explicitly related to a continuous equation, and an upper index "c" if it is a difference equation which is explicitly the contiguity relation of one of the (continuous) Painlevé equations, namely P_{VI} for D_4, P_V for A_3, P_{IV} for A_2, (full) P_{III} for $2A_1$ (which means the direct product of twice A_1 in a self-dual way), P_{II} for the A_1 on the last line and finally the one parameter P_{III} for the A_1 on the line above last. Neither P_I nor the zero-parameter P_{III} appear here, since having no parameter they have no contiguity relations, hence no discrete difference equation related to them. We are not going to present here examples of equations for each Weyl group in the Sakai degeneration pattern. The interested reader may find them in other published work of ours [7].

One important finding of Sakai are the equations related to the group $E_8^{(1)}$. The upper index "e" indicates a new kind of discrete $\mathbb{P}$'s: mappings where the independent variable as well as the parameters enter through the arguments of elliptic functions. The construction of Sakai is a global one. If one wishes to construct explicit examples of the equations associated to specific affine Weyl groups one has to specify a periodically repeated nonclosed pattern in the appropriate space and obtain the corresponding d-$\mathbb{P}$. (A consequence of this last statement is that the potential number of d-$\mathbb{P}$'s is infinite since *any* nonclosed periodically repeated pattern in each of the spaces of the affine Weyl groups of the degeneration pattern would lead to a different d-$\mathbb{P}$.) In [20] we have obtained explicit examples of elliptic d-$\mathbb{P}$'s and despite their lengthy expressions we cannot resist the temptation to present one such example here:

$$\frac{x_{n-1} - \operatorname{sn}^2(\lambda n + \lambda/2 + \phi(n-1) - (-1)^n\psi + \omega(n); m)}{x_{n-1} - \operatorname{sn}^2(\lambda n + \lambda/2 + \phi(n-1) - (-1)^n\psi - \omega(n); m)}$$

$$\times \frac{x_n - \operatorname{sn}^2(-2\lambda n - \phi(n-1) - \phi(n+1) + \omega(n); m)}{x_n - \operatorname{sn}^2(-2\lambda n - \phi(n-1) - \phi(n+1) - \omega(n); m)}$$

$$\times \frac{x_{n+1} - \operatorname{sn}^2(\lambda n - \lambda/2 + \phi(n+1) + (-1)^n\psi + \omega(n); m)}{x_{n+1} - \operatorname{sn}^2(\lambda n - \lambda/2 + \phi(n+1) + (-1)^n\psi - \omega(n); m)}$$

$$= \frac{1 - m^2 \operatorname{sn}^2(\lambda n + \sigma_+ + \omega(n)/2; m)\operatorname{sn}^2(\lambda n + \sigma_- + \omega(n)/2; m)}{1 - m^2\operatorname{sn}^2(\lambda n + \sigma_+ - \omega(n)/2; m)\operatorname{sn}^2(\lambda n + \sigma_- - \omega(n)/2; m)}$$

$$\times \frac{1 - m^2 \operatorname{sn}^2(\lambda n + \sigma_+ - \omega(n)/2; m)\operatorname{sn}^2(\rho + \omega(n)/2; m)}{1 - m^2 \operatorname{sn}^2(\lambda n + \sigma_+ + \omega(n)/2; m)\operatorname{sn}^2(\rho - \omega(n)/2; m)}$$

$$\times \frac{1 - m^2 \operatorname{sn}^2(\lambda n + \sigma_- - \omega(n)/2; m)\operatorname{sn}^2(\rho - \omega(n)/2; m)}{1 - m^2 \operatorname{sn}^2(\lambda n + \sigma_- + \omega(n)/2; m)\operatorname{sn}^2(\rho + \omega(n)/2; m)}. \quad (2.13)$$

Here

$$\sigma_\pm = (2\phi(n \mp 1) + \phi(n \pm 1) \mp (-1)^n\psi)/3 \pm \lambda/6,$$

$$\rho = (\phi(n+1) - \phi(n-1) + 2(-1)^n\psi - \lambda)/3,$$

where ψ is a constant, $\phi(n+3) = \phi(n)$, i.e., ϕ has a period three and $\omega(n+4) = \omega(n)$, i.e., ω has a period four so the whole equation has period 12. The total number of degrees of freedom is 8, including the independent variable.

While the Sakai approach may seem somewhat abstract it is quite useful for the understanding of various aspects of d-$\mathbb{P}$'s and discrete systems in general. Sakai himself provided the link between the property of singularity confinement and the construction of the space of initial conditions. He has shown that all d-$\mathbb{P}$'s have a maximum of 8 confined singularities and can be described by a maximum of 8 blow-ups. The procedure of blowing-up at each singularity is the one first advocated by Kruskal [16]. According to Kruskal one must provide a complete description of the dynamics of the mapping and this means lifting the indeterminacies of each of the singularities. This program of complete description of the dynamics of mappings with confined singularities was carried out by Takenawa [35]. He has studied the discrete $\mathbb{P}$'s of the Sakai classification and reconstituted their dynamics through a series of blow-ups and blow-downs. He has used this approach in order to compute the algebraic entropy of these systems and has shown that all these d-$\mathbb{P}$'s have a degree growth that goes like n^2 (a result previously established, albeit in a nonrigorous way, in [19]).

2.3 A succinct presentation of the bottom-up description of the Painlevé equations

In this section we shall present the derivation of Painlevé equations, both continuous and discrete, based on some integrability criterion. In the case of continuous equations the criterion we shall use is the one based on the Painlevé property. For discrete systems two different criteria will be illustrated, singularity confinement and algebraic entropy. For a detailed presentation of all these methods the interested reader is referred to previous, already published, work of ours.

Derivation of continuous Painlevé equations

The starting point for Painlevé's approach was the observation that critical singularities of second-order equations can be branch points, both algebraic and logarithmic, as well as essential singularities. Since the position of these singularities depends on the initial conditions (this is what the term "movable," customarily used with such singularities, means) the uniformisation procedures used for linear equations cannot be applied in the nonlinear case. The solution of Painlevé to this problem was to consider only equations which are devoid of movable critical singularities.

Painlevé introduced his method (known as α-method) that made it possible to test an equation for the existence of such "bad" singularities in the solution. In order to illustrate the Painlevé's α-method we will examine the derivation of the first transcendental equation that bears his name and consider an equation of the form,

$$x'' = 6x^2 + f(t), \tag{2.14}$$

where $f(t)$ is analytic. Painlevé introduces a small parameter α by a scaling, $x = X/\alpha^2, t = t_0 + \alpha T$. We thus find:

$$\frac{d^2 X}{dT^2} = 6X^2 + \alpha^4 f(t_0) + \alpha^5 f'(t_0) + \frac{1}{2}\alpha^6 f''(t_0) + O(\alpha^7) \tag{2.15}$$

and seek a solution in the form of a power series in α,

$$X(T) = X_0(T) + \alpha^4 X_4(T) + \alpha^5 X_5(T) + \alpha^6 X_6(T) + O(\alpha^7). \tag{2.16}$$

(There is no need to introduce terms proportional to $\alpha, \alpha^2, \alpha^3$ [4].) We find,

$$\frac{d^2 X_0}{dT^2} = 6X_0^2 \tag{2.17}$$

and

$$\frac{d^2 X_{r+4}}{dT^2} - 12X_0 X_{r+4} = \frac{T^r}{r!} \frac{d^r f}{dt^r}(t_0), \tag{2.18}$$

for $r = 0, 1, 2$. The general solution of (2.17) is the Weierstrass elliptic function, $X_0 = \wp(T - T_0; 0, h)$, with h and T_0 as constants of integration. Thus the homogeneous part of (2.18) is a Lamé equation,

$$\frac{d^2 Y}{dT^2} - 12\wp(T - T_0; 0, h)Y = 0, \tag{2.19}$$

and its general solution is

$$Y(T) = a\left(T\frac{d\wp}{dT} + 2\wp\right) + b\frac{d\wp}{dT}, \tag{2.20}$$

with a, b integration constants. The solution of the full (2.18) is obtained by the method of variation of parameters,

$$X_{r+4} = U_{r+4}\left(T\frac{d\wp}{dT} + 2\wp\right) + V_{r+4}\frac{d\wp}{dT}, \tag{2.21}$$

and the coefficients U, V are given by

$$\frac{dU_{r+4}}{dT} = \frac{T^r}{24r!} \frac{d^r f}{dt^r}(t_0)\frac{dX_0}{dT} \tag{2.22}$$

$$\frac{dV_{r+4}}{dT} = \frac{T^r}{24r!} \frac{d^r f}{dt^r}(t_0)\left(T\frac{dX_0}{dT} + 2X_0\right). \tag{2.23}$$

Integrating (2.22) and (2.23) we find that U and V are given in terms of elliptic functions for $r = 0, 1$. For $r = 2$, expanding the solution X_0 around the movable singularity at T_0, where $X_0 \sim (T - T_0)^{-2}$, we find that a logarithm appears. For the solution to be free of movable critical points it is necessary for the coefficient of the logarithm to vanish and the explicit calculation leads to

$$\frac{d^2 f}{dt^2}(t_0) = 0. \tag{2.24}$$

Since t_0 is arbitrary, this means that, for integrability, f must be linear in t. Apart from cases that are integrable in terms of elementary functions, one finds the P_I equation,

$$x'' = 6x^2 + t. \tag{2.25}$$

Painlevé also showed that (2.25) is free of movable essential singularities, thus completing the proof that P_I has no movable critical points. In practice, the Painlevé α-method requires the exact solution of a nonlinear ODE as well as that of inhomogeneous linear ODEs with the same homogeneous part and different inhomogeneous parts at each order. Thus, a particular solution is needed at each order for the integration. As a result the whole approach is somewhat cumbersome.

A much simpler approach was introduced by Painlevé's disciple, Gambier, who applied to this problem a variant of method of Kovalevskaya. Let us illustrate the derivation of P_I by Gambier's method (which we should point out here is much closer in spirit to the modern method of singularity analysis known as the ARS algorithm [1]). Starting with (2.14) we look for the dominant behaviour in the neighbourhood of a singularity t_0. We assume that

$$x \sim a\tau^p, \tag{2.26}$$

where $\tau = t - t_0$. Substituting into (2.14) we find $p = -2$ and $a = 1$, corresponding to x'' and x^2 being dominant. Since p is an integer, we can proceed further and look for the second integration constant (t_0 being the first). We look in particular for the power of τ, called the "index" according to Fuchs, or the resonance in the ARS terminology, at which this second constant appears. We introduce

$$x = \tau^{-2} + \gamma\tau^{r-2} \tag{2.27}$$

into the dominant part of (2.14). Linearizing for γ we find that

$$(r-2)(r-3) - 12 = 0, \tag{2.28}$$

with roots $r = -1$, corresponding to the arbitrariness of t_0, and $r = 6$. Since this second resonance is integer we can proceed to a check of compatibility that will guarantee the absence of logarithmic branch points. We expand

$$x = \tau^{-2} \sum_{r=0}^{6} a_r \tau^r \tag{2.29}$$

with $a_0 = 1$. The calculations are straightforward and we find as a condition $d^2 f / dt^2 = 0$, i.e., f must be linear.

Gambier obtained all the equations of the Painlevé type and in particular produced a list of 24 fundamental ones [6]. Here is the Gambier (t) list, where a, b, c, d, e are constants, q, r are free functions of t, and f_n, ϕ_n, ψ_n are three functions of t satisfying one constraint (that depends on n):

$$x'' = 0 \tag{G1}$$

$$x'' = 6x^2 \tag{G2}$$

$$x'' = 6x^2 - \frac{1}{24} \tag{G3}$$

$$x'' = 6x^2 + t \tag{G4}$$

$$x'' = -3xx' - x^3 + q(x' + x^2) \tag{G5}$$

$$x'' = -2xx' + qx' + q'x \tag{G6}$$

$$x'' = 2x^3 \tag{G7}$$

$$x'' = 2x^3 + ax + b \tag{G8}$$

$$x'' = 2x^3 + tx + a \tag{G9}$$

$$x'' = \frac{x'^2}{x} \tag{G10}$$

$$x'' = \frac{x'^2}{x} + ax^3 + bx^2 + c + \frac{d}{x} \tag{G11}$$

$$x'' = \frac{x'^2}{x} - \frac{x'}{t} + \frac{1}{t}(ax^2 + b) + cx^3 + \frac{d}{x} \tag{G12}$$

$$x'' = \frac{x'^2}{x} + q\frac{x'}{x} - q' + rxx' + r'x^2 \tag{G13}$$

$$x'' = \left(1 - \frac{1}{n}\right)\frac{x'^2}{x} + qxx' - \frac{nq^2}{(n+2)^2}x^3 + \frac{nq'}{n+2}x^2 \tag{G14}$$

$$x'' = \left(1 - \frac{1}{n}\right)\frac{x'^2}{x} + \left(f_n x + \phi_n - \frac{n-2}{nx}\right)x' - \frac{nf_n^2}{(n+2)^2}x^3$$
$$+ \frac{n(f_n' - f_n\phi_n)}{n+2}x^2 + \psi_n x - \phi_n - \frac{1}{nx} \tag{G15}$$

$$x'' = \frac{x'^2}{2x} + \frac{3x^3}{2} \tag{G16}$$

$$x'' = \frac{x'^2}{2x} + \frac{3x^3}{2} + 4ax^2 + 2bx - \frac{c^2}{2x} \tag{G17}$$

$$x'' = \frac{x'^2}{2x} + \frac{3x^3}{2} + 4tx^2 + 2(t^2 - a)x - \frac{b^2}{2x} \tag{G18}$$

$$x'' = \frac{x'^2 - 1}{2x} \tag{G19}$$

$$x'' = x'^2\left(\frac{1}{2x} + \frac{1}{x-1}\right) \tag{G20}$$

$$x'' = x'^2\left(\frac{1}{2x} + \frac{1}{x-1}\right) + (x-1)^2\left(ax + \frac{b}{x}\right) + cx + \frac{dx}{x-1} \tag{G21}$$

$$x'' = x'^2(\frac{1}{2x} + \frac{1}{x-1} \frac{)}{-} \frac{x'}{t} + \frac{(x-1)^2}{t^2}\left(ax + \frac{b}{x}\right) + c\frac{x}{t} + \frac{dx(x+1)}{x-1} \tag{G22}$$

$$x'' = \frac{x'^2}{2}\left(\frac{1}{x} + \frac{1}{x-1} + \frac{1}{x-a}\right) \tag{2.30}$$

$$+ x(x-1)(x-a)\left(b + \frac{c}{x^2} + \frac{d}{(x-1)^2} + \frac{e}{(x-a)^2}\right) \tag{G23}$$

$$x'' = \frac{x'^2}{2}\left(\frac{1}{x} + \frac{1}{x-1} + \frac{1}{x-t}\right) - x'\left(\frac{1}{t} + \frac{1}{t-1} + \frac{1}{x-t}\right)$$
$$+ \frac{x(x-1)(x-t)}{2t^2(t-1)^2}\left(a - \frac{bt}{x^2} + c\frac{t-1}{(x-1)^2} + \frac{(d-1)t(t-1)}{(x-t)^2}\right) \quad \text{(G24)}$$

If one knows the solution of these 24 equations, then one can construct the solution of any other equation of the Painlevé type at order two.

Derivation of discrete Painlevé equations

In order to obtain discrete Painlevé equations through the application of the singularity confinement method we start from the general QRT [26] mapping:

$$x_{n+1} = \frac{f_1(x_n) - x_{n-1}f_2(x_n)}{f_2(x_n) - x_{n-1}f_3(x_n)} \quad (2.31)$$

The rationale of this approach is that since the continuous Painlevé equations are nonautonomous extensions of the elliptic functions one would expect the discrete Painlevé equations to be obtained by deautonomizing the QRT mapping (the solutions of which are elliptic functions).

In order to gain some insight into the choice of the f_i's we rewrite the QRT map as:

$$f_3(x_n)\Pi - f_2(x_n)\Sigma + f_1(x_n) = 0 \quad (2.32)$$

where $\Sigma = x_{n+1} + x_{n-1}$, $\Pi = x_{n+1}x_{n-1}$, the f_i are quartic polynomials and ask that this equation go over to the continuous Painlevé under consideration at the continuous limit. We introduce a lattice parameter δ and obtain:

$$\Sigma = 2x + \delta^2 x'' + O(\delta^4)$$
$$\Pi = x^2 + \delta^2(xx'' - x'^2) + O(\delta^4) \quad (2.33)$$

and when we extract from (2.33) the part involving derivatives, we obtain a continuous limit (as δ goes to zero) of the form:

$$x'' = \frac{f_3(x)}{xf_3(x) - f_2(x)}x'^2 + g(x) \quad (2.34)$$

If we are looking for a specific Painlevé equation we must first choose f_2, f_3 so as to get $f_3(x)/(xf_3(x) - f_2(x))$ to coincide with the factor multiplying x'^2 in that equation.

For P_I and P_{II} we have, clearly, $f_3 = 0$. We are not going to present the derivation for those two d-$\mathbb{P}$'s but illustrate the application of singularity confinement [8] to the derivation of the discrete Painlevé III. In this case we have

$x'' = x'^2/x + g(x)$. Moreover we introduce the the transformation $z \to e^z$ of the continuous variable and work with the more convenient form:

$$w'' = \frac{w'^2}{w} + e^z(aw^2 + b) + e^{2z}\left(cw^3 + \frac{d}{w}\right) \tag{2.35}$$

The leading terms of this form agree with (2.34) with $f_2 = 0$. In that case, the mapping takes the form:

$$x_{n+1}x_{n-1} = \frac{\kappa(n)x_n^2 + \zeta(n)x_n + \mu(n)}{x_n^2 + \beta(n)x_n + \gamma(n)} \tag{2.36}$$

In order to fix the n-dependent coefficients we will study the singularity behaviour as described before. When one solves for x_{n+1} there are two possible sources of singularity for this mapping. Either x_n is a zero of the denominator or x_{n-1} becomes zero. In the first case, the singularity sequence is the following: x_{n+1} diverges, x_{n+2} has a finite value $\kappa(n + 1)/x_n$ and x_{n+3} would in principle be proportional to $1/x_{n+1}$ and thus zero. This would lead to a new divergence. The only way out is to ask that x_{n+2} also be a zero of the appropriate denominator, so that x_{n+3} does not vanish. Expressing x_{n+2} in terms of x_n and taking into account that this must be true for both zeros of x_n, we obtain $\beta(n) = \beta(n+2)\kappa(n+1)/\gamma(n+2)$ and $\gamma(n) = \kappa^2(n+1)/\gamma(n+2)$. Multiplying x_n by an arbitrary function of n does not change the form of (2.21) but only affects the coefficients. This scaling freedom allows us to take a constant value β for $\beta(n)$, resulting to $\kappa(n+1) = \gamma(n+2)$, $\gamma(n) = \gamma(n+2)$. Thus the γ's and κ's must be constants within a given parity: $\gamma(even) = \kappa(odd) = \gamma_+$, $\gamma(odd) = \kappa(even) = \gamma_-$. In order to study the second kind of singularity, we start with x_n such that x_{n+1} vanishes (i.e., $\kappa(n)x_n^2 + \zeta(n)x_n + \mu(n) = 0$). We find then that x_{n+2} has a finite value $\mu(n + 1)/(\gamma(n + 1)x_n)$ and this would lead to a divergent x_{n+3} unless the numerator also vanishes. Substituting the expression for x_{n+2} and using the fact that again this must be true for both zeros of $\kappa(n)x_n^2 + \zeta(n)x_n + \mu(n)$, we obtain $\mu(n) = \zeta(n)\mu(n + 1)/\zeta(n + 2) = \mu^2(n + 1)/\mu(n + 2)$. The solution to these equations is straightforward: $\mu(n) = \mu_0\lambda^{2n}$ and $\zeta(n) = \zeta_{0,\pm}\lambda^n$, where μ_0, $\zeta_{0,\pm}$, are constants, the $\pm$ sign being related to the parity of n. Note that, in that case, there is no second kind of singularity at all! Indeed x_{n+3} is not allowed to diverge even though $x_{n+1} = 0$. (This is reminiscent of the case of continuous equations where, if a denominator appears, one must consider the values of the dependent variable that makes this denominator vanish to ascertain that this does not generate a singularity.) Neglecting the distinction between even and odd we can rewrite d-P$_{\mathrm{III}}$, after a change of the variable as

$$x_{n+1}x_{n-1} = \frac{cd(x_n - aq_n)(x_n - bq_n)}{(x_n - c)(x_n - d)} \tag{2.37}$$

where $q_n = \lambda^n$. Thus the discrete P_{III} just obtained is in fact a multiplicative, q-discrete, one. (Including the even-odd degree of freedom produces a q-discrete form of P_{VI} as shown by Jimbo and Sakai [14].)

We turn now to the algebraic entropy [11], slow growth, approach. Let us start with a simple case. We consider the mapping:

$$x_{n+1} + x_{n-1} = \frac{ax_n + b}{x_n^2} \tag{2.38}$$

where a and b are constants. We start by computing the degrees of the iterates and find, (with $x_0 = p$, $x_1 = q/r$), that the common degree of homogeneity in q and r of the numerator and denominator of the iterates is $0, 1, 2, 5, 8, 13, 18, 25, 32, 41, \ldots$: clearly a polynomial growth. Next we turn to the deautonomisation of the mapping. The singularity confinement result is that a and b must satisfy the conditions $a_{n+1} - 2a_n + a_{n-1} = 0$, $b_{n+1} = b_{n-1}$, i.e., a is linear in n while b is a constant with an even/odd dependence. Assuming now that a and b are arbitrary functions of n we compute the degrees of the iterates of (2). We obtain successively $0, 1, 2, 5, 10, 21, 42, 85, \ldots$. The growth is now exponential, the degrees behaving like $d_{2m-1} = (2^{2m} - 1)/3$ and $d_{2m} = 2d_{2m-1}$, a clear indication that the mapping is not integrable in general. Already at the fourth iteration the degrees differ in the autonomous and nonautonomous cases. Our approach consists in requiring that the degree in the nonautonomous case be *identical* to the one obtained in the autonomous one. If we implement the requirement that d_4 be 8 instead of 10 we find two conditions $a_{n+1} - 2a_n + a_{n-1} = 0$, $b_{n+1} = b_{n-1}$, i.e., precisely the ones obtained through singularity confinement. Moreover, once these two conditions are satisfied, the subsequent degrees of the nonautonomous case coincide with that of the autonomous one. Thus this mapping, leading to polynomial growth, should be integrable, and, in fact, it is. Equation (2.38) with $a(n) = \alpha n + \beta$ and b constant (the even-odd dependence can be gauged out by a parity-dependent rescaling of the variable x) is a discrete form of the Painlevé I equation. In the examples that follow, we shall show that in all cases the nonautonomous form of an integrable mapping obtained through singularity confinement leads to exactly the same degrees of the iterates as the autonomous one.

A second example can be presented using the multiplicative mapping:

$$x_{n+1}x_{n-1} = \frac{a_n x_n + b}{x_n^2} \tag{2.39}$$

where one can put $b = 1$ through an appropriate gauge. In the autonomous case we obtain, starting with $x_0 = p$ and $x_1 = q/r$, successively the degrees: $0, 1, 2, 3, 4, 7, 10, 13, 16, 21, 26, \ldots$, i.e., again a quadratic growth. In fact, if n is of the form $4m+k$, ($k = 0, 1, 2, 3$) the degree is given by $d_n = 4m^2 + (2m+1)k$.

The deautonomisation of (2.39) is straightforward. We compute the successive degrees in the generic case and find: $0, 1, 2, 3, 4, 7, 11, \ldots$. At this stage we require that a factorization occur in order to bring the degree d_6 from 11 to 10. The condition for this is $a_{n+2}a_{n-2} = a_n^2$, i.e., a of the form $a_{e,o}\lambda_{e,o}^n$ with an even-odd dependence which can be easily gauged away. This condition is sufficient in order to bring the degrees of the successive iterates down to the values obtained in the autonomous case. Quite expectedly the condition on a is *precisely* the one obtained by singularity confinement.

2.4 Properties of the, continuous and discrete, Painlevé equations: a parallel presentation

The Painlevé equations possess a host of properties which make them unique. Moreover the properties of the discrete Painlevé equations are in perfect analogy with those of the continuous ones. In what follows we present a (non exhaustive) list of these properties.

2.4.1 Degeneration cascade

As explained in Section 2.2 the continuous Painlevé equations form a degeneration cascade, through coalescence of singularities [28]. We can illustrate this process here considering the degeneration of both P_{IV} and P_{III} to P_{II}.

In fact, starting form P_{IV}

$$X'' = \frac{X'^2}{2X} + \frac{3X^3}{2} + 4TX^2 + 2(T^2 - A)x - \frac{B}{X} \tag{2.40}$$

and putting $X = 2/(ep^3) + x/\epsilon$, $T = -2/(ep^3) + t\epsilon$, $A = -2/(ep^6) + \mu$ and $B = 8/\epsilon^{12}$ we obtain, at the limit $\epsilon \to 0$ the Painlevé II equation

$$x'' = 2x^3 + 8tx + 4\mu \tag{2.41}$$

in a slightly noncanonical form.

Similarly starting from P_{III}

$$X'' = \frac{X'^2}{X} - \frac{X'}{T} + \frac{1}{T}(AX^2 + B) + CX^3 + \frac{D}{X} \tag{2.42}$$

we put $X = 1 + 2\epsilon x$, $T = 1 + \epsilon^2 t$, $A = -1/(2\epsilon^6)$, $B = -A + 2\mu/\epsilon^3$, $C = 1/(\epsilon^6)$, $D = -C$ and obtain again, at the limit $\epsilon \to 0$ the Painlevē II equation for x as a function of t

$$x'' = 2x^3 + tx + \mu \tag{2.43}$$

in canonical form.

The degeneration cascade for discrete Painlevé equations is much more complicated than that of the continuous ones, as explained in Section 2.2. Below we shall illustrate such a degeneration by working out in full detail the case d-P$_{\mathrm{II}}$ → d-P$_{\mathrm{I}}$. We start with the equation

$$X_{n+1} + X_{n-1} = \frac{ZX_n + A}{1 - X_n^2} \tag{2.44}$$

We put $X = 1 + \delta x$, whereupon the equation becomes

$$4 + 2\delta(x_{n+1} + x_{n-1} + x_n) = -\frac{Z(1 + \delta x_n) + A}{\delta x_n} \tag{2.45}$$

Now, clearly, Z must cancel A up to order δ and this suggests the ansatz $Z = -A - 2\delta^2 z$. Moreover, the $O(\delta^0)$ term in the right-hand side must cancel the 4 of the left-hand side and we are thus led to $A = 4 + 2\delta a$. Using these values of Z and A we find (at $\delta \to 0$),

$$x_{n+1} + x_{n-1} + x_n = \frac{z}{x_n} + a \tag{2.46}$$

i.e., precisely d-P$_{\mathrm{I}}$.

Mapping (2.46) is not the only coalescence limit of d-P$_{\mathrm{II}}$. Putting $X = x/\delta$, $Z = -z/\delta^2$, and $c = -\gamma/\delta^3$ we recover an alternate d-P$_{\mathrm{I}}$ at the limit $\delta \to 0$,

$$x_{n+1} + x_{n-1} = \frac{\gamma}{x_n^2} + \frac{z}{x_n} \tag{2.47}$$

On the other hand the equation usually called alternate d-P$_{\mathrm{I}}$ (2.47) does not belong to the same cascade but comes from an alternate d-P$_{\mathrm{II}}$,

$$\frac{z_{n+1}}{x_{n+1} x_n + 1} + \frac{z_n}{x_n x_{n-1} + 1} = -x_n + \frac{1}{x_n} + z_n + \mu \tag{2.48}$$

just as predicted by the Sakai theory.

2.4.2 Lax pairs

As we explained in Section 2.2 the Painlevé equations can be obtained from the compatibility of a linear system of PDEs. Lax pairs are known for all six Painlevé equations. They have the general form:

$$\psi_\zeta = A\psi \tag{2.49}$$

$$\psi_t = B\psi \tag{2.50}$$

where ζ is the spectral parameter and A, B are matrices depending explicitly on ζ and the dependent as well as the independent variables w and t. The continuous Painlevé equation is obtained from the compatibility condition $\psi_{\zeta t} = \psi_{t \zeta}$ leading to:

$$A_t - B_\zeta + AB - BA = 0 \tag{2.51}$$

We illustrate this in the case of the P_{IV} equation. Its Lax pair is [24]:

$$A = \zeta \begin{pmatrix} 1 & 0 \\ 0 & -1 \end{pmatrix} + \begin{pmatrix} t & u \\ 2(v - \theta_0 - \theta_\infty)/u & -t \end{pmatrix}$$

$$+ \zeta^{-1} \begin{pmatrix} \theta_0 - v & -uw/2 \\ 2v(v - 2\theta_0)/uw & -(\theta_0 - v) \end{pmatrix} \tag{2.52}$$

$$B = \zeta \begin{pmatrix} 1 & 0 \\ 0 & -1 \end{pmatrix} + \begin{pmatrix} 0 & u \\ 2(v - \theta_0 - \theta_\infty)/u & 0 \end{pmatrix} \tag{2.53}$$

The compatibility leads to:

$$\frac{dw}{dt} = -4v + w^2 + 2tw + 4\theta_0 \tag{2.54}$$

$$\frac{du}{dt} = -u(w + 2t) \tag{2.55}$$

$$\frac{dv}{dt} = -\frac{2v^2}{w} + \left(\frac{4\theta_0}{w} - w \right)v + (\theta_0 + \theta_\infty)w \tag{2.56}$$

which results to P_{IV}:

$$\frac{d^2 w}{dt^2} = \frac{1}{2w} \left(\frac{dw}{dt} \right)^2 + \frac{3}{2} w^3 + 4tw^2 + 2(t^2 + a)w + \frac{b}{w} \tag{2.57}$$

The parameters a, b are related to the monodromy exponents θ_0, θ_∞ through:

$$a = 1 - 2\theta_\infty, \qquad b = -8\theta_0^2 \tag{2.58}$$

Similar results have been obtained for all the Painlevé equations. For the discrete Painlevé equations not all Lax pairs have been obtained yet. However there exists a class for which the Lax pair can be systematically constructed: it comprises those discrete Painlevé equations which are contiguities of continuous ones.

In general, the Painlevé equation depends on parameters $(\alpha, \beta, \dots)$ which are associated to the monodromy exponents θ_i appearing explicitly in the Lax pair. The Schlesinger transform relates two solutions Ψ and Ψ' of the isomonodromy problem for the equation at hand corresponding to different sets of parameters $(\alpha, \beta, \dots)$ and $(\alpha', \beta', \dots)$. The main characteristic of these transforms is that the monodromy exponents (at the singularities of the associated linear problem), related to the sets $(\alpha, \beta, \dots)$ and $(\alpha', \beta', \dots)$ differ by integers (or half-integers). The general form of a Schlesinger transformation is:

$$\Psi' = R\Psi \tag{2.59}$$

where R is again a matrix depending on ζ, w, z and the monodromy exponents θ_i. The important remark is that (2.49) together with (2.59) constitute *the Lax pair of a discrete equation*. The latter is obtained from the compatibility conditions:

$$R_\zeta + RA - A'R = 0 \tag{2.60}$$

A most interesting result is the isospectral problem associated to q-P$_{\text{III}}$ [24]. Here a q-difference scheme is necessary instead of a differential one,

$$\Phi_n(q\zeta) = L_n(\zeta)\Phi_n(\zeta)$$
$$\Phi_{n+1}(\zeta) = M_n(\zeta)\Phi_n(\zeta), \tag{2.61}$$

leading to

$$M_n(q\zeta)L_n(\zeta) = L_{n+1}(\zeta)M_n(\zeta). \tag{2.62}$$

The resulting Lax pair is written in terms of 4×4 matrices.

2.4.3 Miura and Bäcklund relations

In Section 2.2 we mentioned relations of the solutions of Painlevé equations which either connect two different equations or the same equation for different values of the parameters. In the first case we have a Miura transformation while in the second one we have an auto-Bäcklund (or Schlesinger) transformation.

The (probably) best known Miura among the Painlevé equation is the one relating Painlevé II to the 34th equation in the canonical list of 50 established by Gambier and which is known as P$_{34}$. The starting point is the Miura pair [27]

$$\alpha w = u' + u^2 + t/2$$
$$u = \frac{w' + 1}{2w} \tag{2.63}$$

Eliminating w we obtain the P$_{\text{II}}$ equation

$$u'' = 2u^3 + tu - (\alpha + \tfrac{1}{2}) \tag{2.64}$$

while eliminating u we find P$_{34}$

$$w'' = \frac{w'^2}{2w} + 2\alpha w^2 - tw - \frac{1}{2w} \tag{2.65}$$

Miura transformations are also the starting point for the derivation of the auto-Bäcklund transformations of the Painlevé equations.

We illustrate the auto-Bäcklund transformations by presenting the one of Painlevé V:

$$v'' = \left(\frac{1}{2v} + \frac{1}{v-1}\right)v'^2 - \frac{v'}{z} + \frac{(v-1)^2}{z^2}\left(\alpha v + \frac{\beta}{v}\right) + \frac{\gamma v}{z} + \frac{\delta v(v+1)}{v-1} \qquad (2.66)$$

Before giving the auto-Bäcklund transformations we introduce a new, more convenient parametrisation. First through the appropriate scaling of the independent variable z we put $2\delta = -1$. We write $2\alpha = (n-p)^2$, $2\beta = -(n+p)^2$, $\gamma = -2q$. Furthermore, introducing the two independent signs $\epsilon = \pm 1$, $\eta = \pm 1$, we have $\sqrt{2\alpha} = \epsilon(n-p)$, $\sqrt{-2\beta} = \eta(n+p)$. Thus every instance of P_V is characterised by a triplet (n, p, q). We can now give the auto-Bäcklund [40]:

$$V = 1 - \frac{2zv}{zv' - \epsilon(n-p)v^2 + (\epsilon(n-p) - \eta(n+p) + z)v + \eta(n+p)} \qquad (2.67)$$

which relate $v(n, p, q)$ and $V(N, P, Q)$ where (N, P, Q) are related to (n, p, q) through the following relations:

$$N = \sigma q$$

$$P = \begin{cases} \sigma(\frac{1}{2} - \eta n) & \text{if } \epsilon\eta = 1 \\ \sigma(\frac{1}{2} - \eta p) & \text{if } \epsilon\eta = -1 \end{cases} \qquad (2.68)$$

$$Q = \begin{cases} -\eta p & \text{if } \epsilon\eta = 1 \\ -\eta n & \text{if } \epsilon\eta = -1 \end{cases}$$

where $\sigma = \pm 1$. From (2.67) it is clear that the auto-Bäcklund introduces indeed four transformations depending on the signs ϵ, η.

Let us present now the discrete analogue of the above relations. We start from d-P_{II}, written as

$$x_{n+1} + x_{n-1} = \frac{x_n(z_n + z_{n-1}) + \delta + z_n - z_{n-1}}{1 - x_n^2}. \qquad (2.69)$$

We introduce the Miura transformation [27],

$$y_n = (x_n - 1)(x_{n+1} + 1) + z_n. \qquad (2.70)$$

and we obtain,

$$(y_n + y_{n+1})(y_n + y_{n-1}) = \frac{-4y_n^2 + \delta^2}{y_n - z_n}. \qquad (2.71)$$

Equation (2.71) is d-P_{34}, i.e. the discrete form of equation 34 in the Gambier classification, in perfect analogy to what happens in the continuous case.

An example of auto-Bäcklund transformation will be given in the case of d-P_{IV} [36]. It is written as the pair of equations,

$$y_n = -\frac{x_n x_{n+1} + x_{n+1}(\tilde{z}_n + \kappa) + x_n(\tilde{z}_n - \kappa) + \mu}{x_n + x_{n+1}} \qquad (2.72a)$$

$$x_n = -\frac{y_n y_{n-1} + y_n(z_n - \tilde{\kappa}) + y_{n-1}(z_n + \tilde{\kappa}) + \lambda}{y_n + y_{n-1}}, \qquad (2.72b)$$

where $\tilde{z}_n = z_n + \alpha/2$, $\tilde{\kappa} = \kappa + \alpha/2$ and α is the lattice spacing in the discrete variable n. The meaning of these equations is that, when one eliminates either x or y between the two one ends up with d-P_{IV} in the form

$$(x_{n+1} + x_n)(x_n + x_{n-1}) = \frac{(x_n^2 - \mu)^2 - 4\kappa^2 x_n^2}{(x_n + z_n)^2 - \tilde{\kappa}^2 - \lambda} \qquad (2.73a)$$

$$(y_{n+1} + y_n)(y_n + y_{n-1}) = \frac{(y_n^2 - \lambda)^2 - 4\tilde{\kappa}^2 y_n^2}{(y_n + \tilde{z}_n)^2 - \kappa^2 - \mu}. \qquad (2.73b)$$

The important remark here is that (2.73a) and (2.73b) are *not* on the same lattice (since in (2.73b) the quantity $\tilde{z}$ figures in the denominator, instead of z) but, rather, on 'staggered' lattices.

2.4.4 Particular solutions

The Painlevé equations possess special solutions, which exist *only* for specific values of the parameters and moreover involve at most one integration constant [10].

The simplest such case is that of P_{II}

$$x'' = 2x^3 + tx + \mu \qquad (2.74)$$

It is straightforward to check that when

$$\mu = \epsilon/2 \qquad (2.75)$$

with $\epsilon^2 = 1$ equation (2.74) possess solutions given by the Riccati

$$x' = \epsilon(x^2 + t/2) \qquad (2.76)$$

The latter can be linearized by a Cole–Hopf transformation $x = \epsilon u'/u$ leading to an Airy equation

$$u'' + \frac{t}{2}u = 0 \qquad (2.77)$$

These special function solutions of Painlevé equations involve one integration constant.

Two more types of special solutions do exist. The first type of solution exists for Painlevé equations which have at least two parameters, since they need two constraints. These solutions are expressed in terms of quadratures and involve one integration constant. We illustrate this in the case of the P_{IV} equation:

$$x'' = \frac{x'^2}{2x} + \frac{3x^3}{2} + 4tx^2 + 2x(t^2 + \alpha) - \frac{2\beta^2}{x} \tag{2.78}$$

which has linearizable solutions whenever the constraint:

$$\epsilon_1 \alpha + \epsilon_2 \beta = 1 \tag{2.79}$$

holds. They are given by the solutions of the Riccati:

$$x' = \epsilon_1(x^2 + 2tx) - 2\epsilon_2\beta \tag{2.80}$$

Clearly, if $\beta = 0$, in which case $\alpha = \epsilon_1$, the Riccati becomes a linear equation for $u = 1/x$:

$$u' = -\epsilon_1(2tu + 1) \tag{2.81}$$

The integration of (2.81) is straightforward. We find:

$$u = \left(c - \epsilon_1 \int e^{\epsilon_1 t^2} dt\right) e^{-\epsilon_1 t^2} \tag{2.82}$$

with c an integration constant, i.e., u, or equivalently x, can be expressed in terms of the Error function (of t for $\epsilon_1 = -1$ and of it for $\epsilon_1 = 1$).

Finally, the Painlevé equations have rational solutions, under constraints on the parameters, which involve no integration constant. In the case of P_{II} we find by inspection that if $\mu = 0$, $x = 0$ is a solution. More solutions of this type can be easily constructed. We have for instance when $\mu = \epsilon$ a rational solution $x = -\epsilon/t$.

The solutions we presented above constitute, in some sense, the simplest ones. More particular solutions do exist for values of the parameters simply related to the ones corresponding to these "simplest" solutions. These "higher" solutions can be expressed as Wronskian determinants the elements of which are special functions.

In parallel to the continuous Painlevé equations the discrete $\mathbb{P}$s possess "special function"-type solutions which are solutions of linear difference equations which are discretizations of the corresponding equations for the continuous special functions.

In order to obtain particular solutions of a d-$\mathbb{P}$ we first reduce the equation to a discrete Riccati, i.e. a homographic transformation, which is subsequently linearized and reduced to the equation for some special function. We shall present here the case of q-P_V [37],

$$(x_{n+1}x_n - 1)(x_nx_{n-1} - 1) = \frac{pr(x_n - u)(x_n - 1/u)(x_n - v)(x_n - 1/v)}{(x_nz_n - p)(x_nz_n - r)} \tag{2.83}$$

with $z_n = z_0 \lambda^n$. We propose the following factorization:

$$x_n x_{n+1} - 1 = \frac{p(x_n - u)(x_n - v)}{uv(x_n z_n - p)} \tag{2.84a}$$

$$x_n x_{n-1} - 1 = \frac{uvr(x_n - 1/u)(x_n - 1/v)}{(x_n z_n - r)}. \tag{2.84b}$$

The two equations are compatible only when the following condition holds,

$$uv = p/r\lambda. \tag{2.85}$$

In this case, equation (2.84) can be cast in a more symmetric form that is in fact a discrete Riccati,

$$z_n(x_n x_{n+1} - 1) = px_{n+1} + \lambda r(x_n - u - v) \tag{2.86}$$

We can also show that (2.86) can indeed be linearized. Solving for x_{n+1}, we rewrite it as

$$x_{n+1} = \frac{\lambda r(x_n - u - v) + z_n}{z_n x_n - p} \tag{2.87}$$

We introduce the discrete equivalent of a Cole–Hopf, $x = B/A$, and obtain the system,

$$B_{n+1} = \lambda r B_n + (z_n - \lambda r(u + v))A_n,$$
$$A_{n+1} = z_n B_n - pA_n. \tag{2.88}$$

Eliminating B we get the linear three-point mapping,

$$A_{n+2} + (p - r)A_{n+1} - (z_n z_{n+1} - z_n r(u + v) + pr)A_n = 0, \tag{2.89}$$

which is indeed a discrete form of the confluent hypergeometric equation, up to some straightforward transformations.

Another type of solutions does exist for discrete Painlevé equations just as in the continuous case. They are expressed in terms of special functions under *two* constraints. Here we shall illustrate this in the case of the d-P_{IV} [38]:

$$(x_{n+1} + x_n)(x_n + x_{n-1}) = \frac{(x_n^2 - a^2)(x_n^2 - b^2)}{(x_n - z_n)^2 - c^2} \tag{2.90}$$

where a, b, c are constants and $z_n = \delta n + z_0$. The linearisability condition is:

$$2c - a - b = \delta \tag{2.91}$$

and the corresponding homographic mapping is:

$$x_{n+1} = \frac{x_n(a + b - c - z_n) - ab}{-x_n + c + z_n} \tag{2.92}$$

This mapping can obviously be made linear for $y \equiv 1/x$ provided we take $ab = 0$. Taking, for instance, $b = 0$ and implementing (2.91) we obtain:

$$y_{n+1} = \frac{y_n(c + z_n) - 1}{c - z_{n+1}} \qquad (2.93)$$

The homogeneous part of this equation can be solved simply in terms of Gamma functions whereupon the general solution of (2.93) is given in terms of a discrete quadrature. This special solution can be shown to be the discrete equivalent of the Error function solution of the continuous P_{IV}.

Finally the discrete Painlevé equations have yet another type of solutions, namely rational ones. We shall examine them in the case of q-P_V. One obvious solution of this type is $x = \pm 1$ which exists whenever either u or v takes the value ± 1. Nontrivial solutions also exist. We have, in fact two families of such rational solutions. The first has a most elementary member,

$$x = \pm 1 + (p + r)/z, \qquad (2.94)$$

provided u (or $1/u$) $= \mp 1/\lambda$ and v (or $1/v$) $= \mp p/r$ (or $u \leftrightarrow v$). For the second we find

$$x = (p + r)/z, \qquad (2.95)$$

which exists for $u = \sqrt{\lambda}$, $v = -\sqrt{\lambda}$. These rational solutions exist only on a codimension-two submanifold and, moreover, they do not contain any free integration constants.

The particular solutions of the discrete Painlevé equations can be expressed as Casorati determinants the elements of which are special functions.

2.4.5 Contiguity relations

The contiguity relations of the Painlevé equations are a direct consequence of their auto-Bäcklund transformations. Their interest lies in the fact that if we view them as mappings, they introduce discrete Painlevé equations. Let us give an example, which, if identified at the time, would have opened the domain of the discrete Painlevé equations a decade earlier. The starting point is the auto-Bäcklund of Painlevé II

$$u'' = 2u^3 + tu + m \qquad (2.96)$$

Starting from u we compute the quantity

$$\bar{u} = -u - \frac{2m + 1}{2u' + 2u^2 + t} \qquad (2.97)$$

which is also a solution of P_{II} with parameter $m + 1$ instead of m i.e., $\bar{u} \equiv u(m + 1)$. Using (2.97) and the symmetry of P_{II} $u(m) = -u(-m)$ we can construct also the solution $\underline{u} \equiv u(m - 1)$. It suffices then to eliminate u' which leads to

$$\frac{m + 1/2}{u_{m+1} + u_m} + \frac{m - 1/2}{u_{m-1} + u_m} = -2u_m^2 - t \qquad (2.98)$$

This is the contiguity relation of the solutions of P_{II}, obtained by Jimbo and Miwa in [13]. On the other hand viewed as a mapping, under the evolution of m, this is just a discrete form of P_I.

The obvious question at this point is what do the contiguity relation of the solutions of discrete Painlevé equations lead to. The answer is simple: again discrete Painlevé equations. Moreover while investigating this question we discovered the property of self-duality, i.e., the evolution equation in the discrete independent variable and in the space of the parameters is the same. Let us illustrate what we mean by self-duality in the example where this notion first appeared: the alternate d-P_{II} equation [5]:

$$\frac{z_n}{x_{n+1}x_n + 1} + \frac{z_{n-1}}{x_n x_{n-1} + 1} = -x_n + \frac{1}{x_n} + z_n + a \qquad (2.99)$$

where $z_n = \delta n + z_0$ and a is a parameter. The Schlesinger transforms of (2.99) were presented in [18]. By denoting by $x(a - \delta)$ and $x(a + \delta)$ the solutions of alt-d-P_{II} corresponding to parameters $a - \delta$ and $a + \delta$ respectively, we have:

$$x_n(a - \delta) = \frac{1}{x_n} + \frac{a(1 + x_n x_{n-1})}{1 + x_n x_{n-1} - z_{n-1}x_n} \qquad (2.100)$$

where x_n stands for $x_n(a)$ and

$$x_n(a + \delta) = \left(x_n - \frac{(a + \delta)(1 + x_n x_{n-1})}{1 + x_n x_{n-1} - z_{n-1}x_{n-1}} \right)^{-1} \qquad (2.101)$$

Eliminating x_{n-1} between (2.100) and (2.101) we obtain the dual equation of alt-d-P_{II}, i.e., the equation where the parameter a is now the independent variable. We find:

$$\frac{a + \delta}{x(a)x(a + \delta) - 1} + \frac{a}{x(a)x(a - \delta) - 1} = x(a) + \frac{1}{x(a)} - a - z \qquad (2.102)$$

where we have dropped the index n, and z ($\equiv z_n$) is now just a parameter. We remark that (2.102) is essentially alt-d-P_{II} itself. The only, minor, change is the fact that the x of the dual equation is multiplied by i with respect to the initial one.

Self-duality is a quite general property of the discrete Painlevé equations with *two exceptions*, the equations described by the affine Weyl groups

$(A_2 + A_1)^{(1)}$ and $(A_1 + A_1)^{(1)}$ in the Sakai classification. The self duality property has served as a basis for the proposal of the geometrical description of the discrete Painlevé equations [30], which we introduced under the name of Grand Scheme. While the Grand Scheme approach aimed at providing a geometric description of every known discrete Painlevé equation (and produced many interesting results) the problem was solved in its full generality (and in a most elegant way) proposing the final classification of these integrable discrete systems, by Sakai.

2.5 The ultradiscrete Painlevé equations

In this section we are going to present the extension of Painlevé equations to the ultradiscrete domain. The word ultradiscrete is used to designate systems where the *dependent* variables, as well as the independent ones, take only discrete values. In this respect ultradiscrete systems are generalized cellular automata. The name of ultradiscrete is reserved for systems obtained from discrete ones through a specific limiting procedure introduced in [39], by the Tokyo–Kyoto group.

Before introducing the ultradiscrete limit let us first consider the question of nonlinearity. How simple can a nonlinear system be and still be *genuinely* nonlinear? The nonlinearities we are accustomed to, involving powers, are not necessarily the simplest. It turns out (admittedly with hindsight) that the simplest nonlinear function of x one can think of is $|x|$. It is indeed linear for *both* $x > 0$ and $x < 0$ and the nonlinearity comes only from the different determinations. Thus one would expect the equations involving nonlinearities only in terms of absolute-values to be the simplest. The ultradiscrete limit does just that, i.e., it converts a given (discrete) nonlinear equation to one where only absolute-value nonlinearities appear. The key relation is the following limit:

$$\lim_{\epsilon \to 0^+} \epsilon \log(1 + e^{x/\epsilon}) = \max(0, x) = (x + |x|)/2. \tag{2.103}$$

Other equivalent expressions exist for this limit and the notation that is often used is the truncated power function $(x)_+ \equiv \max(0, x)$. It is easy to show that $\lim_{\epsilon \to 0^+} \epsilon \log(e^{x/\epsilon} + e^{y/\epsilon}) = \max(x, y)$ and the extension to n terms in the argument of the logarithm is straightforward.

Two remarks are in order at this point. First, since the function $(x)_+$ takes only integer values when the argument is integer, the ultradiscrete equations can describe generalized cellular automata, provided one restricts the initial conditions to integer values. Second, the necessary condition for the procedure to be applicable is that the dependent variables be positive, since we are taking a logarithm and we require that the result take values in $\mathbb{Z}$. This means that only some solutions of the discrete equations will survive in the ultradiscretisation.

In order to construct the ultradiscrete analogues of the Painlevé equations we must start with the discrete form that allows the ultradiscrete limit to be taken. The general procedure is to start with an equation for x, introduce X through $x = e^{X/\epsilon}$ and then take appropriately the limit $\epsilon \to 0$. Clearly the substitution $x = e^{X/\epsilon}$ requires x to be positive. This is a stringent requirement that limits the exploitable form of the d-$\mathbb{P}$'s to multiplicative ones. Fortunately many such forms are known for the discrete Painlevé transcendents. We have for instance

d-P$_\text{I}$ $\qquad x_{n+1} x_{n-1} x_n^\sigma = \lambda^n x_n + 1$ $\qquad\qquad$ with $\sigma = 0, 1, 2$ (2.104)

d-P$_\text{II}$ $\qquad x_{n+1} x_{n-1} x_n^\rho = \dfrac{\alpha(x_n + \lambda^n)}{(1 + \lambda^n x_n)}$ $\qquad$ with $\rho = 0, 1$ (2.105)

d-P$_\text{III}$ $\qquad x_{n+1} x_{n-1} = \dfrac{(x_n + \alpha\lambda^n)(x_n + \lambda^n/\alpha)}{(1 + \beta x_n \lambda^n)(1 + x_n \lambda^n/\beta)}$ $\qquad\qquad$ (2.106)

leading to

u-P$_\text{I}$ $\qquad X_{n+1} + X_{n-1} + \sigma X_n = \max(0, n + X_n)$ $\qquad\qquad$ (2.107)

u-P$_\text{II}$ $\qquad X_{n+1} + X_{n-1} + \rho X_n = a + \max(n, X_n) - \max(0, n + X_n)$ $\qquad$ (2.108)

u-P$_\text{III}$ $\qquad X_{n+1} + X_{n-1} = \max(X_n, n + a) + \max(X_n, n - a)$

$$- \max(0, X_n + n + b) - \max(0, X_n + n - b)$$
$$(2.109)$$

Ultradiscrete forms have been derived also for the higher Painlevé equations [9].

2.5.1 Degeneration cascade

Just as their continuous and discrete brethren the ultradiscrete Painlevé equations are characterised by a host of special properties. For instance, they organise themselves into a degeneration cascade. In order to derive this cascade of the u-P's we introduce a large parameter Ω. The coalescence limits are obtained through $\Omega \to +\infty$. Starting from u-P$_\text{III}$ we can obtain u-P$_\text{II}$, for $\rho = 1$, through $a = \Omega + \alpha$, $b = \Omega - \alpha$, $n = m + \Omega$, $X_n = Z_m - \alpha$. Finally from u-P$_\text{II}$ ($\rho = 1$) we obtain u-P$_\text{I}$, for $\sigma = 2$, by putting $a = 4\Omega$, $n = \Omega - m$ and $X_n = Z_m + \Omega$. It turns out that starting from u-P$_\text{II}$, for $\rho = 1$, we can also obtain u-P$_\text{I}$, with $\sigma = 1$. In this case we must take $a = -2\Omega$, $n = -m - \Omega$ and $X_n = (2Z_m - m)/3 - \Omega$.

2.5.2 Lax pairs

Lax pairs can be (and have been) proposed for ultradiscrete Painlevé equations. In particular Joshi and collaborators [15] have shown, on a specific example, how one can apply the ultradiscretisation procedure on the linear system and obtain the ultradiscrete equation as a compatibility condition.

The starting point is a Lax pair for a q-discrete equation

$$\phi(qx, k) = L\phi(x, k)$$
$$\phi(x, qk) = M\phi(x, k)$$
(2.110)

where L, M are matrices with elements involving the dependent variable of the discrete equation and $k = k_0 q^n$, $x = x_0 q^n$. Next the authors of [15] proceed to the ultradiscretisation of (2.110) and moreover introduce the ultradiscrete multiplication of two matrices

$$(A \otimes B)_{ij} = \max_k (A_{ik} + B_{kj})$$
(2.111)

Using this notation the Lax pair at the ultradiscrete limit becomes

$$\phi((n + 1)Q, mQ) = L(nQ, mQ) \otimes \phi(nQ, mQ)$$
(2.112a)
$$\phi(nQ, (m + 1)Q) = M(nQ, mQ) \otimes \phi((n + 1)Q, mQ)$$
(2.112b)

(where Q is related to q through $q = e^{Q/\epsilon}$). The compatibility condition can be written as

$$L(nQ, (m+1)Q) \otimes M(nQ, mQ) = M((n+1)Q, mQ) \otimes L((n+1)Q, mQ) \quad (2.113)$$

and leads indeed to the ultradiscrete Painlevé equation.

While this derivation is formally interesting it is far from clear how one can use it in order to solve the spectral problem (which, after all, is the main usefulness of the Lax pair).

2.5.3 Miura and Bäcklund relations

Miura, Bäcklund and Schlesinger transformations of the solutions of ultradiscrete Painlevé equations do also exist. Let us work out in detail the auto-Bäcklund transformation for u-P_{II} ($\rho = 1$). Let us start with the discrete P_{II}:

$$x_{n-1}x_{n+1} = \frac{\alpha z(x_n + z)}{x_n(1 + x_n)}$$
(2.114)

where $z = \lambda^n$. We readily remark that (2.114) is invariant under the transformation (I) $\alpha \to 1/\alpha$, $x \to z/x$. The Miura transformation (M) relates d-P_{II} to d-P_{34}. It is given as a system:

$$y_n = x_n(x_{n+1} + 1)$$

$$x_n = \frac{y_n y_{n-1} - \alpha z^2}{y_{n-1} + \alpha z} \tag{2.115}$$

Eliminating y we obtain (2.114), while eliminating x we obtain d-P_{34} for y:

$$(y_n y_{n-1} - \alpha z^2)(y_n y_{n+1} - \alpha \lambda^2 z^2) = \alpha z(y_n + z)(y_n + \alpha \lambda z) \tag{2.116}$$

(The form (2.116) is not the canonical one and a gauge transformation $y \to yz\sqrt{\alpha\lambda}$ is needed in order to convert it to canonical form.) Equation (2.116) is invariant under the transformation (J): $y \to \tilde{y} = y/(\alpha\lambda)$, $\alpha \to \tilde{\alpha} = 1/(\alpha\lambda^2)$. In order to obtain the auto-Bäcklund transformation for d-P_{II} (2.114) one must use the Miura to transform to d-P_{34}, use the invariance of the latter and come back through the inverse Miura. The auto-Bäcklund (in fact, the Schlesinger) of d-P_{II} is thus $B = M^{-1}JMI$. It transforms the solution, x, of d-P_{II} with parameter α to one, $\tilde{x}$, corresponding to a parameter $\tilde{\alpha} = \alpha/\lambda^2$. Following the chain of transformation we find:

$$\tilde{x}_n = \frac{z(\lambda x_n x_{n-1} + \alpha(x_n + z))}{\lambda x_n(x_n x_{n-1} + x_n + z)} = \frac{\alpha z(\lambda z + x_{n+1}(x_n + 1))}{\lambda x_n(\alpha z + x_{n+1}(x_n + 1))} \tag{2.117}$$

where the expressions of $\tilde{x}$ are equivalent, as a consequence of (2.114). Using the chain $IM^{-1}JM$ one can compute the inverse Schlesinger leading to $\alpha\lambda^2$.

The ultra-discrete limit of (2.114) is readily obtained:

$$X_{n+1} + X_{n-1} + X_n = a + 2n + (X_n - n)_+ - (X_n)_+ \tag{2.118}$$

(This is not in the canonical form (2.108) we encountered in Section 2.5, but can be easily transformed into it.) The ultra-discretisation procedure cannot be applied to the Miura (2.115) and, in particular, to the second half of it. This means that in the ultra-discrete limit we can compute y_n from given x_n, x_{n+1} but the knowledge of y_n, y_{n-1} does not allow us to compute x_n. Still, the remarkable result is that, while the intermediate step is missing, the end result (2.117) is ultra-discretisable. Thus one can give the ultradiscrete form of the auto-Bäcklund:

$$\tilde{X}_n = n - 1 - X_n + \max(X_n + X_{n-1} + 1, a + X_n, a + n) - \max(X_n + X_{n-1}, X_n, n)$$
$$= a + n - 1 - X_n + \max(X_n + X_{n+1}, X_{n+1}, n + 1) - \max(X_n + X_{n+1}, X_{n+1}, a + n) \tag{2.119}$$

In [34] we have shown that if X_n is a solution of (2.118) corresponding to a parameter a, then $\tilde{X}_n$ is a solution to (2.118) corresponding to parameter $a - 2$.

2.5.4 Particular solutions

Special solutions exist also for the ultradiscrete Painlevé equations (except for u-P_I which is parameter-free). Let us start with u-P_{II} ($\rho = 1$). We find readily that for $a = 0$, a solution $X_n = 0$ exists. This corresponds to the rational solution $x_n = 1$ of d-P_{II} (2.105) for $\alpha = 1$. The next solution for (2.108) is a step-function one. Two solutions, $X_n = \pm 2$ for $n \leq -1$ and $X_n = \pm 1$ for $n \geq 0$, exist for $a = \pm 4$. They are the ultra-discrete limits of the next rational solutions of d-$P_{II} - 1$ (2.108) namely $x_n = (1 + \lambda + \lambda^{1-n})^{\pm 1}(1 + \lambda^{-1} + \lambda^{-1-n})^{\mp 1}$ for $\alpha = \lambda^{\pm 4}$. Higher solutions also exist. When n is large positive, a constant solution exists for $a = 4p$, with integer p, where X is equal to p, while a constant solution with $X = 2p$ exists when n is large negative. The remarkable fact is that these constant "half" solutions do really join to form a solution of (2.108) with p jumps from the value $X = 2p$ to $X = p$. The first jump occurs between n_0 and $n_0 + 1$ for $n_0 = 1 - 2|p|$ and we have successive jumps of $-p/|p|$ at $n_0 + 3k$, $k = 0, 1, 2, \ldots, |p| - 1$. Of course, the higher multistep solutions are also ultradiscrete limits of the higher rational solutions of (2.105).

2.5.5 Contiguity relations

As shown in [32], in order to obtain the contiguity relations of the solutions of some ultradiscrete equation we start from a q-discrete equation which produces the ultradiscrete one at the appropriate limit. In what follows we shall examine the one-parameter q-P_{III} we introduced in [31].

Our starting point is the mapping

$$x_n x_{n+1} = \frac{1 + y_n/z_n}{y_n(1 + y_n/d)} \tag{2.120a}$$

$$y_n y_{n-1} = d\frac{x_n + 1}{x_n^2} \tag{2.120b}$$

where z is the independent variable $z_n = z_0 \lambda^n$. We proceed by obtaining the Schlesinger transformations, which correspond to the changes of the parameter d. We shall denote these changes by an evolution in the direction m, i.e., $d_{m+1} = \lambda d_m$. We first introduce an auxiliary variable v through the relations

$$x_{n+1}yv = xyv_{m+1} = 1 \tag{2.121}$$

where we have omitted the indexes n, m when they are not up-shifted. Using (2.120) and (2.121) we can now establish the equations for the evolution along the m direction:

$$v_m v_{m+1} = \frac{1 + y_m/d_m}{y_m(1 + y_m/z)} \tag{2.122a}$$

$$y_m y_{m-1} = z\frac{v_m + 1}{v_m^2} \tag{2.122b}$$

where z does not vary in the direction m.

However, equation (2.122) is not very nice as a contiguity relation of the solutions of (2.120) since it involves the auxiliary variable v. It is more interesting to write the dual equation in terms of the variables x and y. The simplest relation we can find is the one relating x_{m-1} and y_{m-1} to x_{n+1} and y_n. Using (2.120) and (2.122) we find

$$x_{m-1} = \frac{y_n}{z(1 + y_n x_{n+1})} \tag{2.123a}$$

$$y_{m-1} = z x_{n+1}(1 + y_n x_{n+1}) \tag{2.123b}$$

These relations can be easily inverted. We obtain

$$x_{n+1} = \frac{y_{m-1}}{z + x_{m-1}y_{m-1}} \tag{2.124a}$$

$$y_n = z x_{m-1}(1 + x_{m-1}y_{m-1}) \tag{2.124b}$$

Ultradiscretising the one-parameter q-P$_{\text{III}}$ and its contiguities is straightforward. We introduce the ansatz $x = e^{X/\epsilon}, y = e^{Y/\epsilon}, v = e^{V/\epsilon}, \lambda = e^{1/\epsilon}$ and take the limit of the logarithm of the relations obtained above when $\epsilon \to 0$. We find for the one-parameter q-P$_{\text{III}}$ the ultradiscrete form

$$X_n + X_{n+1} = \max(0, Y_n - n) - Y_n - \max(0, Y_n - m) \tag{2.125a}$$

$$Y_n + Y_{n-1} = m + \max(0, X_n) - 2X_n \tag{2.125b}$$

where we have used explicitly the fact that $z_n = z_0\lambda^n$, $d_m = d_0\lambda^m$ and absorbed the effect of z_0, d_0 in the definition of n, m. The relations involving the auxiliary variable are particularly simple

$$X_{n+1} + Y + V = X + Y + V_{m+1} = 0 \tag{2.126}$$

We can now write the ultradiscrete contiguity relation in terms of V. We find

$$V_m + V_{m+1} = \max(0, Y_m - m) - Y_m - \max(0, Y_m - n) \tag{2.127a}$$

$$Y_m + Y_{m-1} + \max(0, V_m) - 2V_m \tag{2.127b}$$

Finally we give the dual evolution in terms of the variables X and Y. We find

$$X_{m-1} = Y_n - n - \max(0, Y_n + X_{n+1}) \tag{2.128a}$$

$$Y_{m-1} = n + X_{n+1} + \max(0, Y_n + X_{n+1}) \tag{2.128b}$$

or, ultradiscretising (2.124),

$$X_{n+1} = Y_{m-1} - \max(n, Y_{m-1} + X_{m-1}) \tag{2.129a}$$

$$Y_n = n + X_{m-1} + \max(0, Y_{m-1} + X_{m-1}) \tag{2.129b}$$

Again the dual equation is most simple in terms of the auxiliary variable V but one can indeed write contiguity relations in terms of the variable X, Y, which are not more complicated if one chooses the most convenient staggering. It is straightforward to verify that $\{X_{m+1}, Y_{m+1}\}$ satisfy indeed equation (2.125) for parameter $m + 1$.

2.6 Conclusion

This course has been devoted to the presentation of the derivation and properties of Painlevé equations in their three most common avatars, namely continuous, discrete and ultradiscrete. We have shown how the continuous and discrete Painlevé equations can be derived based on some very general theory but also using basic integrability techniques. The latter, in the continuous case, consists in testing for the existence of the Painlevé property while in the discrete case our approach was based on two well-known integrability criteria: singularity confinement and algebraic entropy. We have shown that both continuous and discrete Painlevé equations possess a spate of special properties (bestowed upon them by their integrable character) and moreover that a perfect parallel exists between the properties of the continuous equations and those of the discrete ones. We have complemented our presentation by results on ultradiscrete Painlevé equations. We have shown how one can derive them systematically starting form q-discrete equations and applying the ultradiscretisation procedure. Finally we presented their special properties and we have shown that, quite expectedly, they are also in parallel to those of their continuous and discrete siblings.

References

[1] Ablowitz, M. J., Ramani, A., and Segur, H. 1978. Nonlinear evolution equations and ordinary differential equations of Painlevé type. *Lett. Nuovo Cimento (2)*, **23**(9), 333–338.

[2] Bureau, F. J. 1964. Differential equations with fixed critical points. *Ann. Mat. Pura Appl. (4)*, **64**, 229–364.

[3] Bureau, F. J. 1972. Équations différentielles du second ordre en Y et du second degré en $\ddot{Y}$ dont l'intégrale générale est à points critiques fixes. *Ann. Mat. Pura Appl. (4)*, **91**, 163–281.

[4] Fokas, A. S., and Ablowitz, M. J. 1982. On a unified approach to transformations and elementary solutions of Painlevé equations. *J. Math. Phys.*, **23**, 2033–2042.

[5] Fokas, A. S., Grammaticos, B., and Ramani, A. 1993. From continuous to discrete Painlevé equations. *J. Math. Anal. Appl.*, **180**(2), 342–360.

[6] Gambier, B. 1910. Sur les équations différentielles du second ordre et du premier degré dont l'intégrale générale est à points critiques fixes. *Acta Math.*, **33**(1), 1–55.

[7] Grammaticos, B., and Ramani, A. 2000. The hunting for the discrete Painlevé equations. Sophia Kovalevskaya to the 150th anniversary. *Regul. Chaotic Dyn.*, **5**(1), 53–66.

[8] Grammaticos, B., Ramani, A., and Papageorgiou, V. 1991. Do integrable mappings have the Painlevé property? *Phys. Rev. Lett.*, **67**(14), 1825–1828.

[9] Grammaticos, B., Ohta, Y., Ramani, A., and Takahashi, D. 1998. The ultimate discretisation of the Painlevé equations. *Phys. D*, **114**(3-4), 185–196.

[10] Gromak, V. A., and Lukashevich, N. A. 1990. *Analytic Properties of Solutions of Painlevé Equations*. Minsk: Universitetskoye. in Russian.

[11] Hietarinta, J., and Viallet, C. M. 1998. Singularity confinement and chaos in discrete systems. *Phys. Rev. Lett.*, **81**(2), 325–328.

[12] Ince, E. L. 1944. *Ordinary Differential Equations*. New York: Dover.

[13] Jimbo, M., and Miwa, T. 1981. Monodromy preserving deformation of linear ordinary differential equations with rational coefficients. II. *Phys. D*, **2**(11), 407–448.

[14] Jimbo, M., and Sakai, H. 1996. A q-analog of the sixth Painlevé equation. *Lett. Math. Phys.*, **38**(2), 145–154.

[15] Joshi, N., Nijhoff, F. W., and Ormerod, C. 2004. Lax pairs for ultra-discrete Painlevé cellular automata. *J. Phys. A*, **37**(44), L559–L565.

[16] Kruskal, M. D. 2000. Private communication.

[17] Malmquist, J. 1922. Sur les équations différentielles du second ordre dont l'intégrale générale a ses points critiques fixes. *Ark. Mat. Astr. Fys.*, **17**(8), 1–89.

[18] Nijhoff, F., Satsuma, J., Kajiwara, K., Grammaticos, B., and Ramani, A. 1996. A study of the alternate discrete Painlevé II equation. *Inverse Problems*, **12**(5), 697–716.

[19] Ohta, Y., Tamizhmani, K. M., Grammaticos, B., and Ramani, A. 1999. Singularity confinement and algebraic entropy: the case of the discrete Painlevé equations. *Phys. Lett. A*, **262**(2-3), 152–157.

[20] Ohta, Y., Ramani, A., and Grammaticos, B. 2002. Elliptic discrete Painlevé equations. *J. Phys. A*, **35**(45), L653–L659.

[21] Okamoto, K. 1979. Sur les feuilletages associés aux équations du second ordre à points critiques fixes de P. Painlevé. *Japan. J. Math. (N.S.)*, **5**(1), 1–79.

[22] Okamoto, K. 1981. On the τ-function of the Painlevé equations. *Phys. D*, **2**(3), 525–535.

[23] Painlevé, P. 1888. Sur les équations différentielles du premier ordre. *C. R. Acad. Sci. Paris*, **107**, 221–224, 320–323,724–727.

[24] Papageorgiou, V., Nijhoff, F., Grammaticos, B., and Ramani, A. 1992. Isomonodromic deformation problems for discrete analogues of Painlevé equations. *Phys. Lett. A*, **164**(1), 57–64.

[25] Picard, E. 1889. Mémoire sur la théorie des fonctions algébriques de deux variables. *J. Math. Pures Appl. (4)*, **5**, 135–320.

[26] Quispel, G. R. W., Roberts, J. A. G., and Thompson, C. J. 1989. Integrable mappings and soliton equations. II. *Phys. D*, **34**(1-2), 183–192.

[27] Ramani, A., and Grammaticos, B. 1992. Miura transforms for discrete Painlevé equations. *J. Phys. A*, **25**(14), L633–L637.

[28] Ramani, A., and Grammaticos, B. 1996. Discrete Painlevé equations: coalescences, limits and degeneracies. *Phys. A*, **228**, 160–171.

[29] Ramani, A., Grammaticos, B., and Hietarinta, J. 1991. Discrete versions of the Painlevé equations. *Phys. Rev. Lett.*, **67**(14), 1829–1832.

[30] Ramani, A., Ohta, Y., Satsuma, J., and Grammaticos, B. 1998. Self-duality and Schlesinger chains for the asymmetric d-P_{II} and q-P_{III} equations. *Comm. Math. Phys.*, **192**(1), 67–76.

[31] Ramani, A., Grammaticos, B., Tamizhmani, T., and Tamizhmani, K. M. 2000. On a transcendental equation related to Painlevé III, and its discrete forms. *J. Phys. A*, **33**(3), 579–590.

[32] Ramani, A., Grammaticos, B., and Willox, R. 2008. Contiguity relations for discrete and ultradiscrete Painlevé equations. *J. Nonlinear Math. Phys.*, **15**(4), 353–364.

[33] Sakai, H. 2001. Rational surfaces associated with affine root systems and geometry of the Painlevé equations. *Comm. Math. Phys.*, **220**(1), 165–229.

[34] Takahashi, D., Tokihiro, T., Grammaticos, B., Ohta, Y., and Ramani, A. 1997. Constructing solutions to the ultradiscrete Painlevé equations. *J. Phys. A*, **30**(22), 7953–7966.

[35] Takenawa, T. 2001. A geometric approach to singularity confinement and algebraic entropy. *J. Phys. A*, **34**(10), L95–L102.

[36] Tamizhmani, K. M., Grammaticos, B., and Ramani, A. 1993. Schlesinger transforms for the discrete Painlevé IV equation. *Lett. Math. Phys.*, **29**(1), 49–54.

[37] Tamizhmani, K. M., Ramani, A., Grammaticos, B., and Ohta, Y. 1996. A study of the discrete P_V equation: Miura transformations and particular solutions. *Lett. Math. Phys.*, **38**(3), 289–296.

[38] Tamizhmani, T., Grammaticos, B., Ramani, A., and Tamizhmani, K. M. 2001. On a class of special solutions of the Painlevé equations. *Phys. A*, **295**(3-4), 359–370.

[39] Tokihiro, T., Takahashi, D., Matsukidaira, J., and Satsuma, J. 1996. From soliton equations to integrable cellular automata through a limiting procedure. *Phys. Rev. Lett.*, **76**(18), 3247–3250.

[40] Tokihiro, T., Grammaticos, B., and Ramani, A. 2002. From the continuous P_V to discrete Painlevé equations. *J. Phys. A*, **35**(28), 5943–5950.

3

Definitions and Predictions of Integrability for Difference Equations

Jarmo Hietarinta

Abstract

In these lectures we take a look at various meanings of integrability for difference equations, and the possibility of algorithmic methods to identify (partial) integrability. Analogies with continuum equations are used when possible, but the world of discrete equations is richer and many new things enter.

3.1 Preliminaries

3.1.1 Points of view on integrability

One can approach the integrability of dynamical equations from two opposite directions:

Top down: In this approach one first chooses some *high level mathematical structure* and then derives its consequences/manifestations for dynamical equations. The underlying mathematical structure is there from the start and therefore as a result one gets a method for generating equations with good properties.

Bottom up: In this case the equation is given, for example: "In my application I found this equation, what can you say about it?" In order to answer questions of this type we need a *toolbox of (algorithmic) methods* that can be applied. With them we can, hopefully, identify the equation or at least say whether it is integrable or partially integrable or chaotic. If the equation is at least partially integrable we may be able to say something about its structure, construct some solutions, conserved quantities etc. And although complete integrability itself is structurally unstable, many properties persist in nearby non-integrable systems, and one can try a perturbative expansion around the integrable one.

Here aspects of both points of view are described, with emphasis on providing a toolbox.

3.1.2 Preliminaries on discreteness and discrete integrability

Why should we consider discrete equations? There are several reasons: Perhaps discrete things are more fundamental than continuous ones in the description of nature, perhaps nature itself is discrete at the Planck length scale. In the analysis of many mathematical problems difference equations arise naturally, e.g., in recursion relations or orthogonal polynomials, similarly in some problems of statistical physics it was found that the correlation functions satisfy integrable difference equations. In fact it turns out that there is always interesting mathematics in the background of integrable discrete equations. In any case we need to discretize continuous equations for numerical analysis, because computers work only on discrete things. Finally, continuum integrability is well established and all easy things have already been done; discrete integrability, on the other hand, is relatively new and in that domain there are still new things to be discovered.

What is discrete integrability anyway? Consider an equation of the form

$$y_{n+1} + y_{n-1} = f(y_n). \tag{3.1}$$

Given y_0, y_1 we can compute y_n for all $n \in \mathbb{Z}$. So what's the problem, in particular what does "integrability" mean in this context? Integrability always means that the dynamics is more regular than that described by an arbitrary equation, so let us consider more detailed questions, for example: Can we say anything about y_n without actually computing every intermediate step? Can we find formulae like $y_n = \phi(y_0, y_1; n)$ where ϕ is some reasonable function? How does the error in the initial values propagate with n, for example does the resulting ambiguity grow as n^2, or as 2^n? If the equation is integrable we should be able to tell something about its solutions for arbitrarily long times, which is quite the opposite to chaotic equations.

Map or functional equation? Let us return to equation (3.1) and consider the definitions of the quantities appearing in it. In the above we quietly took the interpretation that $y: \mathbb{Z} \to \mathbb{C}$, i.e, y_n is a sequence of complex numbers, and if y_0, y_1 are given we can compute $y_n, \forall n > 1, n < 0$.

But the same equation can be viewed differently, we may interpret the index n as giving a location on the complex plane and then the equation should be written as

$$y(z + d) + y(z - d) = f(y(z)) \qquad (y: \mathbb{C} \to \mathbb{C}). \tag{3.2}$$

In this interpretation $y(z)$ is a complex analytic function, thus y is required to satisfy the above functional equation for all $z \in \mathbb{C}$.

Formally we can set a point-wise equivalence from (3.2) to (3.1) by $y_n \equiv y(z)$, $y_{n+k} \equiv y(z + dk)$, but not always conversely. Thus the different mathematical settings bring in different properties, tools and results.

Solvability is not integrability This is because integrability is basically about *regularity* or *predictability*, and a closed form explicit solution does not guarantee regularity. The typical example is provided by the logistic map

$$y_{n+1} = 4y_n(1 - y_n),$$

for which one can construct explicit closed form solution for all n:

$$y_n = \tfrac{1}{2}[1 - \cos(2^n c_0)].$$

This solution shows "sensitive dependence on the initial value" which can be seen, e.g., if we calculate the derivative with respect to the initial value c_0:

$$\frac{dy_n}{dc_0} = \frac{1}{2}2^n \sin(2^n c_0).$$

Thus we can see that the error grows exponentially with n, which is one of the indicators associated with "chaos."

Can integrability be preserved during discretization? We mentioned above that one reason we have to consider discrete equations is that they are needed for numerical computations. One requirement for a good discretization is that it should preserve at least the most important properties of the original equation. A natural requirement is that the solution of the discrete version should sample the solution of the original continuous equation, but then a simple discretization is rarely good enough.

As an example consider the following ODE:

$$\frac{du}{dt} = \alpha u(1 - \beta u), \tag{3.3}$$

which can be easily integrated yielding the solution

$$u(t) = \frac{u_0}{\beta u_0 + (1 - \beta u_0)e^{-\alpha t}}. \tag{3.4}$$

How to discretize (3.3) so that the discrete equation implies similar behavior? The naive forward discretization of a derivative is

$$\frac{du}{dt} \approx \frac{u(t + \Delta t) - u(t)}{\Delta t}$$

and would lead to the discretized version

$$u(t + \Delta t) - u(t) = \Delta t\, \alpha u(t)(1 - \beta u(t)). \tag{3.5}$$

If we define the discrete variable y_n by $u(t) = u(t_0 + n\,\Delta t) = [a/(\alpha\beta\,\Delta t)]y_n$ and introduce a new constant a by $a = 1 + \alpha\,\Delta t$, then we can write (3.5) as $y_{n+1} = ay_n(1 - y_n)$. But this is nothing but the logistic equation, which is well known to contain chaotic behavior, thus the above is not a good discretization.

The good discretization is, in fact, the following

$$u(t + \Delta t) - u(t) = \Delta t\, \alpha u(t + \Delta t)(1 - \beta u(t)), \tag{3.6}$$

or after solving for $u(t + \Delta t)$

$$u(t + \Delta t) = \frac{u(t)}{(1 - \alpha\,\Delta t) + \alpha\beta\,\Delta t\, u(t)}.$$

It is difficult to guess this discretization, but one is led to it by analyzing the underlying structure. The important observation is that equation (3.3) can be *linearized*: Using

$$u = \frac{1}{w + \beta}$$

in (3.3) we get

$$\frac{dw}{dt} = -\alpha w,$$

with solution $w = ce^{-\alpha t}$. Now it is safe to discretize the linear equation with forward difference and this leads to

$$w(t + \Delta t) - w(t) = -\alpha\,\Delta t\, w(t) \tag{3.7}$$

and with the inverse transformation $w = -\beta + 1/u$ we get (3.6). Furthermore, the difference equation (3.7) is solved by

$$w(t + n\,\Delta t) = (1 - \alpha\,\Delta t)^n w(t)$$

leading to the solution

$$u(t) \equiv u(n\,\Delta t) = \frac{u_0}{\beta u_0 + (1 - \beta u_0)(1 - \alpha\,\Delta t)^n}. \tag{3.8}$$

The discrete solution (3.8) of (3.6) samples the continuum solution (3.4) of (3.3), and thus the discretization (3.6) is the correct one.

Continuum limits One of the general problems mentioned above was the discretization of a given equation, in fact there can be several discretizations. But conversely, the top-down approach yields difference equations and then we may ask whether they may be considered as discretizations of some continuum equation. To study this question the first step is to take the continuum limit of the discrete equation.

As an example consider the *discrete Painlevé I equation* (d-PI) given by

$$x_{n+1} + x_n + x_{n-1} = \frac{\alpha + \beta n}{x_n} + b. \tag{3.9}$$

We would expect that the continuum limit of this equation is the usual first Painlevé equation, and this will happen, but in the process the parameters have to be scaled in a particular way.

First we introduce a small parameter ϵ related to the lattice spacing

$$\epsilon n = z, \qquad x_n = f(z), \qquad x_{n\pm 1} = f(z \pm \epsilon),$$

and then take the limit

$$\epsilon \to 0, \quad n \to \infty, \qquad \epsilon n \text{ fixed}$$

which yields first

$$3f + \epsilon^2 f'' + O(\epsilon^4) = \frac{\alpha + \beta z/\epsilon}{f} + b.$$

In order to get rid of the denominator (the first Painlevé equation is polynomial) we must take

$$f(z) = c_1 + c_2 \epsilon^\kappa y(z),$$

and expand, with the power $\kappa > 0$ for the moment open. This yields

$$3c_1 + 3c_2 \epsilon^\kappa y(z) + 3c_2 \epsilon^{2+\kappa} y'' + \cdots$$

$$= b + \frac{1}{c_1}\left(\alpha + \beta \frac{z}{\epsilon}\right)\left(1 - \frac{c_2}{c_1}\epsilon^\kappa y + \left(\frac{c_2}{c_1}\right)^2 \epsilon^{2\kappa} y^2 \cdots\right).$$

Since the discrete equation was a three-point equation the continuous one should be second order in derivatives. We see that y'' enters at order $\epsilon^{2+\kappa}$ while the lowest nonlinearity in the RHS enters at $\epsilon^{2\kappa}$. In order to balance them we must take $\kappa = 2$. Thus the first Painlevé equation will appear at order ϵ^4 and the lower order terms should cancel. For this to happen we must also scale the coefficient of z properly, i.e., $\beta \propto \epsilon^5$. Then we have $3c_1 = b + \alpha/c_1$ at order ϵ^0 and $3c_2 = -c_2\alpha/c_1^2$ at order ϵ^2, leading to

$$c_1 = \frac{b}{6}, \qquad \alpha = -\frac{b^2}{12}.$$

Then at ϵ^4 we get the first Painlevé equation

$$y'' = 6y^2 + z,$$

provided that we take

$$c_2 = -\frac{b}{3}, \qquad \beta = -\frac{b^2}{18}\epsilon^5.$$

This example shows that the continuum limit can be rather involved, and that is in fact the normal situation.

3.2 Conserved quantities

One of the fundamental characteristics of integrability is the existence of conserved quantities, they restrict the available phase space and thereby make the motion more predictable, in particular eliminating chaos.

3.2.1 Constants of motion for continuous ODE

Let us first recall the definition of *Liouville integrability*. It is normally given in the Hamiltonian formulation, but for comparison with difference equations we need it in the Lagrangian formulation:

Definition   A Lagrangian $L(\dot{q}, q)$, where q is N-dimensional, is integrable if there are N constants of motion (CM) $I_k(\dot{q}, q)$ (L one of them) such that the I_k

1. are functionally independent,
2. are regular functions of $\dot{q}, q$,
3. $dI/dt = 0$ (using equations of motion).

The requirement of regularity should not be ignored, because formally the initial values are constants of motion for any equation, but usually they cannot be expressed as regular functions of the coordinates and velocities.

The Euler–Lagrange equations generated by $L(\dot{q}, q)$ are

$$\frac{d}{dt}\frac{\partial L}{\partial \dot{q}_i} - \frac{\partial L}{\partial q_i} = 0, \qquad \forall i. \tag{3.10}$$

The relation of CM to the equation is revealed by computing the total derivative:

$$\frac{dI(\dot{q}, q)}{dt} = \sum_i \frac{\partial I}{\partial \dot{q}_i}\ddot{q}_i + \sum_i \frac{\partial I}{\partial q_i}\dot{q}_i. \tag{3.11}$$

Here the RHS should vanish when we impose the equations of motion of the type (3.10). The one-dimensional case is always integrable and the invariant has usually the role of energy, and in this case one can conversely take any $I(\dot{q}, q)$ and derive the corresponding dynamical equation from (3.11). (The correspondence between the first conserved quantity and the equation of motion is clearer in the Hamiltonian formulation where the Hamiltonian itself is the conserved quantity, while here $I \neq L$.)

If one wants to search for higher dimensional integrable Lagrangian systems one can propose ansatze for the Lagrangian and the conserved quantity (typically polynomials in $\dot{q}$, with coefficients depending on q) and then derive equations for the coefficients and solve them. Many integrable systems have been discovered in this way [10].

3.2.2 The standard discrete case

Let us try to see whether the ideas of the continuum case can be extended for discrete equations, at least in the one-dimensional case.

Consider the discrete equivalent of a second order ODE, namely a 3-point OΔE relating u_{n-1}, u_n, u_{n+1}. This equation should be linear in u_{n+1} and u_{n-1} to guarantee well defined evolution in either direction. In this case a conserved quantity must depend on two consecutive variables (corresponding to the continuous $(q, \dot{q})$). Suppose we now take some $K(x, y)$, then by definition it is a conserved quantity if

$$K(u_{n+1}, u_n) - K(u_n, u_{n-1}) = 0. \tag{3.12}$$

How could this produce an equation linear in u_{n+1}, u_{n-1} if K is nonlinear? Recall that in the continuous case the corresponding equation (3.11) contained $\ddot{q}$ linearly. The root problem here (and in many other occasions) is that the Leibniz rule is not valid in the discrete case.

In order to make progress let us assume that $K(x, y) = K(y, x)$. Then the equation (3.12) vanishes when $u_{n+1} = u_{n-1}$, and if K is polynomial this means that it has the factor $u_{n+1} - u_{n-1}$. Let us therefore try a general symmetric biquadratic K:

$$K(x, y) := c_5 x^2 y^2 + c_4 xy(x + y) + c_3 xy + c_2(x^2 + y^2) + c_1(x + y). \tag{3.13}$$

Then we get

$$\frac{K(u_{n+1}, u_n) - K(u_n, u_{n-1})}{u_{n+1} - u_{n-1}} = c_1 + c_2(u_{n+1} + u_{n-1}) + c_3 u_n$$
$$+ c_4 u_n(u_{n+1} + u_n + u_{n-1}) + c_5 u_n^2(u_{n+1} + u_{n-1}),$$

from which we get an equation having (3.13) as CM

$$u_{n+1} + u_{n-1} = \frac{c_4 u_n^2 + c_3 u_n + c_1}{c_5 u_n^2 + c_4 u_n + c_2}.$$

The above can be generalized by taking a rational symmetric biquadratic:

$$K(x,y) = \frac{c_5 x^2 y^2 + c_4 xy(x+y) + c_3 xy + c_2(x^2+y^2) + c_1(x+y)}{d_5 x^2 y^2 + d_4 xy(x+y) + d_3 xy + d_2(x^2+y^2) + d_1(x+y)},$$

having 9 essential parameters. Then direct computation shows that this is a CM for the *symmetric version of the Quispel–Roberts–Thomson (QRT) map* [27]:

$$u_{n+1} = \frac{f_1(u_n) - f_2(u_n)u_{n-1}}{f_2(u_n) - f_3(u_n)u_{n-1}}$$

where f_i are certain specific quartic polynomials. This contains almost all integrable 3-point maps.

3.2.3 The Hirota–Kimura–Yahagi (HKY) generalization

Consider the biquadratic CM

$$K(x,y) = \frac{2xy}{x^2 + y^2 + \beta^2}. \tag{3.14}$$

Then we have

$$K(u_{n+1}, u_n) - K(u_n, u_{n-1}) = \frac{-2u_n(u_{n+1} - u_{n-1})[u_{n+1}u_{n-1} - (u_n^2 + b^2)]}{(u_{n+1}^2 + u_n^2 + b^2)(u_n^2 + u_{n-1}^2 + b^2)},$$

and thus K of (3.14) is the CM for the 3-point equation

$$u_{n+1}u_{n-1} = u_n^2 + b^2.$$

But another computation yields

$$K(u_{n+1}, u_n) + K(u_n, u_{n-1}) = \frac{2u_n(u_{n+1} + u_{n-1})[u_{n+1}u_{n-1} + (u_n^2 + b^2)]}{(u_{n+1}^2 + u_n^2 + b^2)(u_n^2 + u_{n-1}^2 + b^2)},$$

in which we can identify the equation

$$u_{n+1}u_{n-1} = -(u_n^2 + b^2).$$

It seems that in the second case K is conserved "up to sign" and then $K(x,y)^2$, which is biquartic, should be a genuine invariant. Indeed, a direct computation shows that

$$K(u_{n+1}, u_n)^2 - K(u_n, u_{n-1})^2 = \frac{-4u_n^2(u_{n+1} + u_{n-1})(u_{n+1} - u_{n-1})}{(u_{n+1}^2 + u_n^2 + b^2)^2(u_n^2 + u_{n-1}^2 + b^2)^2}$$
$$\times [u_{n+1}u_{n-1} + (u_n^2 + b^2)][u_{n+1}u_{n-1} - (u_n^2 + b^2)],$$

having both equations as factors. Thus

- $u_{n+1}u_{n-1} = u_n^2 + b^2$ has a quadratic invariant,
- $u_{n+1}u_{n-1} = -(u_n^2 + b^2)$ has a quartic invariant.

Alternatively we can say that the second equation has a CM with explicit n-dependence: $K(u_n, u_{n-1}) = (-1)^n 2u_n u_{n-1}/(u_n^2 + u_{n-1}^2 + \beta^2)$.

The above can be generalized by considering other involutions besides $K \rightarrow -K$ [18].

3.3 Singularity confinement and algebraic entropy

One of the goals in studying discrete integrability is to develop *algorithmic ways to identify integrable equations*, which could be analogues of the methods that we already have in the continuous case. Thus we would like to identify equations with regular behavior algorithmically, without actually solving the equation. For ODEs the most popular method has been the Painlevé method, which concentrates on local analysis (for complex time) to check whether solutions have *movable singularities*. This was used as search program by Painlevé, Gambier, etc. and led to the discovery of the six Painlevé equations. Another method, although less used, has been the growth analysis of the solution based on Nevanlinna theory.

What about difference equations? Maybe for a discrete Painlevé test we should again study what happens at a singularity? As for growth analysis, recall that difference equations can trivially be solved step by step, so growth analysis of the solution should give us important information about regularity.

3.3.1 Singularity analysis for difference equations

The idea of singularity analysis for difference equations was first proposed in [9]. The idea behind the *Singularity Confinement Criterion* (SC) was that if the dynamics leads to a singularity then after a few steps one should be able to get out of it (confinement), and this should take place without loss of information. (This in contrast with chaotic systems containing attractors that absorb information.) SC amounts to the requirement of *well defined evolution even near singular points*. This sounds like a reasonable requirement, but it is not clear whether it really is an analogue of the Painlevé test. Anyway, using this principle it has been possible to find discrete analogies of Painlevé equations [28].

In order to illustrate SC let us consider first the autonomous case of d-PI (3.9)

$$x_{n+1} = -x_n - x_{n-1} + \frac{a}{x_n} + b. \tag{3.15}$$

This equation is singular at $x = 0$, and let us assume that we reach the singularity at $x_0 = 0$ with a finite $x_{-1} = u \neq 0$. The sequence x_n continues as follows:

$$x_1 = -0 - u + a/0 + b = \infty,$$
$$x_2 = -\infty - 0 + a/\infty + b = -\infty,$$
$$x_3 = +\infty - \infty - a/\infty + b = ?$$

Here the infinities are not the problem, but rather the ambiguity "$\infty - \infty$". In order to resolve this ambiguity we assume $x_0 = \epsilon$ (small) and redo the calculations:

$$x_{-1} = u,$$
$$x_0 = \epsilon,$$
$$x_1 = \frac{a}{\epsilon} + b - u - \epsilon,$$
$$x_2 = -\frac{a}{\epsilon} + u + \epsilon + \frac{u-b}{a}\epsilon^2 + O(\epsilon^3)$$
$$x_3 = -\left[-\frac{a}{\epsilon} + u + \epsilon + \frac{u-b}{a}\epsilon^2 + O(\epsilon^3) \right] - \left[\frac{a}{\epsilon} + b - u - \epsilon \right]$$
$$\qquad\qquad + \frac{a}{-a/\epsilon + u + O(\epsilon)} + b$$
$$= -\epsilon + \frac{b-2u}{a}\epsilon^2 + O(\epsilon^3),$$
$$x_4 = u + O(\epsilon)$$

Thus the ambiguity is resolved, furthermore we find that singularity is confined and initial information u is finally recovered, although in the intermediate steps it only appeared in the subleading terms. One says that in this case the *singularity pattern* is $\ldots, 0, \infty, -\infty, 0, \ldots$.

Singularity is not always confined, a worst case example is given by

$$x_{n+1} - 2x_n + x_{n-1} = \frac{a}{x_n} + b,$$

for which we obtain

$$x_{-1} = u,$$

$$x_0 = \epsilon,$$

$$x_1 = \frac{a}{\epsilon} + b - u + 2\epsilon,$$

$$x_2 = 2\frac{a}{\epsilon} + 3b - 2u + O(\epsilon),$$

$$x_3 = 3\frac{a}{\epsilon} + 6b - 3u + O(\epsilon),$$

and in general

$$x_k = k\frac{a}{\epsilon} + \ldots,$$

and the singularity is not confined, ever. It is also important to observe that in this nonconfined case there are no ambiguities.

The singularity confinement idea has turned out to be very useful. Its best application has been the *de-autonomization* of discrete equations. In this process one introduces an ansatz for some n-dependency in the free coefficients but insists that the non-autonomous equation has exactly the same singularity pattern and degree growth as the autonomous one [26]. This yields equations for the n-dependent coefficient(s).

Let us consider the previous example (3.9) but allowing a to depend on n. Starting $x_{-1} = u$, $x_0 = \epsilon$, as before, one now finds

$$x_1 = \frac{a_0}{\epsilon} + b - u - \epsilon,$$

$$x_2 = -\frac{a_0}{\epsilon} + u + \frac{a_1}{a_0}\epsilon + \frac{a_1}{a_0}\frac{u-b}{a_0}\epsilon^2 + O(\epsilon^3),$$

$$x_3 = -\frac{a_2 + a_1 - a_0}{a_2}\epsilon + \frac{[a_1/a_0]b - [(a_1 + a_2)/a_0]u}{a_0}\epsilon^2 + O(\epsilon^3),$$

$$x_4 = -\frac{a_3 - a_2 - a_1 + a_0}{a_2 + a_1 - a_0}\frac{a_0}{\epsilon} + \cdots$$

Since we want to keep the old singularity pattern we should get something like $x_4 = u + \cdots$. Thus the condition for singularity confinement at this *same* step implies

$$a_{n+3} - a_{n+2} - a_{n+1} + a_n = 0, \qquad \forall n,$$

with solution

$$a_n = \alpha + \beta n + \gamma(-1)^n. \tag{3.16}$$

If $\gamma = 0$ we recover d-PI of (3.9). In general, with a_n as in (3.16) the singularity is confined, and

$$x_4 := \frac{u(\alpha + \gamma) + 2b\beta}{\alpha + 3\beta - \gamma} + O(\epsilon),$$

in particular, if $\beta = \gamma = 0$ (i.e., $a_n = \alpha$), $x_4 = u + \cdots$. The parameters α, β, γ should be chosen so that $a_n \neq 0$, $\forall n$ (which also implies that $a_2 + a_1 - a_0 \neq 0$).

3.3.2 Singularity confinement in projective space

The singularities reveal their nature best in projective space, where we can follow more closely what happens at ambiguities and what cancellations take place. We will be using $\mathbb{P}^2$ where $(u, v, f) \approx (\lambda u, \lambda v, \lambda f)$, $\lambda \neq 0$

We will consider again the equation $x_{n+1} + x_n + x_{n-1} = a_n/x_n + b$, and first write it as a first order system:

$$\begin{cases} x_{n+1} = -x_n - y_n + \dfrac{a_n}{x_n} + b, \\ y_{n+1} = x_n. \end{cases}$$

Next this is homogenized by substituting $x_n = u_n/f_n$, $y_n = v_n/f_n$:

$$\begin{cases} \dfrac{u_{n+1}}{f_{n+1}} = -\dfrac{u_n}{f_n} - \dfrac{v_n}{f_n} + a_n\dfrac{f_n}{u_n} + b, \\ \dfrac{v_{n+1}}{f_{n+1}} = \dfrac{u_n}{f_n}, \end{cases}$$

and then after clearing denominators we get a *polynomial* map in $\mathbb{P}^2$

$$\begin{cases} u_{n+1} = -u_n(u_n + v_n) + f_n(a_n f_n + b u_n), \\ v_{n+1} = u_n^2, \\ f_{n+1} = f_n u_n. \end{cases} \tag{3.17}$$

Note that the RHS is homogeneous of degree 2, and therefore under iteration the default growth of degree (= *complexity*) is $\deg(u_{n+k}) = 2^k$.

Let us recall from Sect. 3.3.1 the sequence that led to a singularity: $x_{-1} = u$, $x_0 = 0$, $x_1 = \infty$, $x_2 = \infty$, $x_3 = \infty - \infty = ?$ In projective space the corresponding sequence, as derived from (3.17), looks like

$$\begin{pmatrix} u_0 \\ v_0 \\ f_0 \end{pmatrix} = \begin{pmatrix} 0 \\ u \\ 1 \end{pmatrix} \rightarrow \begin{pmatrix} 1 \\ 0 \\ 0 \end{pmatrix} \rightarrow \begin{pmatrix} 1 \\ -1 \\ 0 \end{pmatrix} \rightarrow \begin{pmatrix} 0 \\ 1 \\ 0 \end{pmatrix} \rightarrow \begin{pmatrix} 0 \\ 0 \\ 0 \end{pmatrix}.$$

In projective space infinity is a regular point and therefore only the last term is a true singularity, since it is not in $\mathbb{P}^2$.

For the detailed ϵ study with $x_{-1} = u$, $x_0 = \epsilon$ we have

$$\begin{pmatrix} x_0 \\ x_{-1} \\ 1 \end{pmatrix} \approx \begin{pmatrix} u_0 \\ v_0 \\ f_0 \end{pmatrix} = \begin{pmatrix} \epsilon \\ u \\ 1 \end{pmatrix},$$

$$\begin{pmatrix} x_1 \\ x_0 \\ 1 \end{pmatrix} \approx \begin{pmatrix} u_1 \\ v_1 \\ f_1 \end{pmatrix} = \begin{pmatrix} a_0 + (-u + b)\epsilon + \cdots \\ \epsilon^2 \\ \epsilon \end{pmatrix},$$

$$\begin{pmatrix} x_2 \\ x_1 \\ 1 \end{pmatrix} \approx \begin{pmatrix} u_2 \\ v_2 \\ f_2 \end{pmatrix} = \begin{pmatrix} -a_0^2 + \epsilon a_0(2u - b) + \cdots \\ a_0^2 + 2\epsilon a_0(-u + b) + \cdots \\ \epsilon a_0 + \epsilon^2(-u + b) + \cdots \end{pmatrix},$$

$$\begin{pmatrix} x_3 \\ x_2 \\ 1 \end{pmatrix} \approx \begin{pmatrix} u_3 \\ v_3 \\ f_3 \end{pmatrix} = \begin{pmatrix} \epsilon^2 a_0^2(-a_0 + a_1 + a_2) + \cdots \\ a_0^4 + 2\epsilon a_0^3(-2u + b) \cdots \\ -\epsilon a_0^3 + \epsilon^2 a_0^2(3u - 2b) + \cdots \end{pmatrix},$$

and finally

$$\begin{pmatrix} x_4 \\ x_3 \\ 1 \end{pmatrix} \approx \begin{pmatrix} u_4 \\ v_4 \\ f_4 \end{pmatrix} = \begin{pmatrix} \epsilon^2 a_0^6 A_3 + \epsilon^3 a_0^5[(4b - 6u)A_3 + a_0(u - b) + a_2 b] + \cdots \\ \epsilon^4 a_0^4 A_2^2 + \cdots \\ -\epsilon^3 a_0^5 A_2 + \cdots \end{pmatrix},$$

where $A_2 := a_2 + a_1 - a_0$, $A_3 := a_0 - a_1 - a_2 + a_3$.

This is the crucial point of singularity confinement: If $A_3 = 0$ and $A_2 \neq 0$ then ϵ^3 is a *common factor* and can be divided out and then the $\epsilon \to 0$ limit is no longer singular, but yields

$$\begin{pmatrix} x_4 \\ x_3 \\ 1 \end{pmatrix} \approx \begin{pmatrix} u_4 \\ v_4 \\ f_4 \end{pmatrix} = \begin{pmatrix} a_0(u - b) + a_2 b \\ 0 \\ a_3 \end{pmatrix}.$$

Thus we have emerged from the singularity and in particular recovered the initial data u. In fact, two things happen at the same time:

- The cancellation of the common factor ϵ^3 *removes the singularity*.
- The cancellation also *reduces growth of complexity*, as defined by the degree of the iterate.

These are two sides of the same phenomenon: Singularity implies reduction in growth by cancellation, but on the other hand, cancellation is only possible if we would otherwise end up in the singular point $(0, 0, 0)$.

For integrability the precise amount of cancellation will be crucial: the cancellations have to be so strong that the default exponential growth of the degree of the iterate is reduced to polynomial growth. Over the years the following classification has been shown to work very well:

- growth is linear in $n \Rightarrow$ equation is linearizable.
- growth is *polynomial* in $n \Rightarrow$ equation is *integrable*.
- growth is exponential in $n \Rightarrow$ equation is chaotic.

3.3.3 Singularity confinement is not sufficient

Consider the following three-point map [13]

$$x_{n+1} + x_{n-1} = x_n + \frac{1}{x_n^2}. \tag{3.18}$$

Let us do the epsilon analysis of singularity confinement, assuming $x_{-1} = u$, $x_0 = \epsilon$, we get

$$x_1 = \epsilon^{-2} - u + \epsilon,$$
$$x_2 = \epsilon^{-2} - u + \epsilon^4 + O(\epsilon^6),$$
$$x_3 = -\epsilon + 2\epsilon^4 + O(\epsilon^6),$$
$$x_4 = u + 3\epsilon + O(\epsilon^3).$$

Thus singularity is confined with pattern $\dots, 0, \infty, \infty, 0, \dots$. Furthermore the initial information u is recovered in x_4, so from the singularity confinement point of view everything looks fine. The problem is that (3.18) shows numerical chaos [13]. The existence of a singularity implies some cancellations, but apparently the cancellations are not sufficiently strong. A more detailed computation shows that the degrees of the iterates of (3.18) grow as

$$1, 3, 9, 27, 73, 195, 513, 1347, 3529, \dots$$

which grows asymptotically as $d_n \approx [(3 + \sqrt{5})/2]^n$. In contrast to this, the degrees of the integrable dP-I (3.9) grow as

$$1, 2, 4, 8, 13, 20, 28, 38, 49, 62, 76, \dots$$

which is fitted by $d_n = \frac{1}{8}(9 + 6n^2 - (-1)^n)$ [14].

In summary we can say that singularity confinement is *necessary* for a well-defined evolution, it is easy to verify and it can be used effectively to de-autonomize a given map. But it is not sufficient for integrable evolution. Thus one needs improvements or other tests, one of which is the requirement of slow growth of complexity or low algebraic entropy [8]. Another method of interest is the application of Nevanlinna theory for difference equations [1].

3.4 Integrability in 2D

3.4.1 Definitions and examples

For two-dimensional dynamics we must first assume some discrete structure in the underlying space. The most popular choice is to assume that everything takes place on an infinite rectangular lattice, where the values of the dynamical variable u are given at the points (n, m) with value $u_{n,m}$. However, other models have also been proposed.

As examples of constructs in the square lattice we have the discrete KdV (dKdV), which can be given as

$$\alpha(y_{n+2,m-1} - y_{n,m}) = \frac{1}{y_{n+1,m-1}} - \frac{1}{y_{n+1,m}}, \tag{3.19}$$

or after the transformation $(n, m) \mapsto (n + m, m + 1)$ as

$$\alpha(y_{n+1,m} - y_{n,m+1}) = \frac{1}{y_{n,m}} - \frac{1}{y_{n+1,m+1}}. \tag{3.20}$$

and the discrete "potential" KdV (dpKdV)

$$(u_{n,m+1} - u_{n+1,m})(u_{n,m} - u_{n+1,m+1}) = p^2 - q^2, \tag{3.21}$$

The equation of "similarity constraint" for KdV is given by

$$\left(\lambda(-1)^{n+m} + \frac{1}{2}\right)u_{n,m} + \frac{np^2}{u_{n-1,m} - u_{n+1,m}} + \frac{mq^2}{u_{n,m-1} - u_{n,m+1}} = 0. \tag{3.22}$$

Figure 3.1 shows the lattice points that are involved in these equations.

It is interesting to observe that several numerical acceleration algorithms

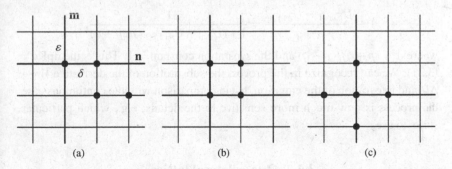

(a)　　　　　　　　(b)　　　　　　　　(c)

Figure 3.1 A difference equation may involve a different configuration of points in the lattice. Case (a) corresponds to (3.19), case (b) to (3.20) and (3.21) and case (c) to (3.22).

(for partial sums) are integrable lattice equations. For example in the Shanks–Wynn ϵ-algorithm one assumes the initial sequences $\epsilon_0^{(m)} = 0$, $\epsilon_1^{(m)} = S_m$, and then generates new sequences $\epsilon_n^{(m)}$ (that approach the limit S_∞ faster) by

$$(\epsilon_{n+1}^{(m)} - \epsilon_{n-1}^{(m+1)})(\epsilon_n^{(m+1)} - \epsilon_n^{(m)}) = 1.$$

This is the integrable discrete potential KdV equation (3.21). Similarly, Bauer's η-algorithm $(X_k^{(m)} = [\eta_k^{(m)}]^{(-1)^{k+1}})$

$$X_{n+1}^{(m)} - X_{n-1}^{(m+1)} = \frac{1}{X_n^{(m+1)}} - \frac{1}{X_n^{(m)}}$$

is the integrable discrete KdV equation (3.19).

It was noted above that many things are more complicated in the discrete case due to the lack of Leibniz rule. In the continuous case the relationship between the KdV equation and its potential form is obtained by substituting $u = \partial_x v$ and integrating the equation. Let us see what the relationship is between the discrete versions (3.19), (3.21).

Let us introduce $y_{n,m} = 1 + W_{n,m}$ into dKdV (3.20) which then becomes

$$\alpha(W_{n,m+1} - W_{n+1,m}) = \frac{1}{1 + W_{n,m}} - \frac{1}{1 + W_{n+1,m+1}}. \tag{3.23}$$

Next let $W_{n,m} = (U_{n-1,m-1} - U_{n,m})/(p+q)$ and then after arranging the terms suitably we can write (3.23) as

$$\frac{\alpha}{p+q}(U_{n-1,m} - U_{n,m-1}) - \frac{1}{1 + (U_{n-1,m-1} - U_{n,m})/(p+q)}$$

$$= \frac{\alpha}{p+q}(U_{n,m+1} - U_{n+1,m}) - \frac{1}{1 + (U_{n,m} - U_{n+1,m+1})/(p+q)}.$$

Now the RHS is a double shift of the LHS and they can be separated as

$$1 + \frac{U_{n,m+1} - U_{n+1,m}}{p-q} = \frac{1}{1 + (U_{n,m} - U_{n+1,m+1})/(p+q)},$$

where $\alpha = (p+q)/(p-q)$ and the separation constant $= 1$. This is the dpKdV (3.21). We can recognize in the process the substitution of the derivative $W \to \Delta U$ and integration of the equation, but in comparison with the continuous case the process is now much more sensitive to the details, e.g., which particular shifts to use.

3.4.2 Quadrilateral lattices

In the following we will only consider quadrilateral lattices, i.e., those in which the dynamics is given by a relation between the four corners on an elementary square, as shown in Figure 3.1(b). In the literature several different shorthand notations are used for the corner elements, e.g., $x_{n,m} = x_{00} = x$, $x_{n+1,m} = x_{10} = x_{[1]} = \tilde{x}$, $x_{n,m+1} = x_{01} = x_{[2]} = \hat{x}$, $x_{n+1,m+1} = x_{11} = x_{[12]} = \hat{\tilde{x}}$, we use all notations (almost randomly) in the following. The four corner values are assumed to be related by a multilinear equation:

$$k\,xx_{[1]}x_{[2]}x_{[12]} + l_1\,xx_{[1]}x_{[2]} + l_2\,xx_{[1]}x_{[12]} + l_3\,xx_{[2]}x_{[12]} + l_4\,x_{[1]}x_{[2]}x_{[12]}$$

$$+ s_1\,xx_{[1]} + s_2\,x_{[1]}x_{[2]} + s_3\,x_{[2]}x_{[12]} + s_4\,x_{[12]}x + s_5\,xx_{[2]} + s_6\,x_{[1]}x_{[12]}$$

$$+ q_1\,x + q_2\,x_{[1]} + q_3\,x_{[2]} + q_4.x_{[12]} + u$$

$$\equiv Q(x, x_{[1]}, x_{[2]}, x_{[12]}; p_1, p_2) = 0.$$

Multilinearity is important, because then we can compute any remaining corner value once the other three are known. The p_i are some parameters associated with shift directions $[i]$, they may appear in the coefficients k, l_j, s_j, q_j, u. This definition allows a well-defined evolution from any staircase-like initial condition, up or down.

Examples of this type include the lattice (potential) KdV

$$(p_1 - p_2 + x_{n,m+1} - x_{n+1,m})(p_1 + p_2 + x_{n,m} - x_{n+1,m+1}) = p_1^2 - p_2^2, \quad (3.24)$$

which goes into (3.21) after the translation $x_{n,m} = u_{n,m} + p_1 n + p_2 m$, lattice modified KdV (dmKdV) given by

$$p_1(x_{n,m}x_{n,m+1} - x_{n+1,m}x_{n+1,m+1}) = p_2(x_{n,m}x_{n+1,m} - x_{n,m+1}x_{n+1,m+1}),$$

and the lattice Schwarzian KdV (dSKdV)

$$(x - \tilde{x})(\hat{x} - \hat{\tilde{x}})p_2^2 = (x - \hat{x})(\tilde{x} - \hat{\tilde{x}})p_1^2.$$

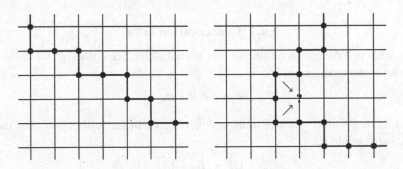

Figure 3.2 Evolution from any step-like arrangement of initial values is possible in the quadrilateral lattice, but any overhang could lead to a contradiction.

3.4.3 Continuum limit

Above it was mentioned that (3.24) is the discrete potential KdV and therefore we would expect that its continuum limit should be the potential Korteweg–de Vries equation

$$v_t = v_{xxx} + 3v_x^2. \tag{3.25}$$

Note, however, that (3.24) is fully n, m-symmetric while (3.25) is not x, t-symmetric, so the relation is not obvious. Apparently in the continuum limits some specific scaling must be used. Since in (3.25) ∂_t scales as ∂_x^3 and v scales as ∂_x we are led to try the following

$$x_{nm} = \epsilon v(x + \epsilon(c_1 n + c_2 m), t + \epsilon^3(d_1 n + d_2 m)), \tag{3.26}$$

where ϵ is a small parameter. Substituting this into (3.24) and expanding in powers of ϵ yields $c_1 p = c_2 q$ at levels ϵ^2, ϵ^3 and then at ϵ^4 we get (3.25), e.g., if $c_1 = 2/p$, $c_2 = 2/q$, $d_1 = 2/(3p^3)$, $d_2 = 2/(3q^3)$. (Note also that (3.24) is invariant under simultaneous scaling of u, p, q.)

On can also take the limits in the two coordinates n, m in sequence, with intermediate semi-discrete equations. There are two ways to do the first limit: we can either take a "straight" continuum limit by

$$x(n, m) = y_n(\xi + \epsilon m),$$

or first change variables by $x_{nm} = w_{Nm}$, where $N = n + m$, and then take the continuum limit in m. This second case can be called the "skew" continuum limit, cf. Figures 3.1(a) and (b). After taking the continuum limit on the remaining discrete variable one again arrives to (3.25), after choosing/scaling the independent variables suitably.

3.4.4 Conservation laws

In order to have conservation laws for partial difference equations we should look for the analogue of

$$\partial_x A + \partial_t B = 0.$$

Let T_n be the shift operator in the n direction, T_m in the m direction, then the analogue could be

$$(T_n - 1)F + (T_m - 1)G = 0.$$

The vanishing of the LHS should now follow from the quadrilateral equation and therefore in order to keep the indices within a specific square we will assume

$$F = F(u_{n,m}, u_{n,m+1}); \qquad G = G(u_{n,m}, u_{n+1,m}).$$

An example of this construction is

$$F = (u_{n,m}u_{n,m+1} - 1)(u_{n,m+1} - u_{n,m}), \qquad G = u_{n,m}u_{n+1,m}(u_{n,m} - u_{n+1,m}),$$

which yields a conservation law for the discrete potential KdV [29].

3.5 Singularity confinement in 2D

Previously we considered singularity confinement for the 1D case but it works for 2D as well [9].

As an example let us consider the dKdV in the form

$$w_{n+1,m+1} = w_{n,m} + \frac{1}{w_{n+1,m}} - \frac{1}{w_{n,m+1}}.$$

The initial data is given by $a, b, 0, c, d, f, g$ in Figure 3.3 (in which the coordinates have been rotated so that the staircase is horizontal). The single zero leads to infinities at two places, then to a zero and finally to ambiguities in two places.

A more detailed analysis with the initial value $0_1 = \varepsilon$ (small) yields the following values at the subsequent iterations

$$\infty_1 = b + \frac{1}{\varepsilon} - \frac{1}{a}, \qquad \infty_2 = c + \frac{1}{d} - \frac{1}{\varepsilon},$$

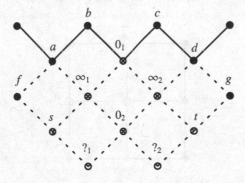

Figure 3.3 Propagation of singularities in a 2D map. Here n grows in the SE direction and m in the SW direction.

at the first stage, and on the next

$$s = a + \frac{1}{\infty_1} - \frac{1}{f}, \qquad t = d + \frac{1}{g} - \frac{1}{\infty_2},$$

$$0_2 = \varepsilon + \frac{1}{\infty_2} - \frac{1}{\infty_1} = -\varepsilon + \left(b - c - \frac{1}{a} - \frac{1}{d}\right)\varepsilon^2 + \cdots$$

Then at the next step we can resolve the ambiguities:

$$?_1 = \infty_1 + \frac{1}{0_2} - \frac{1}{s} = c + \frac{1}{d} - \frac{1}{a - 1/f} + O(\varepsilon),$$

$$?_2 = \infty_2 + \frac{1}{t} - \frac{1}{0_2} = b - \frac{1}{a} + \frac{1}{d + 1/g} + O(\varepsilon),$$

and thus the singularity is confined.

The difficulty in the 2D singularity confinement is the great freedom of inducing infinities and singularities in various places and this makes the systematic comprehensive analysis quite difficult. One possibility is to fix beforehand the 2D configuration and its singularities. In [30] the "ultra-local SC" was proposed: The idea is to consider a 3×3 square of points as in Figure 3.4, with initial values given at black disks, chosen so that $u_{11} = \infty$. The requirement is that u_{12}, u_{21} are finite and ambiguity at u_{22} can be resolved using ϵ-analysis.

As an example consider dpKdV

$$u_{n+1,m+1} = u_{n,m} - \frac{1}{u_{n,m+1} - u_{n+1,m}}.$$

The initial value $u_{01} = u_{10}$ implies

$$u_{11} = \infty, \qquad u_{12} = u_{01} = u_{21}, \qquad u_{22} = \infty - \infty?$$

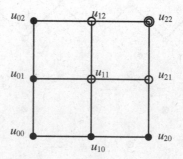

Figure 3.4 Arrangement for the ultra-local singularity confinement study.

A detailed analysis reveals that if $u_{10} = u_{01} + \epsilon$ then

$$u_{11} = \frac{1}{\epsilon} + u_{00},$$

$$u_{12} = u_{01} + \epsilon + \epsilon^2(u_{02} - u_{00}) + O(\epsilon^3),$$

$$u_{21} = u_{01} + 0 + \epsilon^2(-u_{20} + u_{00}) + O(\epsilon^3),$$

$$u_{22} = \frac{1}{\epsilon} + u_{00} - \frac{1}{\epsilon + \epsilon^2(u_{02} + u_{02} - 2u_{00}) + O(\epsilon^3)} = u_{02} + u_{20} - u_{00} + O(\epsilon).$$

This resolves the $\infty - \infty$ singularity, and recovers the initial value u_{00}.

Thus we have seen that the singularity confinement idea generalizes to 2D lattices, but as noted before, it is only a necessary condition. It may still be the first tool to apply but something more definitive is needed.

3.6 Algebraic entropy for 2D lattices

We will now consider the generalization of degree growth for the study of integrability in lattices.

3.6.1 Default growth of degree and factorization

From 1D studies we recall that the default growth of complexity (=degree of the iterate) is usually exponential. Reduced growth is obtained by cancellations which are associated with singularity confinement, and sufficient cancellations can lead to polynomial growth of complexity associated with integrability.

In the 2D case we will here only consider quadratic maps in a quadrilateral lattice,

$$p_1 \, xx_{[1]} + p_2 \, x_{[1]}x_{[2]} + p_3 \, x_{[2]}x_{[12]} + p_4 \, x_{[12]}x + p_5 \, xx_{[2]} + p_6 \, x_{[1]}x_{[12]}$$
$$+ q_1 \, x + q_2 \, x_{[1]} + q_3 \, x_{[2]} + q_4 \, x_{[12]} + u = 0. \quad (3.27)$$

As usual we write the map in the projective plane with $x = v/f$:

$$\begin{cases} v_{[12]} = p_1 \, v \, v_{[1]}f_{[2]} + p_2 \, v_{[1]}v_{[2]}f + p_5 \, v \, v_{[2]}f_{[1]} \\ \qquad\quad + q_1 \, v \, f_{[1]}f_{[2]} + q_2 \, v_{[1]}f_{[2]}f + q_3 \, v_{[2]}f_{[1]}f + u \, f \, f_{[1]}f_{[2]}, \\ f_{[12]} = p_3 \, v_{[2]}f_{[1]}f + p_4 \, v \, f_{[1]}f_{[2]} + p_6 \, v_{[1]}f_{[2]}f + q_4 f \, f_{[1]}f_{[2]}. \end{cases}$$

We noted earlier that a well-defined evolution can start from a staircase-like initial configuration. We will now specialize either to a regular staircase or to a corner, as described in Figure 3.5, initial values are given on the points marked

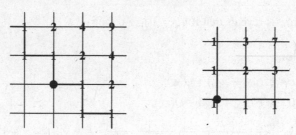

Figure 3.5 Default degree growth for staircase or corner initial data in the case of a quadratic map (3.27). The origin is marked by a black disk.

with "1," on those points the v's are free, but f's should be the same, hence the default degree growth is given by

$$\deg(z_{[12]}) = \deg(z) + \deg(z_{[1]}) + \deg(z_{[2]}) - 1,$$

where the extra -1 is because the map is quadratic and a common f can be cancelled. Here $z = v$ or $= f$, because they have the same degree. Asymptotically the default degree growth for the staircase initial data is $\frac{1}{2}(1 + \sqrt{2})^n$.

Some well-known models were studied from this point of view in [31]. For the dpKdV the following degrees were obtained

⋮	⋮	⋮	⋮	⋮	⋮			⋰	⋰	⋰	⋰	⋰	⋰
1	4	7	10	13	16	...		1	2	4	7	11	16	...
1	3	5	7	9	11	...		1	1	2	4	7	11	...
1	2	3	4	5	6	...			1	1	2	4	7	...
1	1	1	1	1	1	...				1	1	2	4	...

In the corner case the degrees follow the rule $d_{n,m} = nm + 1$, while in the staircase $d_{n,m} = 1 + (n + m)(n + m - 1)/2$, thus in each case the growth is polynomial.

Note that the first interesting factorizations take place at the point $(2, 2)$ of Figure 3.5, where in the staircase the degree 9 reduces to 7 and in the corner where degree 7 reduces to 5. A detailed computation reveals the factors that cancel: For the staircase on finds that the variables factorize as

$$v_{22}, f_{22} = (\text{main parts of degree 7}) \times (v_{01} - v_{10})^2,$$

and for the corner

$$v_{22}, f_{22} = (\text{main parts of degree 5}) \times (v_{01} - v_{10})^2.$$

Thus in each case v_{22} and f_{22} have some different main parts but their greatest common divisor is $(v_{01} - v_{10})^2$, which cancels.

3.6.2 Search based on factorization

Since integrable maps seem to generate a quadratic common factor at $(2, 2)$, this can be taken as the first requirement, supplemented later by other considerations.

In the simplest case (including the one mentioned above) the quadratic factor is a product of two linear factors. A search for new integrable maps reported in [15] was based on the requirement that at least *one linear factor* can be extracted at the point $(2, 2)$. Figure 3.6 indicates once more the default degrees. The equation $Q = 0$ can be used to solve for x_{22}, which is given by the rational expression v_{22}/f_{22}, from which one should be able to factor out a linear factor. Another way of saying the same thing is that when one substitutes the values computed at $(1, 1), (1, 2), (2, 1)$ into the polynomial Q, it should factorize with a linear factor. This last problem was solved in [15] for a completely arbitrary quadratic map (3.27). The solution process branched a lot and even after eliminating the results for which Q factorized, or did not depend on all corners, 125 candidate maps remained.

For the candidate maps the final test was done by studying their growth of complexity. The growth analysis was done using staircase configurations with propagation to all directions (NE,NW,SE,SW). Among the 125 tentative results 23 different degree growth patterns were observed, and all of them could be fitted with rational generating functions. There were linear patterns $1, 2, 3, 4, 5, 6, 7, 8, 9, 10, \ldots$ with generating functions $1/(1 - s)^2$ and $1, 2, 3, 5, 6, 8, 9, 11, 12, 14, \ldots$, with $(1 + s + s^3)/[(s + 1)(1 - s)^2]$, several different patterns with quadratic growth, including $1, 2, 3, 5, 7, 11, 14, 20, 24, 32, \ldots$ with $(1 + s - s^2 + s^4 + s^5)/[(s + 1)^2(1 - s)^3]$ and $1, 2, 4, 7, 11, 16, 22, 29, 37, 46, \ldots$ with $(1 - s + s^2)/(1 - s)^3$. The remaining patterns were

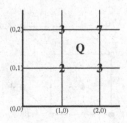

Figure 3.6 The situation at first factorization with corner initial data.

exponential; these included $1, 2, 3, 5, 8, 13, 21, 34, 55, 89, \ldots$ with generating function $(1 + s)/(1 - s - s^2)$ and asymptotic growth with exponent $1.618\ldots$ and $1, 2, 4, 8, 16, 33, 69, 145, 305, 642, \ldots$ with $(1 - s - s^4)/[(1 - s)(1 - 2s - s^4)]$ and exponent $2.107\ldots$.

From the integrable equations found in this search [15] we would like to mention two:

$$x_{00}x_{10}c_1 + x_{00}x_{01}c_5 + (x_{00}x_{11} + x_{10}x_{01})c_2 + x_{10}x_{11}c_6 + x_{01}x_{11}c_3 = 0 \quad (3.28)$$

which is homogeneous, and integrable for all values of the five parameters c_i, (growth pattern $1, 2, 4, 7, 11, 16, 22, 29, 37, 46, \ldots$ in all directions) and

$$(x_{00} - x_{01})(x_{10} - x_{11}) + (x_{00} - x_{11})a + (x_{01} - x_{10})b + c = 0 \quad (3.29)$$

(growth pattern $1, 2, 3, 5, 7, 11, 14, 20, 24, 32, \ldots$ in all directions).

3.7 Consistency around a cube

3.7.1 Definition

An important concept in lattice integrability is the idea of "Consistency Around a Cube" (CAC). It means that the map on the plane can be consistently extended to higher dimensions. From 2D to 3D this goes as follows: Adjoin a third direction $x_{n,m} \to x_{n,m,k}$ and construct a cube as in Figure 3.7, with a quadrilinear map with corresponding parameters on each side:

on the bottom	$Q_{12}(x_{000}, x_{100}, x_{010}, x_{110}; p, q) = 0,$
on the top	$Q_{12}(x_{001}, x_{101}, x_{011}, x_{111}; p, q) = 0,$
on the back	$Q_{23}(x_{000}, x_{010}, x_{001}, x_{011}; q, r) = 0,$
on the front	$Q_{23}(x_{100}, x_{110}, x_{101}, x_{111}; q, r) = 0,$
on the left side	$Q_{31}(x_{000}, x_{001}, x_{100}, x_{101}; r, p) = 0,$
on the right side	$Q_{31}(x_{010}, x_{011}, x_{110}, x_{111}; r, p) = 0.$

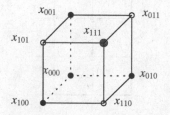

Figure 3.7 The consistency cube

The problem of consistency is the following: Given values $x_{000}, x_{100}, x_{010}, x_{001}$ (at black disks in Figure 3.7) we can compute values at open disks $x_{110}, x_{101}, x_{011}$ uniquely using the corresponding equations on the bottom, left and back sides, respectively. Then x_{111} can be computed from the equations on the top, left and front side equations, and they should all agree. This is the consistency condition.

The CAC concept represents a rather high level of integrability. It is a kind of Bianchi identity, and was observed to hold for a sequence of Moutard transforms [25], but it was first proposed as a property of maps in [24]. One consequence of CAC is that it allows the immediate construction of Lax presentation [23].

3.7.2 Lax pair

We follow the recipe given in [21]. The idea is to take the third direction as the spectral direction, and then generate the auxiliary functions from the variables x_{**1} as follows: One solves Q_{13} for x_{101} and Q_{23} for x_{011} and the dependence on these variables is linearized by introducing f, g: $x_{001} = f/g$, $x_{101} = f_{[1]}/g_{[1]}$, $x_{011} = f_{[2]}/g_{[2]}$, furthermore let us denote the parameter of the spectral direction by $\lambda = r$.

For the discrete KdV $(x_{n,m+1} - x_{n+1,m})(x_{n,m} - x_{n+1,m+1}) = p^2 - q^2$, we have $Q_{13} \equiv (x_{001} - x_{100})(x_{000} - x_{101}) = p^2 - r^2$, and therefore

$$\frac{f_{[1]}}{g_{[1]}} = \frac{xf + (\lambda^2 - p^2 - \tilde{x}x)g}{f - \tilde{x}g},$$

$$\frac{f_{[2]}}{g_{[2]}} = \frac{xf + (\lambda^2 - q^2 - \hat{x}x)g}{f - \hat{x}g}.$$

We define $\phi = \binom{f}{g}$ and then write the above as

$$\phi_{[1]} = L\phi, \qquad \phi_{[2]} = M\phi$$

where

$$L = \gamma \begin{pmatrix} x & \lambda^2 - p^2 - x\tilde{x} \\ 1 & -\tilde{x} \end{pmatrix}, \qquad M = \gamma' \begin{pmatrix} x & \lambda^2 - q^2 - x\hat{x} \\ 1 & -\hat{x} \end{pmatrix}.$$

Here γ, γ' are separation constants. The consistency condition is $\phi_{[12]} = \phi_{[21]}$, i.e., $L_{[2]}M = M_{[1]}L$. This yields a condition on the constants γ, γ', which in this case is solved by $\gamma = \gamma' = 1$, and also yields the map $(\hat{x} - \tilde{x})(x - \hat{\tilde{x}}) = p^2 - q^2$.

3.7.3 CAC as a search method

CAC has also been used as a method to search and classify lattice equations see [5, 6]. In [5] two additional assumptions were made

- symmetry ($\varepsilon, \sigma = \pm 1$):

$$Q(x_{000}, x_{100}, x_{010}, x_{110}; p_1, p_2) = \varepsilon Q(x_{000}, x_{010}, x_{100}, x_{110}; p_2, p_1)$$
$$= \sigma Q(x_{100}, x_{000}, x_{110}, x_{010}; p_1, p_2),$$

- "tetrahedron property": x_{111} does not depend on x_{000}.

With these assumptions a complete classification was obtained. Here we just mention three models of the list [5]

(H1) $(x - \hat{\tilde{x}})(\tilde{x} - \hat{x}) + q - p = 0,$

(Q3) $(q^2 - p^2)(x\hat{\tilde{x}} + \tilde{x}\hat{x}) + q(p^2 - 1)(x\tilde{x} + \hat{x}\hat{\tilde{x}}) - p(q^2 - 1)(x\hat{x} + \tilde{x}\hat{\tilde{x}})$
$$- \delta^2(p^2 - q^2)(p^2 - 1)(q^2 - 1)/(4pq) = 0,$$

(Q4)   which is the root model from which others follow:

$$a_0 x\tilde{x}\hat{x}\hat{\tilde{x}} + a_1(x\tilde{x}\hat{x} + \tilde{x}\hat{x}\hat{\tilde{x}} + \hat{x}\hat{\tilde{x}}x + \hat{\tilde{x}}x\tilde{x} + a_2(x\hat{\tilde{x}} + \tilde{x}\hat{x})$$
$$+ \bar{a}_2(x\tilde{x} + \hat{x}\hat{\tilde{x}}) + \tilde{a}_2(x\hat{x} + \tilde{x}\hat{\tilde{x}}) + a_3(x + \tilde{x} + \hat{x} + \hat{\tilde{x}}) + a_4 = 0,$$

in which the parameters a_i are given by Weierstrass elliptic functions.

H1 is the potential KdV equation (3.21), and Q3 contains various KdV-type equations as special limits. Q4 was first derived by Adler as a superposition rule of BT's for the Krichever–Novikov equation [2]. It is worth noting that the discrete KdV-equation (3.23) does not appear in the list.

Another search based on the CAC-idea, but without the tetrahedron assumption was made in [12]. The new non-tetrahedron results were

- $x + x_{[1]} + x_{[2]} + x_{[12]} = 0$
- $xx_{[12]} + x_{[1]}x_{[2]} = 0$
- $(xx_{[1]}x_{[2]} + xx_{[1]}x_{[12]} + xx_{[2]}x_{[12]} + x_{[1]}x_{[2]}x_{[12]}) + (x + x_{[1]} + x_{[12]} + x_{[2]}) = 0.$

They are all without spectral parameters and furthermore linearizable. As a side-product of this search exercise a simpler Jacobi form for (Q4) was found:

$$(h_1 f_2 - h_2 f_1)[(xx_{[1]}x_{[12]}x_{[2]} + 1)f_1 f_2 - (xx_{[12]} + x_{[1]}x_{[2]})]$$
$$+ (f_1^2 f_2^2 - 1)[(xx_{[1]} + x_{[12]}x_{[2]})f_1 - (xx_{[2]} + x_{[1]}x_{[12]})f_2] = 0, \quad (3.30)$$

$h_i^2 = f_i^4 + \delta f_i^2 + 1$, parameterizable by Jacobi elliptic functions. Note also that this is even in the variables. This and the original (Q4) are related by a Möbius transform [4].

A further result with CAC but without symmetry or tetrahedron property was found in [11]

$$\frac{x + e_2}{x + e_1}\frac{x_{[12]} + o_2}{x_{[12]} + o_1} = \frac{x_{[1]} + e_2}{x_{[1]} + o_1}\frac{x_{[2]} + o_2}{x_{[2]} + e_1}.$$

Note that the parameters and variables appear symmetrically. This model has an interesting geometric interpretation as it describes some special relation between eight points on a conic [3]. However, also this model is linearizable.

3.8 Soliton solutions

One consequence of the the CAC property is that soliton solutions can be constructed using the third directions in various ways. We will basically follow "Hirota's direct method" which consists of the following steps:

1. find a background or vacuum solution,
2. find a 1-soliton-solution (1SS) built on top of the background,
3. use this info to guess a *dependent variable transformation into Hirota bilinear form*,
4. construct the first few soliton solutions perturbatively,
5. guess the general form (usually a determinant: Wronskian, Pfaffian etc.) and prove it.

Hirota's bilinear form is well suited for constructing continuous soliton solutions, because the dependent variable is then a polynomial of exponentials in which the independent variables appear linearly. Something similar should work for the discrete case.

Below we go through these steps in the case of the model H1,

$$(u - \hat{\tilde{u}})(\tilde{u} - \hat{u}) - (p - q) = 0,$$

following [19]. (In this section we use tilde for the n-shift, hat for the m-shift and bar for the k-shift).

3.8.1 Background solutions

First problem in the perturbative approach is to find the background solution. From the form of H1 we see that the natural guess $u_{nm} = 0$ is not a solution. The idea proposed in [7] is to take the CAC cube and insist that the solution is a fixed point of the shift in the third direction. In that case the "side"-equations of the cube are

$$Q(u, \tilde{u}, u, \tilde{u}; p, r) = 0, \qquad Q(u, \hat{u}, u, \hat{u}; q, r) = 0,$$

which for (H1) read

$$(\tilde{u} - u)^2 = r - p, \qquad (\hat{u} - u)^2 = r - q,$$

For convenience we reparametrize $(p, q) \to (a, b)$ by

$$p = r - a^2, \qquad q = r - b^2.$$

and then the side-equations factorize as

$$(\tilde{u} - u - a)(\tilde{u} - u + a) = 0, \qquad (\hat{u} - u - b)(\hat{u} - u + b) = 0,$$

Since the factor that vanishes may depend on n, m we actually have to solve

$$\tilde{u} - u = (-1)^\theta a, \qquad \hat{u} - u = (-1)^\chi b,$$

where $\theta, \chi \in \mathbb{Z}$ may depend on n, m. From consistency of the two equations we get $\theta \in \{n, 0\}$, $\chi \in \{m, 0\}$, and the set of possible background solution turns out to be $an + bm + \gamma$, $\frac{1}{2}(-1)^n a + bm + \gamma$, $an + \frac{1}{2}(-1)^m b + \gamma$, or $\frac{1}{2}(-1)^n a + \frac{1}{2}(-1)^m b + \gamma$. In the following we only consider the background solution $u_{nm}^{OSS} := an + bm + \gamma$.

3.8.2 1SS

Next we use the cube again, but now assuming that the bottom square is for the background solution and the top square for a one-soliton solution (1SS). Thus we have to solve the side equations

$$(u - \bar{\tilde{u}})(\tilde{u} - \bar{u}) = p - \varkappa, \tag{3.31a}$$

$$(u - \bar{\hat{u}})(\hat{u} - \bar{u}) = \varkappa - q. \tag{3.31b}$$

Here u is the background solution $an + bm + \gamma$, $\bar{u}$ is the new 1SS, and $\varkappa$ is the soliton parameter (the parameter in the bar-direction). Following the procedure of [19] we search for a new solution $\bar{u}$ of the form

$$\bar{u} = \bar{u}_0 + v,$$

where $\bar{u}_0$ is the bar-shifted background solution

$$\bar{u}_0 = an + bm + k + \lambda.$$

For v we then find

$$\dot{v} = \frac{Ev}{v + F}, \qquad \hat{v} = \frac{Gv}{v + H},$$

where

$$E = -(a + k), \qquad F = -(a - k), \qquad G = -(b + k), \qquad H = -(b - k),$$

and k is related to $\varkappa$ by $\varkappa = r - k^2$. As in the treatment of the Lax pair we introduce $v = f/g$ and $\Phi = (g, f)^T$ and then we can write the v-equations as matrix equations

$$\Phi(n + 1, m) = \mathcal{N}(n, m)\Phi(n, m), \qquad \Phi(n, m + 1) = \mathcal{M}(n, m)\Phi(n, m),$$

where

$$\mathcal{N}(n, m) = \Lambda \begin{pmatrix} E & 0 \\ 1 & F \end{pmatrix}, \qquad \mathcal{M}(n, m) = \Lambda' \begin{pmatrix} G & 0 \\ 1 & H \end{pmatrix}.$$

In this case E, F, G, H are constants and therefore we can choose the separation constants as $\Lambda = \Lambda' = 1$.

Since the matrices $\mathcal{N}, \mathcal{M}$ commute it is easy to find

$$\Phi(n, m) = \begin{pmatrix} E^n G^m & 0 \\ \dfrac{E^n G^m - F^n H^m}{-2k} & F^n H^m \end{pmatrix} \Phi(0, 0),$$

and if we let

$$\rho_{n,m} = \left(\frac{E}{F}\right)^n \left(\frac{G}{H}\right)^m \rho_{0,0} = \left(\frac{a + k}{a - k}\right)^n \left(\frac{b + k}{b - k}\right)^m \rho_{0,0},$$

then we obtain

$$v_{n,m} = \frac{-2k\rho_{n,m}}{1 + \rho_{n,m}}.$$

Finally we obtain the 1SS for H1:

$$u_{n,m}^{1SS} = an + bm + \lambda + k + \frac{-2k\rho_{n,m}}{1 + \rho_{n,m}}.$$

The function ρ is the analogue of the continuous e^η in Hirota's approach.

3.8.3 NSS

On the basis of the above result we propose a bilinearizing dependent variable transformation

$$u_{n,m} = an + bm + \lambda - \frac{g_{n,m}}{f_{n,m}}. \tag{3.32}$$

When this is substituted into H1 we find

$$\begin{aligned} \text{H1} &\equiv (u - \hat{\tilde{u}})(\tilde{u} - \hat{u}) - p + q \\ &= -[\mathcal{H}_1 + (a - b)f\hat{\tilde{f}}][\mathcal{H}_2 + (a + b)\hat{f}\tilde{f}]/(f\hat{f}\tilde{f}\hat{\tilde{f}}) + (a^2 - b^2), \end{aligned}$$

where

$$\mathcal{H}_1 \equiv \hat{g}\tilde{f} - \tilde{g}\hat{f} + (a - b)(\hat{f}\tilde{f} - f\hat{\tilde{f}}) = 0, \tag{3.33a}$$

$$\mathcal{H}_2 \equiv g\hat{\tilde{f}} - \hat{\tilde{g}}f + (a + b)(f\hat{\tilde{f}} - \hat{f}\tilde{f}) = 0. \tag{3.33b}$$

Thus (3.33) can be considered as the bilinearization of H1.

Discrete bilinear equations are often solved by Casoratians (Casorati determinants) defined as follows: Given functions $\varphi_i(n, m, h)$ we define the column vectors

$$\varphi(n, m, h) = (\varphi_1(n, m, h), \varphi_2(n, m, h), \dots, \varphi_N(n, m, h))^T,$$

and then compose the $N \times N$ Casorati matrix from columns with different shifts h_i, and then the determinant

$$C_{n,m}(\varphi; \{h_i\}) = |\varphi(n, m, h_1), \varphi(n, m, h_2), \cdots, \varphi(n, m, h_N)|.$$

For example

$$C_{n,m}^1(\varphi) := |\varphi(n, m, 0), \varphi(n, m, 1), \cdots, \varphi(n, m, N - 1)|$$
$$\equiv |0, 1, \cdots, N - 1| \equiv |\widehat{N - 1}|,$$
$$C_{n,m}^2(\varphi) := |\varphi(n, m, 0), \cdots, \varphi(n, m, N - 2), \varphi(n, m, N)|$$
$$\equiv |0, 1, \cdots, N - 2, N| \equiv |\widehat{N - 2}, N|,$$

where we have also introduced some shorthand notations.

The main result for H1 is the following: The bilinear equations $\mathcal{H}_i$ are solved by Casoratians $f = |\widehat{N - 1}|$, $g = |\widehat{N - 2}, N|$, with φ given by

$$\varphi_i(n, m, h) = \varrho_i^+ k_i^h (a + k_i)^n (b + k_i)^m + \varrho_i^- (-k_i)^h (a - k_i)^n (b - k_i)^m,$$

and thus with these f, g we get the NSS from (3.32).

Similar results exist for other equations in the ABS list [16, 19, 22]. In each case the structure of the soliton solution is similar to those of the Hirota–Miwa equation [17, 20].

3.9 Conclusions

In these lectures we have considered many forms of integrability for difference equations. In particular we have discussed

Singularity confinement, which is simple to apply and in practice rather efficient, especially for deautonomization. But it is only necessary, not sufficient.

Algebraic entropy, which is perhaps more complicated to apply but also more precise and more generic. In practice it turns out that

- linear growth = linearizability
- polynomial growth = integrability
- exponential growth = chaos

Consistency-Around-Cube, which is only applicable to maps defined on a square lattice, but on the other hand it is a quite strong concept. It leads to Lax pairs in a straightforward way and using this concept one can construct soliton solutions systematically.

Acknowledgements

Many of the results presented here are from recent collaborations with C. Viallet, F. Nijhoff, J. Atkinson and D.-J. Zhang.

References

[1] Ablowitz, M. J., Halburd, R., and Herbst, B. 2000. On the extension of the Painlevé property to difference equations. *Nonlinearity*, **13**(3), 889–905.

[2] Adler, V. E. 1998. Bäcklund transformation for the Krichever–Novikov equation. *Internat. Math. Res. Notices*, **1998**(1), 1–4.

[3] Adler, V. E. 2006. Some incidence theorems and integrable discrete equations. *Discrete Comput. Geom.*, **36**(3), 489–498.

[4] Adler, V. E., and Suris, Yu. B. Private communication.

[5] Adler, V. E., Bobenko, A. I., and Suris, Yu. B. 2003. Classification of integrable equations on quad-graphs. The consistency approach. *Comm. Math. Phys.*, **233**(3), 513–543.

[6] Adler, V. E., Bobenko, A. I., and Suris, Yu. B. 2009. Discrete nonlinear hyperbolic equations; classification of integrable cases. *Funct. Anal. Appl.*, **43**(1), 3–17.

[7] Atkinson, J., Hietarinta, J., and Nijhoff, F. 2007. Seed and soliton solutions of Adler's lattice equation. *J. Phys. A*, **40**(1), F1–F8.

[8] Falqui, G., and Viallet, C.-M. 1993. Singularity, complexity, and quasi-integrability of rational maps. *Comm. Math. Phys.*, **154**(1), 111–125.

[9] Grammaticos, B., Ramani, A., and Papageorgiou, V. 1991. Do integrable mappings have the Painlevé property? *Phys. Rev. Lett.*, **67**(14), 1825–1828.

[10] Hietarinta, J. 1987. Direct methods for the search of the second invariant. *Phys. Rep.*, **147**(2), 87–154.

[11] Hietarinta, J. 2004. A new two-dimensional lattice model that is 'consistent around the cube'. *J. Phys. A*, **37**(6), L67–L73.

[12] Hietarinta, J. 2005. Searching for CAC-maps. *J. Nonlinear Math. Phys.*, **12**(suppl. 2), 223–230.

[13] Hietarinta, J., and Viallet, C. 1998. Singularity confinement and chaos in discrete systems. *Phys. Rev. Lett.*, **81**(2), 325–328.

[14] Hietarinta, J., and Viallet, C. 2000. Discrete Painlevé I and singularity confinement in projective space. *Chaos Solitons Fractals*, **11**(1), 29–32.

[15] Hietarinta, J., and Viallet, C. 2007. Searching for integrable lattice maps using factorization. *J. Phys. A*, **40**(42), 12629–12643.

[16] Hietarinta, J., and Zhang, D.-j. 2009. Soliton solutions for ABS lattice equations II: Casoratians and bilinearization. arXiv:0903.1717.

[17] Hirota, R. 1981. Discrete analogue of a generalized Toda equation. *J. Phys. Soc. Japan*, **50**(11), 3785–3791.

[18] Hirota, R., Kimura, K., and Yahagi, H. 2001. How to find the conserved quantities of nonlinear discrete equations. *J. Phys. A*, **34**(48), 10377–10386.

[19] J. Atkinson, J. Hietarinta, and Nijhoff, F. 2008. Soliton solutions for Q3. *J. Phys. A*, **41**(14), 142001.

[20] Miwa, T. 1982. On Hirota's difference equations. *Proc. Japan Acad. Ser. A Math. Sci.*, **58**(1), 9–12.

[21] Nijhoff, F. 2002. Lax pair for the Adler (lattice Krichever–Novikov) system. *Phys. Lett. A*, **297**(1-2), 49–58.

[22] Nijhoff, F., Atkinson, J., and Hietarinta, J. 2009. Soliton solutions for ABS lattice equations I: Cauchy matrix approach. arXiv:0902.4873.

[23] Nijhoff, F. W., and Walker, A. J. 2001. The discrete and continuous Painlevé VI hierarchy and the Garnier systems. *Glasg. Math. J.*, **43A**, 109–123.

[24] Nijhoff, F. W., Ramani, A., Grammaticos, B., and Ohta, Y. 2001. On discrete Painlevé equations associated with the lattice KdV aystems and the Painlevé VI equation. *Stud. Appl. Math.*, **106**(3), 261–314.

[25] Nimmo, J. J. C., and Schief, W. K. 1997. Superposition principles associated with the Moutard transformation: an integrable discretization of a (2 + 1)-dimensional sine-Gordon system. *Proc. Roy. Soc. London Ser. A*, **453**(1957), 255–279.

[26] Ohta, Y., Tamizhmani, K. M., Grammaticos, B., and Ramani, A. 1999. Singularity confinement and algebraic entropy: the case of the discrete Painlevé equations. *Phys. Lett. A*, **262**(2-3), 152–157.

[27] Quispel, G. R. W., Roberts, J. A. G., and Thompson, C. J. 1988. Integrable mappings and soliton equations. *Phys. Lett. A*, **126**(7), 419–421.

[28] Ramani, A., Grammaticos, B., and Hietarinta, J. 1991. Discrete versions of the Painlevé equations. *Phys. Rev. Lett.*, **67**(14), 1829–1831.

[29] Rasin, O. G., and Hydon, P. E. 2007. Conservation laws for integrable difference equations. *J. Phys. A*, **40**(42), 12763–12773.

[30] Sahadevan, R., and Capel, H. W. 2003. Complete integrability and singularity confinement of nonautonomous modified Korteweg–de Vries and sine-Gordon mappings. *Phys. A*, **330**(3-4), 373–390.

[31] Tremblay, S., Grammaticos, B., and Ramani, A. 2001. Integrable lattice equations and their growth properties. *Phys. Lett. A*, **278**(6), 319–324.

4

Orthogonal Polynomials, their Recursions, and Functional Equations

Mourad E. H. Ismail[a]

Abstract

These lectures contain a brief introduction to orthogonal polynomials with some applications to nonlinear recurrence relation. We discuss the spectral theory of orthogonal polynomials. We also mention other applications which may not be well known to the integrable systems community. For example we treat birth and death processes, the J-matrix method, and inversions of large Hankel matrices. The spectral theory of the Askey–Wilson operators on bounded and unbounded intervals is briefly mentioned.

4.1 Introduction

Many mathematicians and physicists think orthogonal polynomials are polynomial solutions of Sturm–Liouville differential equations. This is usually emphasized in textbooks written with an applied flavor. Although the Jacobi, Hermite, Laguerre polynomials and their special cases are indeed solutions to such equations, all orthogonal polynomials arise from difference equations. In addition to satisfying second order difference equations in the degree they may also satisfy an operator equation in a continuous variable.

In this article we include a brief account of certain aspects of orthogonal polynomials which may appeal to specialists in difference equations. In particular we give many instances in the theory of orthogonal polynomials where difference equations play a fundamental role. We also discuss the spectral theory of orthogonal polynomials and review the direct and inverse problems for orthogonal polynomials, which is the theory of linear symmetric second order difference equations. We also show how structure relations for orthogonal

[a] This research was supported by King Saud University, Riyadh through grant DSFP/Math 01

polynomials lead to nonlinear difference equations satisfied by the recurrence coefficients of the orthogonal polynomials under very special initial conditions.

Definition 4.1 Let $f(x_n, x_{n+1}, \ldots, x_{n+m}) = 0$, $n = 1, 2, \ldots$ be a difference equation. We say that the equation has the discrete Painlevé property (singularity confinement) if x_n is such that it results in a singularity for x_{n+1}, then there is a natural number p such that this singularity is confined to $x_{n+1}, x_{n+2}, \ldots, x_{n+p}$. Furthermore x_{n+p+1} depends only on $x_n, x_{n-1}, \ldots$.

4.2 Orthogonal polynomials

We start with a given positive probability measure with infinite support and finite moments of all orders. The monomials form a basis for the space of polynomials hence by the Gram–Schmidt orthogonalization procedure there exists a unique set of monic polynomials orthogonal with respect to μ. We write the monic polynomials as

$$P_n(x) = x^n + \sum_{k=0}^{n-1} c_{n,n-k} x^k. \tag{4.1}$$

Following the notation in [24] we write

$$\int_{\mathbb{R}} P_m(x) P_n(x) \, d\mu(x) = \zeta_n \delta_{m,n}. \tag{4.2}$$

Let

$$m_n = \int_{\mathbb{R}} x^n \, d\mu(x). \tag{4.3}$$

Define the Hankel determinants

$$D_n = \begin{vmatrix} m_0 & m_1 & \cdots & m_n \\ m_1 & m_2 & \cdots & m_{n+1} \\ \vdots & \vdots & & \vdots \\ m_n & m_{n+1} & \cdots & m_{2n} \end{vmatrix}. \tag{4.4}$$

It is not difficult to see that the quadratic form associated with the matrix in (4.4) is positive definite for all n, $n = 0, 1, \ldots$. The uniqueness of P_n implies the representation

$$P_n(x) = \frac{1}{D_{n-1}} \begin{vmatrix} m_0 & m_1 & \cdots & m_n \\ m_1 & m_2 & \cdots & m_{n+1} \\ \vdots & \vdots & & \vdots \\ m_{n-1} & m_n & \cdots & m_{2n-1} \\ 1 & x & \cdots & x^n \end{vmatrix}. \tag{4.5}$$

The above representation implies

$$\zeta_n = \frac{D_n}{D_{n-1}}. \tag{4.6}$$

To see this note that when $m = n$ the left-hand side of (4.2) becomes $\int_{\mathbb{R}} x^n P_n(x)\,d\mu(x)$ which, in view of (4.5) is D_n/D_{n-1}.

Every monic sequence of orthogonal polynomials satisfies a three term recurrence relation of the form

$$xP_n(x) = P_{n+1}(x) + \alpha_n P_n(x) + \beta_n P_{n-1}(x). \tag{4.7}$$

The reason is that $\{P_k(x)\}$ forms a basis for the space of polynomials and the coefficient of $P_k(x)$ in the expansion of $xP_n(x)$ is a positive multiple of $\int_{\mathbb{R}} P_n(x)xP_k(x)\,d\mu(x)$ which is zero when $k < m-1$ by the orthogonality of the P_n's. One can use (4.7) to show that the norms $\{\zeta_n\}$ are related to the recursion coefficients through,

$$\zeta_n = \beta_1\beta_2\cdots\beta_n. \tag{4.8}$$

Multiply (4.7) by $P_n(y)$ and subtract the result from the same formula after interchanging x and y and after some manipulations we establish the Christoffel–Darboux identity

$$\sum_{k=0}^{n-1} \frac{P_k(x)P_k(y)}{\zeta_k} = \frac{P_n(x)P_{n-1}(y) - P_n(y)P_{n-1}(x)}{\zeta_{n-1}(x-y)}. \tag{4.9}$$

As $y \to x$ we find

$$\sum_{k=0}^{n-1} \frac{P_k(x)^2}{\zeta_k} = \frac{P'_n(x)P_{n-1}(x) - P_n(x)P'_{n-1}(x)}{\zeta_{n-1}}. \tag{4.10}$$

The orthonormal polynomials will be denoted by $\{p_n(x)\}$, that is

$$p_n(x) = P_n(x)/\sqrt{\zeta_n}. \tag{4.11}$$

In the orthonormal polynomial notation the recursion (4.7) becomes

$$xp_n(x) = a_{n+1}p_{n+1}(x) + \alpha_n p_n(x) + a_n p_{n-1}(x), \qquad a_n^2 = \beta_n. \tag{4.12}$$

Let

$$A_n = \begin{pmatrix} \alpha_0 & a_1 & 0 & \cdots & 0 & 0 & 0 \\ a_1 & \alpha_1 & a_2 & \cdots & 0 & 0 & 0 \\ \vdots & \ddots & \ddots & \ddots & \vdots & \vdots & \vdots \\ 0 & 0 & \cdots & & a_{n-2} & \alpha_{n-2} & a_{n-1} \\ 0 & 0 & 0 & \cdots & 0 & a_{n-1} & \alpha_{n-1} \end{pmatrix}. \tag{4.13}$$

The recurrence relation (4.12) can be written as

$$x \begin{pmatrix} p_0(x) \\ p_1(x) \\ \vdots \\ p_n(x) \\ \vdots \end{pmatrix} = A_\infty \begin{pmatrix} p_0(x) \\ p_1(x) \\ \vdots \\ p_n(x) \\ \vdots \end{pmatrix}. \tag{4.14}$$

The matrix A_∞ has nonzero elements in the $j, j \pm 1$ positions and zeros in the j, k position, if $|j - k| > 1$. It is called a *Jacobi* or *tridiagonal* matrix.

Let

$$P_n(x) = \sum_{k=0}^{n} c_{n,k} x^{n-k}, \qquad c_{n,0} := 1. \tag{4.15}$$

By substituting for the polynomials in (4.7) we see that

$$c_{n,1} = -\sum_{k=0}^{n-1} \alpha_k, \qquad n > 0, \quad c_{0,1} := 0. \tag{4.16}$$

Moreover we have

$$c_{n,2} = -\sum_{k=1}^{n-1} \beta_k - \sum_{k=1}^{n-1} \alpha_k c_{k,1}, n > 1, \quad c_{1,2} = 0. \tag{4.17}$$

When we study the q-difference equations we will repeatedly use the algebraic relation

$$P_n(qy) = q^n P_n(y) + (1 - q)q^{n-1} c_{n,1} P_{n-1}(y) + \text{lower order terms.} \tag{4.18}$$

When the P_n's are symmetric polynomials, that is $P_n(-x) = (-1)^n P_n(x)$, we need one more term in (4.18). In this case $c_{n,1} = \alpha_n = 0$ for all n and

$$P_n(qy) = q^n P_n(y) + (1 - q^2)q^{n-2} c_{n,2} P_{n-2}(y) + \text{lower order terms} \tag{4.19}$$

and $c_{n,2} = -\sum_{k=1}^{n-1} \beta_k, n > 1$.

4.3 The spectral theorem

First we note that the polynomials P_n are characteristic polynomials of real symmetric matrices.

Theorem 4.2 *The monic polynomials have the determinant representation*

$$P_N(x) = determinant\ of\ (\lambda I_N - A_N), \tag{4.20}$$

where I_N is an $N \times N$ identity matrix.

Proof Let $S_n(\lambda)$ be the characteristic polynomial of A_n, $n = 1, 2, \ldots$. By expanding the determinant expression for S_{N+1} about the last row it follows that $\{S_n(x)\}$ satisfies the recurrence relation (4.7). On the other hand $S_1(x) = x - \alpha_0$, $S_2(x) = (x - \alpha_0)(x - \alpha_1) - \beta_1$, so $S_1 = P_1$ and $S_2 = P_2$. Therefore S_n and P_n agree for all n. □

Theorem 4.3 *The zeros of P_n are real and simple for all $n > 0$. Moreover the zeros of $P_n(x)$ and $P_{n-1}(x)$ interlace.*

Proof The reality of the zeros follows from Theorem 4.2. It is clear from (4.10) that $P_n(x)$ and $P'_n(x)$, $n > 0$, do not vanish at the same point. Moreover (4.10) shows that $P'_n(\xi)P_{n-1}(\xi) > 0$ when $P_n(\xi) = 0$. Thus $P_{n-1}(x)$ changes sign an odd number of times on any interval formed by consecutive zeros of $P_n(x)$. This establishes the interlacing property. □

Let

$$x_{N,1} > x_{N,2} > \cdots > x_{N,N}, \tag{4.21}$$

be the zeros of $P_N(x)$. It is easy to see from (4.12) that

$$v_k := (p_0(x_{N,k}), p_1(x_{N,2}), \ldots, p_{N-1}(x_{N,2}))^T \tag{4.22}$$

is an eigenvector of A_N. Moreover $\|v_k\|^2 = P'_N(x_{N,k})P_{N-1}(x_{N,k})/\zeta_{N-1}$ follows from (4.10). Define $\rho_k(N)$, $1 \le k \le N$ by

$$\frac{1}{\rho_k(N)} = \zeta_{N-1} P'_N(x_{N,k})P_{N-1}(x_{N,k}). \tag{4.23}$$

We summarize our findings so far in a theorem.

Theorem 4.4 *The matrix A_N has eigenvalues $x_{N,k}$, $1 \le k \le N$ and the corresponding normalized eigenvectors are $\sqrt{\rho_k(N)}\, v_k$, $1 \le k \le N$.*

The next result is the spectral theorem for orthogonal polynomials which is also known as Favard's theorem. The theorem was known long before Favard and in fact is stated and proved in Stone's book [39] which appeared three years before Favard's paper [16].

Theorem 4.5 (The spectral theorem)   *Let $\{P_n(x)\}$ be a sequence of monic polynomials generated by $P_0(x) = 1$, $P_1(x) = x - \alpha_0$ and the recurrence relation (4.7). If $\alpha_n \in \mathbb{R}$, $n = 0, 1, \ldots$ and $\beta_n > 0$, $n = 1, 2, \ldots$ then there is a probability measure μ such that the orthogonality relation (4.2) holds with $\zeta_0 = 1$, $\zeta_n = \beta_1 \beta_2 \cdots \beta_n$.*

Proof The matrix whose row vectors are $v_k \sqrt{\zeta_{N-1}/\rho_k}$ is an orthogonal matrix because the v_k is orthogonal to v_j when $j \neq k$ and the rows are unit vectors. Consequently the columns are mutually orthogonal unit vectors. In other words $\sum_{i=1}^{N} p_n(x_{N,i}) p_m(x_{N,i}) \rho_i(N) = \delta_{m,n}$. Equivalently

$$\sum_{i=1}^{N} P_n(x_{N,i}) P_m(x_{N,i}) \rho_i(N) = \zeta_n \delta_{m,n}, \qquad 0 \leq m, n < N. \qquad (4.24)$$

It is clear from the above equation that $\sum_{k=1}^{N} \rho_k(N) = 1$. Let μ_n be a right continuous step function with $\mu_N(x) = 0$ if $x < x_{N,N}$, has jump $\rho_k(N)$ at $x_{N,k}$, and $= 1$ for $x \geq x_{N,1}$. Since the variation of μ_N is 1 for all N, by Helley's selection principle a subsequence $\{\mu_{N_k}\}$ of $\{\mu_N\}$ will converge to a nondecreasing function μ and we now have the orthogonality relation (4.2) for all $m, n \geq 0$.   □

Note that (4.7) determines the polynomials in terms of the moments. The converse is also true, that is the orthogonal polynomials determine the moments uniquely. To see this use (4.14) and the spectral theorem to see that x^n is the first component of the vector

$$A_{\infty}^n (P_0(x), P_1(x), \ldots, P_m(x), \ldots)^T,$$

hence m_n is the first component of the vector

$$\int_{\mathbb{R}} A_{\infty}^n (P_0(x), P_1(x), \ldots, P_m(x), \ldots)^T \, d\mu(x).$$

The measure μ produced by the spectral theorem may not be unique. Moreover the spectral theorem is an existence theorem and does not provide a tool to construct the orthogonality measure(s). The spectral theorem for orthogonal polynomials also solves an associated problem, namely the problem of recovering μ from its moments $\{m_n\}$.

The spectral theorem can be viewed as the spectral theorem for a semi-infinite tridiagonal (Jacobi) matrix operator. This point of view is explained in [39] and a more modern treatment is in [1]. A very recent and up-to-date treatment of the moment problem is in Simon's excellent survey article [38]. Brown and Christiansen [9] study the maximal and minimal extensions of the Jacobi operator corresponding to three term recurrence relations. In §4.7 we will revisit the topic of determining the spectral measures.

Remark 4.6 When $\alpha_n = 0$ then the largest zero of $P_n(x)$ is $< 2\sqrt{\beta_n}$, see Theorem 7.2.7 in [24] (with $c_n = 1/4$. It seems plausible that when $\alpha_n = 0$ and β_n increases then the largest zero of $P_n(x)$ is $\approx 2\sqrt{\beta_n}$.

4.4 The Freud nonlinear recursions

Freud [17] realized that the recursion coefficients of polynomials orthogonal with respect to an exponential weight function $e^{-v(x)}$ satisfy nonlinear recurrence relation. As an example of his technique we consider the weight function

$$w(x) = c\exp(-Ntx^4/4 - aNx^2/2), \qquad x \in \mathbb{R}. \tag{4.25}$$

Theorem 4.7 *The recursion coefficients of the polynomials orthogonal with respect to the weight function w in (4.25) satisfy the discrete Painlevé-I equation*

$$\frac{n}{N} = a_n^2[a + t(a_{n+1}^2 + a_n^2 + a_{n-1}^2)]. \tag{4.26}$$

Proof Consider the integral $I_n := \int_{\mathbb{R}}(p_{n-1}(y)p_n(y))'w(y)\,dy$. The orthogonality of the p_n's implies

$$I_n = \int_{\mathbb{R}} p_n(y)'p_{n-1}(y)w(y)\,dy = \frac{n}{a_n},$$

since the coefficient of x^n in $p_n(x)$ is $1/(a_1a_2\cdots a_n)$. We now evaluate I_n by integration by parts and obtain

$$
\begin{aligned}
I_n/N &= \int_{\mathbb{R}} \frac{v'(y)}{N} p_n(y)p_{n-1}(y)w(y)\,dy = \int_{\mathbb{R}} [ty^3 + ay]p_n(y)p_{n-1}(y)w(y)\,dy \\
&= \int_{\mathbb{R}} [ty^2 + a][a_{n+1}p_{n+1}(y) + a_np_{n-1}(y)]p_{n-1}(y)w(y)\,dy \\
&= aa_n + t\int_{\mathbb{R}} [a_{n+1}a_{n+2}p_{n+2}(y) + (a_n^2 + a_{n+1}^2)p_n(y) + a_na_{n-1}p_{n-2}(y)] \\
&\qquad\qquad\qquad\qquad \times [a_np_n(y) + a_{n-1}p_{n-2}(y)]w(y)\,dy \\
&= aa_n + ta_n[a_{n+1}^2 + a_n^2 + a_{n-1}^2],
\end{aligned}
$$

and the proof is complete. $\qquad\qquad\qquad\qquad\qquad\qquad\qquad\qquad\square$

Freud established (4.26) in [17] when $a = 0$, $N = 1$. Freud also considered the case $v(x) = x^6$ and was interested in the large n asymptotics of a_n for the exponential weights. It clear from (4.26), with $a = 0$, $N = 1$, that one can make the ansatz $a_n \approx bn^\alpha$ and proceed to conclude that $\alpha = \frac{1}{4}$ and $b^4 = \frac{1}{3}$. Freud used clever upper and lower bounds to prove that $a_n = (n/3)^{1/4}[1 + o(1)]$.

A treatment of Freud's techniques is in Van Assche's very informative article [41].

Equation (4.26) when written in the monic recursion coefficients takes the form

$$\frac{n/N}{\beta_n} = a + t[\beta_{n+1} + \beta_n + \beta_{n-1}]. \tag{4.27}$$

The nonlinear recursion (4.27) is discrete Painlevé 1 (dP1) in the current classification of difference equations [19]. It is also called the string equation in the physics literature. Of course the solution when the $\{\beta_n\}$ is the sequence of recursion coefficients of the corresponding monic orthogonal polynomials then $\beta_n > 0$. In general when we consider (4.27) with general initial conditions then some β_n's may vanish causing a singularity in the solution. The singularity however is confined. This can be seen as follows. First absorb N in a and t, so N is now $= 1$. Let $a = tA$ so that (4.27), after replacing β_n by x_n, becomes

$$x_{n+1} = \frac{n}{tx_n} - A - x_n - x_{n-1} \tag{4.28}$$

Let $x_m = \epsilon$, and we will then let $\epsilon \to 0$. Thus

$$x_{m+1} = \frac{m}{t\epsilon} - A - \epsilon - x_{m-1},$$

$$x_{m+2} = -\frac{m}{t\epsilon} + x_{m-1} + \frac{m+1}{m}\epsilon + O(\epsilon^2),$$

$$x_{m+3} = -\frac{m+3}{m}\epsilon + O(\epsilon^2)$$

$$x_{m+4} = -A - x_{m-1} + \frac{2}{m}\epsilon + O(\epsilon^2).$$

When $A + x_{m-1} \neq 0$ then the next element, namely x_{m+5} does not tend to zero as $\epsilon \to 0$. This shows that the singularity is indeed confined.

4.5 Differential equations

In this section we derive differential equations and raising and lowering operators for general orthogonal polynomials. The differential equations were derived in [6, 8, 11, 34]. Discrete and q-analogue are in [22, 29].

Theorem 4.8 *Assume v' is continuous on (a, b) and*

$$\int_a^b \frac{v'(x) - v'(y)}{x - y} y^n w(y) \, dy$$

exists for nonnegative integers n. Define $A_n(x)$, $B_n(x)$ by

$$\zeta_n A_n(x) = \frac{w(y)P_n^2(y)}{y - x}\Big]_{y=a^+}^{y=b^-} + \int_a^b \frac{v'(x) - v'(y)}{x - y}P_n^2(y)w(y)\,dy, \qquad (4.29)$$

$$\zeta_{n-1}B_n(x) = \frac{w(y)P_n(y)P_{n-1}(y)}{y - x}\Big]_{y=a^+}^{y=b^-}$$
$$+ \int_a^b \frac{v'(x) - v'(y)}{x - y}P_n(y)P_{n-1}(y)w(y)\,dy. \qquad (4.30)$$

and assume that the boundary terms in (4.29)–(4.30) exist. Then the structure relation

$$\frac{d}{dx}P_n(x) = \beta_n A_n(x)P_{n-1}(x) - B_n(x)P_n(x), \qquad (4.31)$$

holds.

The first example is to consider the weight function $w(x) = c\exp(-tx^4/4 - ax^2/2)$ considered also in Alexander Its' lectures in this conference. The constant c is chosen to make $\int_{\mathbb{R}} w(x)\,dx = 1$. Thus $v'(x) = tx^3 + ax$. Polynomials orthogonal to even weight functions always have the symmetry property $p_n(-x) = (-1)^n p_n(x)$. Hence

$$A_n(x) - \frac{1}{\zeta_n}\int_{\mathbb{R}}[a + t(x^2 + xy + y^2)]|P_n^2(y)w(y)\,dy$$

$$= a + tx^2 + \frac{t}{\zeta_n}\int_{\mathbb{R}}[P_{n+1}(y) + \beta_n P_{n-1}(y)]^2\,w(y)\,dy$$

$$= a + t[x^2 + \beta_{n+1} + \beta_n],$$

where we used (4.12) and $\zeta_{n+1} = \beta_{n+1}\zeta_n$. Similarly

$$B_n(x) = \frac{1}{\zeta_{n-1}}\int_{\mathbb{R}} txyP_n(y)P_{n-1}(y)w(y)\,dy = xt\beta_n.$$

To summarize, we have

$$A_n(x) = a + t[x^2 + \beta_{n+1} + \beta_n], \qquad B_n(x) = xt\beta_n. \qquad (4.32)$$

Another proof of Theorem 4.7 It is clear that (4.17) implies

$$c_{n+1,2} - c_{n,2} = -\beta_n. \qquad (4.33)$$

Substitute for P_n from (4.15) in equation (4.31) and take into account the evaluations of $A_n(x)$ and $B_n(x)$ in (4.32) (after replacing t and a by Nt and Na then equate coefficients of x^{n+1} and x^{n-1}. The first gives $0 = 0$ but the second gives

$$\frac{n}{N\beta_n} = a + t[\beta_n + \beta_{n+1}] + tc_{n-1,2} - tc_{n,2},$$

which simplifies to (4.27) after applying (4.33). $\qquad\square$

The A_n's and B_n's satisfy linear recurrence relations. The recursions are

$$B_n(x) + B_{n+1}(x) = (x - \alpha_n)A_n(x) - v'(x), \tag{4.34}$$

and

$$B_{n+1}(x) - B_n(x) = \frac{\beta_{n+1}A_{n+1}(x)}{x - \alpha_n} - \frac{\beta_n A_{n-1}(x)}{(x - \alpha_n)} - \frac{1}{x - \alpha_n}. \tag{4.35}$$

Our thesis is that equation (4.31) and the recurrence relations (4.34)–(4.35) embody the properties of the recursion coefficients. In particular they imply nonlinear relations among the recursion coefficients including the string equation. To see this consider the weight $w(x) = c \exp(-tx^4/4 - ax^2/2)$. In this case $\alpha_n = 0$ and equation (4.35) becomes

$$1 = \beta_{n+1}[a + t(\beta_{n+1} + \beta_{n+2})] - \beta_n[a + t(\beta_n + \beta_{n-1})]. \tag{4.36}$$

To get a telescoping series add $t\beta_n$ in the first square bracket and compensate by adding $t\beta_{n+1}$ in the second square bracket. Now summing the above equations gives the Freud equation (4.27), after replacing a and t by Na and Nt, respectively and checking the initial condition. Actually we may take $\beta_0 = 0$.

Remark 4.9   It is important to note that (4.36) is a third order difference equation which integrates (by summation) to the second order difference equation (4.27). On the other hand equating coefficients of x^{n-1} in (4.31) leads directly to the integrated equation (4.27). Also note that the initial conditions have been used in deriving (4.27)

Remark 4.10   Lew and Quarles showed that the equation (4.27) with $a = 0$, that is

$$x_n(x_n + x_{n+1} + x_{n-1}) = n, \tag{4.37}$$

has a unique positive solution with the initial condition $x_0 = 0$. Of course this solution is $x_n = \beta_n$.

Let us revisit the paragraph before (4.3) and consider a Freud weight function $w(x) = \exp(-|x|^\alpha)$ on $\mathbb{R}$, so the orthonormal polynomials $\{p_n(x)\}$ satisfy

$$\frac{2}{\alpha} \int_0^\infty p_m(y^{1/\alpha})p_n(y^{1/\alpha})e^{-y}y^{-1+1/\alpha}\,\mathrm{d}y = \delta_{m,n},$$

when $m - n$ is even, and $m \neq n$. The Laguerre polynomials $\{L_n^{(\beta)}(x)\}$ are orthogonal with respect to the weight function $x^\beta e^{-x}$ on $(0, \infty)$. Their largest zero $x_{n,1}^{(L)}$ satisfies $x_{n,1}^{(L)}/n \to 4$. This suggests that the largest zero of $p_n(x)$ is possibly $\approx c[x_{n,1}^{(L)}]^{1/\alpha}$, for a constant c and $\beta = -1 + 1/\alpha$. Thus, by Remark 1 at the end of §4.3, $\beta_n = an^{2/\alpha}[1 + o(1)]$ for some constant a. This argument is very heuristic since $p_n(y^{1/\alpha})$ is not even a polynomial. Freud made the very precise conjecture that

$$\lim_{n \to \infty} \frac{\beta_n}{n^{2/\alpha}} = \left[\frac{\Gamma(\alpha + 1)}{\Gamma(\alpha/2)\Gamma(1 + \alpha/2)} \right]^{-2/\alpha}. \tag{4.38}$$

This conjecture led to many developments in orthogonal polynomials and approximation theory and was proved in [33, 36] using potential theoretic and weighted polynomial approximation techniques.

4.6 *q*-difference equations

Recall the definition of the q-difference operator

$$D_q f(x) = \frac{f(qx) - f(x)}{qx - x}. \tag{4.39}$$

As we saw in §4.5 a weight function introduces a potential function v which is then used to derive the differential relations of the orthogonal polynomials. Ismail and Simeonov [27] realized that a weight function w leads to two potential functions u and v defined by

$$u(x) = -\frac{D_{q^{-1}} w(x)}{w(x)}, \qquad v(qx) = -\frac{D_q w(x)}{w(x)}. \tag{4.40}$$

The relation between the two potential functions is

$$v(x) = \frac{u(x)}{1 + (1 - 1/q)xu(x)}. \tag{4.41}$$

Earlier Chen and Ismail introduced the u potential function in [12]. Chen and Ismail [12], and Ismail and Simeonov [27] assumed that $\int_0^\infty p(x)w(x)\,dx/x$ converges for polynomials $p(x)$. The next theorem [28] is an extension of the results in [27] where the above assumption is removed.

Theorem 4.11 *Let v be as in (4.40) and define*

$$C_n(x) := \frac{\ln q \, P_n^2(0)w(0)}{(1 - q)\zeta_n \, x} + \frac{q}{\zeta_n} \int_0^\infty \frac{v(x) - v(qy)}{x - qy} P_n(y)P_n(qy)w(y)\,dy, \tag{4.42}$$

$$D_n(x) := \frac{\ln q \, P_n(0)P_{n-1}(0)w(0)}{(1 - q)\zeta_{n-1} \, x}$$

$$+ \frac{q}{\zeta_{n-1}} \int_0^\infty \frac{v(x) - v(qy)}{x - qy} P_n(y)P_{n-1}(qy)w(y)\,dy, \tag{4.43}$$

if w is defined on $(0, \infty)$ and continuous at $x = 0$ from the right. If w is defined *on $\mathbb{R}$ and is continuous at $x = 0$ then we replace $\int_0^\infty$ by $\int_{\mathbb{R}}$ and delete the boundary terms in (4.42) and (4.43). We then have the structure relation*

$$D_{q^{-1}}P_n(x) = \beta_n C_n(x)P_{n-1}(x) - D_n(x)P_n(x). \tag{4.44}$$

Moreover the functions $C_n(x)$ and $D_n(x)$ satisfy the recursions

$$D_{n+1}(x) + D_n(x) = (x - \alpha_n)C_n(x) + x(1/q - 1)\sum_{j=0}^n C_j(x) - v(x), \tag{4.45}$$

$$(x - \alpha_n)D_{n+1}(x) - (x/q - \alpha_n)D_n(x) = -1 + \beta_{n+1}C_{n+1}(x) - \beta_n C_{n-1}(x), \tag{4.46}$$

where α_n and β_n are as in (4.7).

Example 1   Consider the weight function

$$w(x) = C\prod_{n=0}^\infty \frac{1}{1 + a(1-q)q^{2n+1}x^2}, \qquad x \in \mathbb{R}, \tag{4.47}$$

where C is a normalization constant to make $\int_{\mathbb{R}} w(x)\,dx = 1$. Thus $v(x) = ax$ and $P_n(-x) = (-1)^n P_n(x)$, which implies $P_{2n+1}(0) = 0$. A simple calculation implies

$$D_n(x) = 0, \qquad C_n(x) = aq^{n+1} \tag{4.48}$$

Equation (4.46) implies

$$1 = \beta_{n+1}C_{n+1}(x) - \beta_n C_{n-1}(x). \tag{4.49}$$

Now equation (4.49) establishes the difference equation $1 = aq^{n+2}\beta_{n+1} - aq^n\beta_n$, whose solution subject to $\beta_0 = 0$ is

$$\beta_n = \frac{q^{-2n}(1-q^n)}{a(1-q)}. \tag{4.50}$$

The polynomials in this case are the discrete q-Hermite polynomials. Their recurrence relation is

$$xP_n(x) = P_{n+1}(x) + q^{-n}(q^{-n} - 1)P_{n-1}(x). \tag{4.51}$$

One can establish the generating function

$$\sum_{n=0}^\infty \frac{(-1)^n q^{\binom{n}{2}}}{(q;q)_n}P_n(x)t^n = \prod_{n=0}^\infty \frac{1 - q^n xt}{1 + q^{2n}t^2/a}, \tag{4.52}$$

where $(q;q)_n = \prod_{j=1}^n (1 - q^j)$ with $(q;q)_0 := 1$. The generating function gives the explicit formula

$$P_n(x) = \sum_{k=0}^n \frac{(q;q)_n}{(q^2;q^2)_k}(-1)^k x^{n-2k}a^{-k}q^{k(2k+1-2n)}. \tag{4.53}$$

These polynomials are discrete q-analogues of the Hermite polynomials, see [32, §3.29].

Example 2 Consider the weight function

$$w(x) = C \prod_{n=0}^{\infty} \frac{1}{1 + a(1-q)q^{4n+3}x^4}, \qquad x \in \mathbb{R}, \tag{4.54}$$

and $\int_{\mathbb{R}} w(x)\,dx = 1$. Thus $v(x) = ax^3$ and it is straightforward to find

$$D_n(x) = aq^{n+1}\beta_n x. \tag{4.55}$$

Moreover

$$C_n(x) = \frac{aq}{\zeta_n} \int_{\mathbb{R}} [x^2 + q^2 y^2] P_n(y) P_n(qy) w(y)\,dy$$

$$= aq^{n+1}x^2 + \frac{aq^2}{\zeta_n} \int_{\mathbb{R}} [P_{n+1}(y) + \beta_n P_{n-1}(y)]$$
$$\times [P_{n+1}(qy) + \beta_n P_{n-1}(qy)]\, w(y)\,dy$$

$$= aq^{n+1}x^2 + aq^{n+3}\beta_{n+1} + aq^{n+1}\beta_n + \frac{aq^2}{\zeta_{n-1}} \int_{\mathbb{R}} P_{n-1}(y) P_{n+1}(qy) w(y)\,dy.$$

Applying (4.19) we reduce the above to

$$C_n(x) = aq^{n+1}[x^2 + \beta_n + q^2\beta_{n+1} + (1-q^2)c_{n+1,2}]. \tag{4.56}$$

The coefficients of x^{n+1} on both sides of (4.44) match. Now equate coefficients of x^{n-1} in equation (4.44) to see that

$$\frac{1-q^n}{1-q}\frac{q^{-2n}}{a\beta_n} = q^2[\beta_n + \beta_{n+1}] + \beta_{n-1} + (1-q^2)c_{n,2}. \tag{4.57}$$

Equivalently

$$q^2\beta_{n+2} = \beta_{n-1} + \frac{1-q^{n+1}}{1-q}\frac{q^{-2n-2}}{a\beta_{n+1}} - \frac{1-q^n}{1-q}\frac{q^{-2n}}{a\beta_n}. \tag{4.58}$$

In terms of the $c_{n,2}$'s the above difference equations are

$$\frac{1-q^n}{1-q}\frac{q^{-2n}}{a} = [c_{n+1,2} - c_{n,2}][q^2 c_{n+2,2} - c_{n-1,2}]. \tag{4.59}$$

In other words

$$q^2 c_{n+2,2} = c_{n-1,2} + \frac{q^{-2n}(1-q^n)}{a(1-q)}\frac{1}{c_{n+1,2} - c_{n,2}} \tag{4.60}$$

If in (4.58) we make the ansatz $\beta_n = cq^{-\lambda n}[1 + o(1)]$ we find that

$$\beta_n = \frac{q^{-n}}{\sqrt{aq(1-q)}}[1 + o(1)]. \tag{4.61}$$

Examples 1 and 2 suggest considering the weight functions

$$w_m(x;q) = C_m \prod_{n=0}^{\infty} \frac{1}{1 + a(1-q)q^{2mn+2m-1}x^{2m}}, \qquad x \in \mathbb{R}, \qquad (4.62)$$

where $a > 0$ and $m = 1, 2, \ldots$. The normalization constant C_m makes $\int_{\mathbb{R}} w_m(x;q)\,dx = 1$. In Example 1 we saw that $\beta_n = q^{-n}(q^{-n} - 1)$, so, by Remark 4.9 at the end of §4.3, the largest zero of P_n is $2q^{-n}[1 + o(1)]$. This is the model for the weights $w_m(x;q)$. To guess the correct asymptotics we need to say some thing about the q-Laguerre polynomials $\{L_n^{(\alpha)}(x;q)\}$. They satisfy the relation

$$\int_0^{\infty} L_k^{(\alpha)}(x;q)L_j^{(\alpha)}(x;q)\frac{x^{\alpha}\,dx}{\prod_{n=0}^{\infty}(1 + xq^n)} = 0, \qquad j \neq k, \qquad (4.63)$$

[18, 32]. Equation (4.63) implies

$$\int_{\mathbb{R}} L_k^{(\alpha)}(x^{2m};q)L_j^{(\alpha)}(x^{2m};q)\frac{|x|^{2m\alpha+2m-1}\,dx}{\prod_{n=0}^{\infty}(1 + x^{2m}q^n)} = 0, \qquad j \neq k. \qquad (4.64)$$

Let $\xi_n(\alpha, q)$ be the largest zero of $L_n^{(\alpha)}(x;q)$. In [23] we proved that

$$\xi_n(\alpha, q) = \frac{q^{-2n-\alpha}}{i_1(q)}[1 + o(1)], \text{ as } n \to \infty,$$

where $i_1(q)$ is the smallest zero of the Ramanujan function $A_q(x)$. Comparing the orthogonality relation (4.64) and the following orthogonality relation of the polynomials $\{P_n(x;m)\}$

$$\int_R P_r(x;m)P_s(x;m)w_m(x;q)\,dx = 0, \qquad r \neq s,$$

it follows that when $r = 2mk$, the largest zero of $P_r(x;m)$ is $[\xi_k(-1 + 1/2m; q^{2m})]^{1/2m}$, that is $O(q^{-2k})$, which is $O(q^{-r/m})$. The largest zero of an orthogonal polynomial is $\approx 2\sqrt{\beta_n}$, hence β_n of the polynomials $\{P_n(x;m)\}$ will have the limiting property

$$\lim_{n\to\infty} \beta_n q^{2n/m} = b, \qquad (4.65)$$

for some positive constant b. This result, (4.65), together with the analysis of polynomials orthogonal with respect to the weights $w_m(x;q)$ is in the forthcoming article [26]. The above heuristic argument is from [28].

4.7 The inverse problem

The direct problem for orthogonal polynomials is given the orthogonality measure determine the polynomials or the recursion coefficients in the three term recurrence relation. The inverse problem is "given the recursion coefficients in (4.12) with $a_n \neq 0$, $n > 0$ and α_n real for $n \geq 0$, determine the orthogonality measure(s) guaranteed by the spectral theorem (Theorem 4.5)." The technique outlined in this section determines the orthogonality measure when it is unique. The very interesting case of nonuniqueness is more involved and we do not have the space to describe it but the interested reader may consult [1, 37, 39]. Specific examples of orthogonal polynomials with nonunique measures are in [24, chapter 21], where explicit formulas are given for three families of orthogonal polynomials whose orthogonality measures are not unique and all the orthogonality measures are given through their Stieltjes transforms and in one case many explicit measures are found.

We first give necessary and sufficient conditions for the uniqueness of the orthogonality measure.

Theorem 4.12 *If the orthogonality measure is unique and $\{p_n(z)\}$ are the orthonormal polynomials, then the series $\sum_{n=0}^{\infty} |p_n(z)|^2$ diverges for all complex z. Conversely if $\sum_{n=0}^{\infty} |p_n(z)|^2$ diverges for one complex z then the orthogonality measure is unique.*

We next show a way to reduce the problem of finding μ to a problem of determining the large degree asymptotics of polynomials.

Recall that $\{P_n(x)\}$ solves the recurrence relation

$$xy_n = y_{n+1} + \alpha_n y_n + \beta_n y_{n-1} \tag{4.66}$$

with the initial conditions $P_0(x) = 1$, $P_1(x) = x + \alpha_1$. Define $\{P_n^*(x)\}$, a second solution to the same difference equation, (4.66), with the initial conditions $P_0^*(x) := 0$, $P_1^*(x) = 1$.

Theorem 4.13 *If the orthogonality measure μ is unique then*

$$\lim_{n \to \infty} \frac{P_n^*(z)}{P_n(z)} = \int_{\mathbb{R}} \frac{1}{z - t} \, d\mu(t), \qquad z \notin \mathbb{R}. \tag{4.67}$$

Proofs of Theorems 4.12–4.13 can be found in [1, 37]. References [1, 39] also show how the measure(s) of orthogonality is related to the spectral measure(s) of the operator A_∞ acting on the sequence space l^2.

One can find μ by inverting the Stieltjes transform $\int_{\mathbb{R}} 1/(z - t) \, d\mu(t)$. Indeed if $F(z) = \int_{\mathbb{R}} 1/(z - t) \, d\mu(t)$, $z \notin \mathbb{R}$ then

$$\mu(s) - \mu(t) = \int_s^t \frac{F(v - i\epsilon) - F(v + i\epsilon)}{2\pi i} \, dv. \tag{4:68}$$

In particular at a point of continuity of μ, we have

$$\mu'(x) = \lim_{\epsilon \to 0^+} \frac{F(t - i\epsilon) - F(t + i\epsilon)}{2\pi i}.$$

Moreover every isolated pole $z = u$ of $F(z)$ contributes a discrete mass of μ at $x = u$ and the mass equals the residue of $F(z)$ at $z = u$. The nonisolated masses of μ are found from the following theorem, which also hold for isolated masses.

Theorem 4.14 *If the orthogonality measure μ is unique and $\mu(\{u\}) = \rho > 0$ then*

$$1/\rho = \sum_{n=0}^{\infty} P_n^2(u)/\zeta_n.$$

The orthogonality measures of several important systems of orthogonal polynomials were found by computing the large n asymptotics of $P_n(x)$ and $P_n^*(x)$ then applying Theorem 4.13. Examples are in [3, 24]. The reference [24] also contains several examples of polynomials orthogonal with respect to nonunique measures.

4.8 Applications

I am sure the reader knows of many instances where special functions and orthogonal polynomials appear in the solutions of boundary value problems or Sturm–Liouville problems. Discrete orthogonal polynomials appear in combinatorics and coding theory. In this section we mention three applications which do not seem to be widely known but we feel they are interesting and useful.

The first concerns inverting Hankel matrices and uses the Christoffel–Darboux kernel

$$K_n(x, y) := \sum_{k=0}^{n} P_k(x) P_k(y)/\zeta_k. \tag{4.69}$$

Theorem 4.15 *Let μ be a positive measure whose support contains at least N points and set $m_n = \int_{\mathbb{R}} x^n d\mu(x)$. Let H_n be the Hankel matrix whose j, k entry is m_{j+k}, $0 \le j, k \le n \le N/2$. Assume that*

$$K_n(x, y) = \sum_{j,k=0}^{n} a_{j,k}(n) \, x^j y^k. \tag{4.70}$$

Then $a_{j,k}(n) = a_{k,j}(n)$ and the matrix A_n whose entries are $\{a_{j,k}(n)\}$ is the inverse of H_n.

Theorem 4.15 is very useful in inverting large Hankel matrices. One example is the Hilbert matrix [13],

$$\mathbb{H}_n := (h_{j,k} : 0 \le j, k \le n), \qquad h_{j,k} := 1/(1 + j + k). \tag{4.71}$$

Since $1/(j + 1) = \int_0^1 x^j \, dx$, we see that the Hilbert matrix is the Hankel matrix associated with a contant weight function on $[0, 1]$. The corresponding orthogonal polynomials are Legendre polynomials scaled to be orthogonal on $[0, 1]$. The orthonormal polynomias are

$$p_n(x) = (-1)^n \sqrt{2n + 1} \sum_{k=0}^{n} \binom{n}{k}\binom{n + k}{k}(-1)^k x^k. \tag{4.72}$$

Thus

$$K_n(x, y) = \sum_{m=0}^{n}(2m + 1) \sum_{j,k=0}^{m} \binom{m}{j}\binom{m + j}{j}\binom{m}{k}\binom{m + k}{k}(-1)^{j+k} x^j y^k. \tag{4.73}$$

Theorem 4.16 *The determinant of $\mathbb{H}_n$ is given by*

$$\det \mathbb{H}_n = \left[\prod_{k=1}^{n} \frac{(2k + 1)! \, (2k)!}{k! \, k!} \right]^{-1} \tag{4.74}$$

Let $\mathbb{H}_n^{-1} = \tilde{h}_{j,k}$. Then

$$\tilde{h}_{j,k} = (-1)^{j+k} \sum_{m=\max(j,k)} \binom{m}{j}\binom{m + j}{j}\binom{m}{k}\binom{m + k}{k}. \tag{4.75}$$

This shows for example that the elements of $\mathbb{H}_n^{-1}$ are all integers, a highly nontrivial fact since all the entries of $\mathbb{H}_n$ are small fractions. Numerical inversion of $\mathbb{H}_n$ is very difficult and unstable because the elements of $\mathbb{H}_n$ are small.

Theorem 4.16 has been extended to the matrices whose elements are reciprocals of Fibonacci numbers and their q-analogues.

The second problem is the study of birth and death processes. A birth and death processes is a stationary Markov chain whose state space is the nonnegative integers. Let $p_{m,n}(t)$ be the transition probability to go from state m to state n in time t. Here $p_{m,n}(t)$ does not depend on the initial time. We assume that

$$p_{m,n}(t) = \begin{cases} \lambda_m t + o(t), & n = m + 1, \\ \mu_m t + o(t), & n = m - 1, \\ 1 - (\lambda_m + \mu_m)t + o(t) & n = m, \end{cases} \tag{4.76}$$

and $p_{m,n}(t) = o(t)$, if $|m - n| > 1$. The birth rates λ_m and death rates μ_m satisfy $\mu_0 \geq 0, \mu_m > 0, m > 0$, and $\lambda_m > 0$ for $m \geq 0$.

The transition probabilities satisfy the Chapman–Kolmogorov equations

$$\dot{p}_{m,n}(t) = \lambda_{n-1}p_{m,n-1}(t) + \mu_{n+1}p_{m,n+1}(t) - (\lambda_n + \mu_n)p_{m,n}(t), \tag{4.77}$$

$$\dot{p}_{m,n}(t) = \lambda_m p_{m+1,n}(t) + \mu_m p_{m-1,n}(t) - (\lambda_m + \mu_m)p_{m,n}(t), \tag{4.78}$$

where $\dot{y}$ means dy/dt. Equations (4.77)–(4.78) may be solved by the separation of variables, so we set $p_{m,n}(t) = f(t)Q_m F_n$. Therefore $f'(t)/f(t)$ is independent of t and we set it equal to $-x$. We write F_n as $F_n(x)$ to indicate its dependence on x. From (4.77) we get

$$-xF_n(x) = \lambda_{n-1}F_{n-1}(x) - (\lambda_n + \mu_n)F_n(x) + \mu_{n+1}F_{n+1}(x), \tag{4.79}$$

with $F_0(x)$ arbitrary, so we take $F_{-1}(x) := 0$ and $F_0(x) = 1$. A similar calculation gives a recurrence relation for $\{Q_n(x)\}$ from which we conclude that

$$F_n(x) = \frac{\lambda_0 \lambda_1 \cdots \lambda_{n-1}}{\mu_1 \mu_2 \cdots \mu_n} Q_n(x). \tag{4.80}$$

Thus we have determined a solution and we multiply by x-dependent separation constants (functions) and sum or integrate over all choices of x. Therefore

$$p_{m,n}(t) = \frac{1}{\zeta_m} \int e^{-xt} F_m(x)F_n(x) \, d\mu(x),$$

where

$$\zeta_n = \frac{\lambda_0 \lambda_1 \cdots \lambda_{n-1}}{\mu_1 \mu_2 \cdots \mu_n}. \tag{4.81}$$

As $t \to 0^+$, $p_{m,n}(t) \to \delta_{m,n}$. Thus

$$\int F_m(x)F_n(x) \, d\mu(x) = \zeta_m \delta_{m,n},$$

and conclude that the $\{F_n\}$'s are orthogonal with respect to μ. The boundedness of $p_{m,n}(t)$ implies $x \notin (-\infty, 0)$, and we obtain

$$p_{m,n}(t) = \frac{1}{\zeta_m} \int_0^\infty e^{-xt} F_m(x)F_n(x) \, d\mu(x). \tag{4.82}$$

Observe that (4.82) gives the solution in a factored form and it is not difficult to analyze the large t behavior of $p_{m,n}(t)$ from (4.82).

The integral representation (4.82) was established by Karlin and McGregor in [30, 31].

Another problem which leads to orthogonal polynomials is the so called J-matrix method. The idea is to start with a linear operator T which is densely defined on $L^2(\mathbb{R})$ and is symmetric. An example is the Schrödinger operator,

$$T := -\frac{d^2}{dx^2} + V(x), \qquad x \in \mathbb{R}. \tag{4.83}$$

The idea is to find a complete orthonormal basis in the domain of T such that the matrix representation of T in $\{\varphi_n(x)\}$ is tridiagonal, that is

$$\int_{\mathbb{R}} \bar{\varphi}_m T \varphi_n \, dx = 0, \qquad \text{if } |m - n| > 1.$$

We now diagonalize T, that is let $T\psi_E = E\psi_E$ and assume

$$\psi_E(x) \sim \sum_{n=0}^{\infty} \varphi_n(x)\psi_n(E). \tag{4.84}$$

Observe that

$$(E\psi_E, \varphi_n) = (T\psi_E, \varphi_n) = (T\varphi_{n-1}\psi_{n-1} + T\varphi_n\psi_n + T\varphi_{n+1}\psi_{n+1}\varphi_n, \varphi_n).$$

Therefore

$$E\psi_n(E) = \psi_{n+1}(E)(T\varphi_{n+1}, \varphi_n) + \psi_n(E)(T\varphi_n, \varphi_n) + \psi_{n-1}(E)(T\varphi_{n-1}, \varphi_n). \tag{4.85}$$

If $(T\varphi_n, \varphi_{n\pm1}) \neq 0$, then (4.85) is a recurrence relation for a sequence of polynomials $\{\psi_n(E)\}$. It is easy to see that (4.85) can be reduced to (4.7) if and only if

$$(T\varphi, \varphi_{n-1})(T\varphi_{n-1}, \varphi_n) > 0.$$

This is indeed the case since $(T\psi_n, \psi_{n-1}) = (\varphi_n, T\varphi_{n-1}) = \overline{(T\varphi_{n-1}, \varphi_n)}$. One would then expect that the original operator T and the tridiagonal matrix operator $(\phi_m, T\phi_n)$ have the same spectrum. Heller, Reinhardt and Yamani [20] introduced this method with the expectation that T and the tridiagonal matrix operator have the same spectrum. Heller, Reinhardt and Yamani applied their technique to several physical problems and identified the orthogonal polynomials $\{\psi_n(E)\}$. Their examples include the harmonic oscillator, the Morse oscillator, and the hydrogen atom. The polynomials $\{\psi_n(E)\}$ are Meixner polynomials in the harmonic oscillator model. The hydrogen atom with a Coulomb potential was more challenging. In the case when the Coulomb potential is repulsive, i.e., the electron and nucleus have charges of the same sign, $\psi_n(E)$ are the polynomials of Szegő and Pollaczek, see [40]. In the physical case when the Coulomb potential is attractive, the $\psi_n(E)$ are the Bank–Ismail polynomials, [5], see [24]. The same polynomials appeared in the Dirac equation for the hydrogen atom, see A. D. Alhaidari [2] and C. T. Munger [35].

Only recently it was proved rigorously that the J-matrix technique produces a tridiagonal matrix with the same spectrum as the original operator T.

Ismail and Koelink proved this in [25], where they also proved that the spectral measures of T and the matrix operator with entries $\{(\phi_m, T\phi_n)\}$ are the same.

4.9 The Askey–Wilson polynomials

One important development in the theory of special functions is the discovery of the Askey–Wilson polynomials and operators in the late 1970s. The results appeared in an AMS memoir [4]. The subject went through a very rapid development and is treated in detail in [24], see also [18].

Given a polynomial f we set $\breve{f}(e^{i\theta}) := f(x)$, $x = \cos\theta$, that is

$$\breve{f}(z) = f((z + 1/z)/2), \quad z = e^{\pm i\theta}. \tag{4.86}$$

In other words we think of $f(\cos\theta)$ as a function of $e^{i\theta}$. The Askey–Wilson divided difference operator $\mathcal{D}_q$ is

$$(\mathcal{D}_q f)(x) := \frac{\breve{f}(q^{1/2}e^{i\theta}) - \breve{f}(q^{-1/2}e^{i\theta})}{(q^{1/2} - q^{-1/2})i\sin\theta}, \quad x = \cos\theta. \tag{4.87}$$

It is important to observe that the denominator in (4.87) is

$$\breve{e}(q^{1/2}e^{i\theta}) - \breve{e}(q^{-1/2}e^{i\theta}), \quad \text{with } e(x) = x.$$

Recall the definition of the Chebyshev polynomials

$$T_n(\cos\theta) := \cos(n\theta), \qquad U_n(\cos\theta) := \frac{\sin((n+1)\theta)}{\sin\theta}. \tag{4.88}$$

Both T_n and U_n have degree n. Thus $\breve{T}_n(z) = [z^n + z^{-n}]/2$. A calculation gives

$$\mathcal{D}_q T_n(x) = \frac{q^{n/2} - q^{-n/2}}{q^{1/2} - q^{-1/2}} U_{n-1}(x). \tag{4.89}$$

Therefore

$$\lim_{q \to 1}(\mathcal{D}_q f)(x) = f'(x), \tag{4.90}$$

holds for $f = T_n$, hence for all polynomials, since $\{T_n(x)\}_0^\infty$ is a basis for the vector space of all polynomials and $\mathcal{D}_q$ is a linear operator.

In the problem of coupling of three angular momenta it was observed the coupling coefficients, known as the Wigner 6-J symbols, have many symmetries. It turned out that the 6-J symbols are orthogonal polynomials in a certain variable. The 6-J symbols contain 5 parameter and as orthogonal polynomials they contain four parameters and one variable. Their weight function is

$$W(x; \mathbf{t}) = \frac{\prod_{j=1}^{4}[\Gamma(t_j + i\sqrt{x})\Gamma(t_j - i\sqrt{x})]}{\Gamma(2i\sqrt{x})\Gamma(-2i\sqrt{x})}, \qquad x > 0, \tag{4.91}$$

and **t** stands for the vector (t_1, t_2, t_3, t_4). The polynomials are called the Wilson polynomials, [7, 24]. A one parameter generalization is the Askey–Wilson polynomials, whose weight function is

$$w(x; \mathbf{t}) := \frac{\prod_{n=0}^{\infty}(1 - 2(2x^2 - 1)q^n + q^{2n})}{\prod_{k=1}^{4} \prod_{m=0}^{\infty}[1 - 2xt_k q^m + t_k^2 q^{2m}]} \frac{1}{\sqrt{1 - x^2}}, \tag{4.92}$$

for $x \in (-1, 1)$.

The Askey–Wilson polynomials are the polynomial solutions to q-Sturm–Liouville

$$\frac{1}{w(x; \mathbf{t})} \mathcal{D}_q w(x; q^{1/2}\mathbf{t}) \mathcal{D}_q y = \lambda y. \tag{4.93}$$

Indeed (4.93) has a polynomial solution of degree n if and only if

$$\lambda = \lambda_n := \frac{4q}{(1 - q)^2}(1 - q^{-n})(1 - t_1 t_2 t_3 t_4 q^{n-1}), \quad n = 1, 2, \ldots, \tag{4.94}$$

for $x \in [-1, 1]$. The second order operator on the right side of (4.93) is symmetric with respect to the inner product

$$(f, g) = \int_{-1}^{1} f(x)g(x) \frac{dx}{\sqrt{1 - x^2}}. \tag{4.95}$$

In general the analogue of the Sturm-Liouville equation is

$$-\frac{1}{w(x)} \mathcal{D}_q p(x) \mathcal{D}_q y = \lambda y. \tag{4.96}$$

This more general problem under suitable assumptions on w and p has been investigated in [10].

The analogue of the Askey–Wilson operators and the inner products on unbounded domains was introduced by the author in [21]. The parametrization is $x = (\zeta - 1/\zeta)/2$ and we let $\zeta = e^{\xi}$. The analogue of the Askey–Wilson operator is

$$(\mathcal{D}_q f)(x) := \frac{\check{f}(q^{1/2}e^{\xi}) - \check{f}(q^{-1/2}e^{\xi})}{(q^{1/2} - q^{-1/2})\cosh \xi}, \quad x = \sinh \xi, \tag{4.97}$$

where $f(x) = f(\sinh \xi) =: \check{f}(e^{\xi})$. The analogue of the q-Sturm–Liouville is

$$\frac{1}{w(x; \mathbf{t})} \mathcal{D}_q[w(x; q^{1/2}\mathbf{t}) \mathcal{D}_q y] = \lambda y. \tag{4.98}$$

Recent developments on this spectral problem are in [14, 15].

Acknowledgments

I wish to thank the organizers for inviting me to the summer school program and the CRM staff for their hospitality during the summer school program. Special thanks to Pavel Winternitz and Decio Levi for their help and encouragement. Sarah Jane Johnston and Zeinab Mansour read the manuscript and made several corrections.

References

[1] Akhiezer, N. I. 1965. *The Classical Moment Problem and Some Related Questions in Analysis.* New York: Hafner.

[2] Alhaidari, A. D. 2004. Exact L^2 series solution of the Dirac–Coulomb problem for all energies. *Ann. Physics*, **312**(1), 144–160.

[3] Askey, R., and Ismail, M. 1984. Recurrence relations, continued fractions, and orthogonal polynomials. *Mem. Amer. Math. Soc.*, **49**(300).

[4] Askey, R., and Wilson, J. 1985. Some basic hypergeometric orthogonal polynomials that generalize Jacobi polynomials. *Mem. Amer. Math. Soc.*, **54**(319).

[5] Bank, E., and Ismail, M. E. H. 1985. The attractive Coulomb potential polynomials. *Constr. Approx.*, **1**(2), 103–119.

[6] Bauldry, W. C. 1990. Estimates of asymptotic Freud polynomials on the real line. *J. Approx. Theory*, **63**(2), 225–237.

[7] Biedenharn, L. C., and Louck, J. D. 1981. *The Racah–Wigner Algebra in Quantum Theory.* Encyclopedia Math. Appl., vol. 9. Reading, MA: Addison-Wesley.

[8] Bonan, S. S., and Clark, D. S. 1990. Estimates of the Hermite and Freud polynomials. *J. Approx. Theory*, **63**(2), 210–224.

[9] Brown, B. M., and Christiansen, J. S. 2005. On the Krein and Friedrichs extensions of a positive Jacobi operator. *Expo. Math.*, **23**(2), 179–186.

[10] Brown, B. M., Evans, W. D., and Ismail, M. E. H. 1996. The Askey–Wilson polynomials and q-Sturm–Liouville problems. *Math. Proc. Cambridge Philos. Soc.*, **119**(1), 1–16.

[11] Chen, Y., and Ismail, M. E. H. 1997. Ladder operators and differential equations for orthogonal polynomials. *J. Phys. A*, **30**(22), 7818–7829.

[12] Chen, Y., and Ismail, M. E. H. 2008. Ladder operators for q-orthogonal polynomials. *J. Math. Anal. Appl.*, **345**(1), 1–10.

[13] Choi, M. D. 1983. Tricks or treats with the Hilbert matrix. *Amer. Math. Monthly*, **90**(5), 301–312.

[14] Christiansen, J. S., and Ismail, M. E. H. 2006. A moment problem and a family of integral evaluations. *Trans. Amer. Math. Soc.*, **358**(9), 4071–4097.

[15] Christiansen, J. S., and Koelink, E. 2006. Self-adjoint difference operators and classical solutions to the Stieltjes–Wigert moment problem. *J. Approx. Theory*, **140**(1), 1–26.

[16] Favard, J. 1935. Sur les polynômes de Tchebicheff. *C. R. Acad. Sci. Paris*, **200**, 2052–2053.

[17] Freud, G. 1976. On the coefficients in the recursion formulae of orthogonal polynomials. *Proc. Roy. Irish Acad. Sect. A*, **76**(1), 1–6.

[18] Gasper, G., and Rahman, M. 2004. *Basic Hypergeometric Series*. 2nd edn. Encyclopedia Math. Appl., vol. 96. Cambridge: Cambridge Univ. Press.

[19] Grammaticos, B., and Ramani, A. 2004. Discrete Painlevé equations: a review. Pages 245–321 of: *Discrete Integrable Systems*. Lecture Notes in Math., vol. 644. Berlin: Springer.

[20] Heller, E. J., Reinhardt, W. P., and Yamani, H. A. 1973. On an "equivalent quadrature" calculation of matrix elements of $(z - p^2/2m)$ using an L^2-expansion technique. *J. Comput. Phys.*, **13**(4), 536–549.

[21] Ismail, M. E. H. 1993. Ladder operators for q^{-1}-Hermite polynomials. *C. R. Math. Rep. Acad. Sci. Canada*, **15**(6), 261–266.

[22] Ismail, M. E. H. 2003. Difference equations and quantized discriminants for q-orthogonal polynomials. *Adv. in Appl. Math.*, **30**(3), 562–589.

[23] Ismail, M. E. H. 2005a. Asymptotics of q-orthogonal polynomials and a q-Airy function. *Int. Math. Res. Not.*, **2005**(18), 1063–1088.

[24] Ismail, M. E. H. 2005b. *Classical and Quantum Orthogonal Polynomials in One Variable*. Encyclopedia Math. Appl., vol. 98. Cambridge: Cambridge Univ. Press.

[25] Ismail, M. E. H., and Koelink, E. The J-matrix method. *Adv. Appl. Math* to appear.

[26] Ismail, M. E. H., and Mansour, Z. S. q-analogues of Freud weights and nonlinear difference equations.

[27] Ismail, M. E. H., and Simeonov, P. 2009. q-difference operators for orthogonal polynomials. *J. Comput. Appl. Math.*, **233**(3), 749–761.

[28] Ismail, M. E. H., Johnston, S., and Mansour, Z. S. Structure relations or q-polynomials and some applications. *Applicable Anal.* to appear.

[29] Ismail, M. E. H., Nikolova, I., and Simeonov, P. 2004. Difference equations and discriminants for discrete orthogonal polynomials. *Ramanujan J.*, **8**(4), 475–502.

[30] Karlin, S., and McGregor, J. 1957a. The classification of birth and death processes. *Trans. Amer. Math. Soc.*, **86**, 366–400.

[31] Karlin, S., and McGregor, J. 1957b. The differential equations of birth-and-death processes and the Stieltjes moment problem. *Trans. Amer. Math. Soc.*, **85**, 489–546.

[32] Koekoek, R., and R. F, Swarttouw. 1998. *The Askey-scheme of hypergeometric orthogonal polynomials and its q-analogue*. Reports of the Faculty of Technical Mathematics and Informatics 98–17. Delft University of Technology, Delft.

[33] Lubinsky, D. S., Mhaskar, H. N., and Saff, E. B. 1988. A proof of Freud's conjecture for exponential weights. *Constr. Approx.*, **4**(1), 65–83.

[34] Mhaskar, H. N. 1990. Bounds for certain Freud polynomials. *J. Approx. Theory*, **63**(2), 238–254.

[35] Munger, C. T. 2007. Ideal basis sets for the Dirac Coulomb problem: eigenvalue bounds and convergence proofs. *J. Math. Phys.*, **48**(2), 022301.

[36] Rakhmanov, E. A. 1982. Asymptotic properties of orthogonal polynomials on the real axis. *Mat. Sb. (N. S.)*, **119(161)**(2), 163–203. in Russian.

[37] Shohat, J. A., and Tamarkin, J. D. 1950. *The Problem of Moments*. Math. Surveys, vol. 1. Providence, RI: Amer. Math. Soc.

[38] Simon, B. 1998. The classical moment as a self-adjoint finite difference operator. *Adv. Math.*, **137**(1), 82–203.

[39] Stone, M. H. 1932. *Linear Transformations in Hilbert Space and their Application to Analysis*. New York: Amer. Math. Soc.

[40] Szegő, G. 1975. *Orthogonal Polynomials*. 4th edn. Amer. Math. Soc. Colloq. Publ., vol. 23. Providence, RI: Amer. Math. Soc.

[41] Van Assche, W. 2007. Discrete Painlevé equations for recurrence coefficients of orthogonal polynomials. Pages 687–725 of: *Difference Equations, Special Functions and Orthogonal Polynomials*. Hackensack, NJ: World Sci. Publ.

5

Discrete Painlevé Equations and Orthogonal Polynomials

Alexander Its

Abstract

Random matrices and orthogonal polynomials have been, for more than a decade, one of the principal sources of the important analytical ideas and exciting problems in the theory of discrete Painlevé equations. In the orthogonal polynomial setting, the discrete Painlevé equations appear in the form of the nonlinear difference relations satisfied by the relevant recurrence coefficients. The principal analytical question is the analysis of certain double scaling limits of the solutions of the discrete Painlevé equations. In these notes we will present a review on the subject using the Riemann–Hilbert formalism as a main analytic tool.

5.1 General setting

These notes are devoted to the orthogonal polynomials and Painlevé equations: both continuous and discrete. In the theory of orthogonal polynomials, the Painlevé equations, both continuous and discrete, appear as the equations satisfied by the recurrence coefficients of orthogonal polynomials. Our main goal is to discuss some of the results concerned with the global asymptotic analysis of the solutions of discrete Painlevé equations generated by the recurrence coefficients. We shall start with the setting of the Riemann–Hilbert formalism for orthogonal polynomials which has been used to achieve these results. Simultaneously, this formalism will allow us to introduce the discrete Painlevé equations in a very natural way. There will be no new facts in this part of the notes, except, perhaps, the way in which the accents between the different aspects of the subject are distributed.

5.1.1 Orthogonal polynomials

Let $\{P_n(z)\}_{n=0}^{\infty}$ be the system of orthogonal polynomials on the line with respect to an exponential weight,

$$\int_{-\infty}^{\infty} P_n(z)P_m(z)w(z)\,dz = h_n\delta_{nm}, \tag{5.1}$$

$$P_n(z) = z^n + \cdots, \tag{5.2}$$

$$w(z) = e^{-NV(z)}, \qquad V(z) = \sum_{i=1}^{2m} t_j z^j, \qquad t_{2m} > 0. \tag{5.3}$$

They satisfy the tree-term recurrence equation,

$$zP_n(z) = P_{n+1}(z) + Q_n P_n(z) + R_n P_{n-1}(z). \tag{5.4}$$

The recurrence coefficients R_n and Q_n, together with the polynomials themselves, are the main objects of the interest in the theory. The following basic identities take place,

$$R_n = \frac{h_n}{h_{n-1}} \qquad Q_n = b_1^{(n)} - b_1^{(n+1)} \qquad P_n(z) = z^n + b_1^{(n)} z^{n-1} \cdots \tag{5.5}$$

We refer to the classical monograph [36] as the basic reference on the subject.

The principal analytic question of the theory of orthogonal polynomials is the large n asymptotic behavior of the polynomials, and the corresponding recurrence coefficients. A more general version of this question is the following:

$$P_n(z), R_n, Q_n \mapsto? \quad \text{as } n, N \to \infty, \tag{5.6}$$

and

$$(t_1, \ldots, t_{2m}) \equiv \vec{t} \to \vec{t}_0.$$

The principal challenge in attacking this problem is that, unless $V(z) = z^2$, the only available explicit representations for the orthogonal polynomials are the representations either as $n \times n$ determinants or as n-fold integrals.

Remark 5.1   The system $\{P_n\}$ exists iff the Hankel determinant,

$$D_n \equiv \det\left\{ \int_{-\infty}^{\infty} w(z)z^{j+k}\,dz \right\}_{j,k=0,\ldots,n-1} \tag{5.7}$$

is not zero. The study of $\lim_{n\to\infty} D_n$ is of its own great interest.

5.1.2 Connections to integrable systems

Orthogonal polynomials have profound connections with several other areas of classical analysis:

- Special functions
- Moments problems
- Hankel and Toeplitz determinants
- Spectral theory of Jacobi matrices
- Random matrices

The more recent connection, and the one we will be focusing in these notes, is to the integrable systems. Strictly speaking, this is not a very recent connection. Through Hankel and Toeplitz determinants, orthogonal polynomials have been playing a prominent role in exactly solvable statistical mechanics for several decades already. In Soliton Theory, the orthogonal polynomials appeared at the moment when the Toda-lattice equations joined (mid 70s) the list of fundamental integrable equations. In fact, there is one very remarkable example of interactions between Soliton Theory and orthogonal polynomials which I want to mention specifically.

In the late 1960s, after the seminal works of Gardner–Gereen–Kruscal–Miura, Lax, Faddeev and Zakharov on the solution of the KdV equation on the real line via IST, one of the new challenges of the young theory was an extension of the IST to the periodic Cauchy problem. Among other crucial steps in building of the relevant formalism – what we call now the finite-gap or algebrogeometric method in the theory of integrable PDEs, was a realization (V. Matveev) that the question is intimately related to the classical works of Akhiezer on the orthogonal polynomials on the system of intervals.

Starting in the early 1990s, the connection of the orthogonal polynomials to the integrable systems has been increasingly appreciated and used by the orthogonal polynomial community. Many long-standing problems in the area have been solved using this connection and several new exciting sides of this connection have been discovered. One of the most recent is concerned with the discrete Painlevé equations. In fact, the orthogonal polynomials constitute one of the three general contexts within which the discrete Painlevé equations first appeared [20, 21, 35], and within which some of the important difference equations arising in the 2D quantum gravity (Brézin, Kazakov; Gross, Migdal; Duglas, Shenker) were identified as discrete Painlevé equations. The other two fields where the discrete Painlevé equations emerged from are integrable mappings and discrete Lie-symmetries (Grammaticus, Nijhoff, Papageorgiou, Ramani; Levi, Winternitz; see [24] for more on the history of the area).

In the rest of this section, we will explain the relation between orthogonal polynomials and discrete Painlevé equations using the Riemann–Hilbert formalism for orthogonal polynomials [22]. It is worth mentioning that this formalism was the one of the first frameworks that allowed one to introduce discrete Painlevé equations in a systematic way. Moreover, the Riemann–Hilbert scheme also provide a very powerful tool for dealing with the asymptotic analysis of discrete Painlevé equations.

5.1.3 The Riemann–Hilbert representation of the orthogonal polynomials

Put

$$Y(z) = \begin{pmatrix} P_n(z) & \dfrac{1}{2\pi i} \displaystyle\int_{-\infty}^{\infty} \dfrac{P_n(z')w(z')}{z'-z}\,dz' \\ -\dfrac{2\pi i}{h_{n-1}}P_{n-1}(z) & -\dfrac{1}{h_{n-1}} \displaystyle\int_{-\infty}^{\infty} \dfrac{P_{n-1}(z')w(z')}{z'-z}\,dz' \end{pmatrix} \tag{5.8}$$

$$w(z) = e^{-NV(z)}$$

Then (cf. [22])

1. $Y(z) \in H(\mathbb{C} \setminus \mathbb{R})$
2. $Y_+(z) = Y_-(z)\begin{pmatrix} 1 & w(z) \\ 0 & 1 \end{pmatrix}, z \in \mathbb{R}$
3. $Y(z) = (I + O(1/z))z^{n\sigma_3} \; z \to \infty, \sigma_3 = \begin{pmatrix} 1 & 0 \\ 0 & -1 \end{pmatrix}.$

Observation 1   Properties 1–3 determine $Y(z)$ uniquely.

Observation 2   Properties 1–3 are a particular case of the *Riemann–Hilbert factorization problem*:

Given an oriented contour Γ on the z-plane and the map,

$$G: \Gamma \to GL(N, \mathbb{C}),$$

find the matrix valued function $Y(z)$ such that:

- $Y(z) \in H(\mathbb{C} \setminus \mathbb{R})$
- $Y_+(z) = Y_-(z)G(z)$
- $Y(z) \to I, z \to \infty$

The first observation reduces the analysis of the orthogonal polynomials to the analysis of the Riemann–Hilbert problem 1–3. Indeed, knowing $Y(z)$, the all objects of interest, i.e., $P_n(z)$, h_n, R_n, and Q_n can be evaluated via the relations:

$$P_n(\lambda) = Y_{11}(\lambda) \tag{5.9}$$

$$h_n = -2ni(m_1)_{12} \qquad h_{n-1}^{-1} = -\frac{1}{2\pi i}(m_1)_{21} \tag{5.10}$$

$$R_n = (m_1)_{12}(m_1)_{21} \tag{5.11}$$

$$Q_n = (m_1^{(n)})_{11} - (m_1^{(n+1)})_{11}, \tag{5.12}$$

We use the notation $O^{(n)}$ alongside with the notation O_n when we need to indicate the dependence of the object O of n.

where m_1 is defined as the first matrix coefficient in the expansion,

$$Y(z) \cong \left(I + \frac{m_1}{z} + \cdots\right) z^{n\sigma_3}, \qquad z \to \infty. \tag{5.13}$$

The second observation allows one to use for solving 1–3 the classical apparatus of the analytic factorization of matrix functions in conjunction with the new Rieman-Hilbert techniques which have been developed in Soliton Theory, starting from the pioneering works of Zakharov, Shabat and Manakov and culminating in the *Nonliner Steepest Descent* method of Deift and Zhou (see survey [14] for a detail historic review).

In these notes we won't discuss the Rieman–Hilbert approach to the asymptotic analysis of the orthogonal polynomials. For that, and for the historic review, we refer the reader to book [13]. In the next subsection, we will explain, however, how the Riemann–Hilbert problem 1–3 yields the discrete Painlevé equations. Moreover, simultaneously we will demonstrate their integrability by providing relevant Lax pairs.

5.1.4 Discrete Painlevé equations

Following the standard procedure of the theory of integrable systems (cf. [19, 31]; see also [23]), we can, in a canonical way, associate with the Riemann–Hilbert problem 1–3 certain linear differential equations. To this end, we put

$$\Psi(z) = Y(z) e^{-(NV(z)/2)\sigma_3},$$

and notice that in terms of the Ψ-function, the Riemann–Hilbert problem 1–3 transforms into the problem with the *constant* jump matrix,

- $\Psi(z) \in H(\mathbb{C} \setminus \mathbb{R})$
- $\Psi_+(z) = \Psi_-(z) \left(\begin{smallmatrix} 1 & 1 \\ 0 & 1 \end{smallmatrix}\right), z \in \mathbb{R}$
- $\Psi(z) = (I + O(1/z)) e^{-(NV(z)/2)\sigma_3 + n \ln z \sigma_3}, z \to \infty$

Therefore, by the usual reference to Liouville's theorem, the following linear system appears.

$$\begin{cases} \Psi^{(n+1)}(z) = U(z)\Psi^{(n)}(z) \\ \dfrac{\partial \Psi(z)}{\partial z} = A(z)\Psi(z) \\ \dfrac{\partial \Psi(z)}{\partial t_j} = V_j(z)\Psi(z) \end{cases} \tag{5.14}$$

where

$$U(z) = zU_1 + U_0 \tag{5.15}$$

$$A(z) = \sum_{k=0}^{2m-1} A_k z^k \tag{5.16}$$

$$V_j(z) = \sum_{k=0}^{j} V_{jk} z^k \tag{5.17}$$

and the matrix coefficients A_k, V_{jk}, U_1, U_0 are polynomials of the entries $(m_k)_{jl}$ of the matrix coefficients of the asymptotic series,

$$\Psi(z) \cong \left(I + \sum_{k=1}^{\infty} \frac{m_k}{z} \right) e^{-NV(z)/z\sigma_3 + n \ln z\sigma_3}, \qquad z \to \infty. \tag{5.18}$$

It can be shown (cf. examples below) that ultimately all the matrix coefficients of system (5.14) can be expressed in terms of the basic quantities h_n, R_n, Q_n, which in turn are recovered with the help of (5.9)–(5.12) from the first coefficient $m_1^{(n)}$. In fact, it is always, that

$$U(z) = \begin{pmatrix} z - Q_n & \frac{h_n}{2\pi i} \\ -\frac{2\pi i}{h_n} & 0 \end{pmatrix}, \tag{5.19}$$

and the first equation in (5.14) is equivalent to the three term recurrent equation (5.4). The matrix polynomials $A(z), V_j(z)$ depend on the choice of the polynomial $V(z)$ in (5.3); they can be written explicitly in terms of the matrix function $V'(L)$ where

$$L_{nm} = R_{n+1}^{1/2} \delta_{n+1\,m} + Q_n \delta_{nm} + R_n^{1/2} \delta_{n-1\,m} \tag{5.20}$$

is the Jacobi matrix corresponding to the system $\{P_n(z)\}$.

. The compatablity condition of the first two equations in (5.14) yields the *discrete string equation*

$$A^{(n+1)}(z)U(z) - U(z)A^{(n)}(z) = \frac{\partial U(z)}{\partial z} \equiv \begin{pmatrix} 1 & 0 \\ 0 & 0 \end{pmatrix}, \tag{5.21}$$

which is equivalent to the relations (see [22, 35] for the case of even $V(z)$; see [5, 8, 18] for general $V(z)$),

$$R_n^{1/2}[V'(L)]_{n\,n-1} = \frac{n}{N} \tag{5.22}$$

$$[V'(l)]_{nn} = 0. \tag{5.23}$$

System (5.22)–(5.23) is also called the *Freud equation* [35]. This is a nonlinear difference system on R_n, Q_n, and it is an *integrable system*; indeed, equation

(5.21) is its *Lax representation*, and the first two equations of system (5.14) form the *Lax pair* for system (5.22)–(5.23).

From the above consideration, we arrive at the following important conclusion: *The orthogonal polynomials Riemann–Hilbert problem, through the Lax pair* (5.14), *becomes a source of integrable nonlinear difference equations.* The important fact is that to produce (5.21) we do not need $V(z)$ to be necessarily a polynomial. Any rational function $V(z)$ would do. Indeed, even this requirement can be relaxed. We actually only need $V'(z)$ to be rational, i.e., the *semiclassical* orthogonal polynomials [34]. Moreover, as we will see in Section 5.2.3 even this requirement can be lifted (see also [8] where the differential equations for the recurrence coefficients are written for a very general type of the weight).

The difference systems corresponding to the first nontrivial choices of $V(z)$ are all difference equations of the second order. These are the equations which are identified as the *discrete Painlevé equations*. Indeed they

- are integrable
- describe the discrete isomonodromy deformations of the second equation in (5.14)
- satisfy the confinement property which was introduced by Grammaticus, Nijhoff and Ramani as the characteristic property of d-Painlevé equations [24].
- have the usual, differential Painlevé equations as their continuous limit.

Remark   A more detailed and systematic description of the d-Painlevé equations as the difference equations satisfied by the recurrence coefficients of the semiclassical orthogonal polynomials is given in [39].

The principal analytical question related to (5.21)–(5.23) is the asymptotic analysis of their solutions as $n, N \to \infty$. This is where the Riemann-Hilbert formalism and the nonlinear steepest descent method play a key role. In the classical language, this is again the main analytic question of the theory of orthogonal polynomials. In the rest of the notes, we will present some of the results concerning this analysis considering one old example and one new example. Before we do this, however, we would like to make two general remarks concerning system (5.14).

1. Compatibility of the second and the third equations of system (5.14) yields the ordinary differential equation of the Painlevé type; that is, the continuous Painlevé equations appear in the orthogonal polynomial theory even *before* the large N limit is taken. In fact, in the context of the continuous Painlevé theory, the discrete Painlevé equation (5.22) represents the

Bäcklund-Darboux transformation of the continuous Painlevé equation describing the isomonodromy deformations of the second linear equation in (5.14).

2. Compatibility of the first and the third equations of system (5.14) yields the Toda-type hierarchy of the differential-difference equations, while the compatibility of the third equations with different t_j—the integrable PDEs of the KdV type.

Remark For more on integrable hierarchies of differential equations appearing in orthogonal polynomial theory, we refer the reader to the works [38] and [40] and to the references therein.

5.2 Examples

5.2.1 Gaussian weight

Choosing,

$$V(z) = z^2, \qquad N = 1, \tag{5.24}$$

we obtain that

$$A(z) = -z\sigma_3 + A_0, \tag{5.25}$$

where (see (5.10), (5.13))

$$A_0 = [\sigma_3, m_1] = \begin{pmatrix} 0 & -\frac{h_n}{\pi i} \\ \frac{4\pi i}{h_{n-1}} & 0 \end{pmatrix}. \tag{5.26}$$

The evenness of the potential $V(z)$, implies that

$$Q_n = 0, \quad \forall n. \tag{5.27}$$

Taking this into account, and substituting (5.25) and (5.19) into the Lax equation (5.21) we see that the latter degenerates to a simple linear relation,

$$R_{n+1} - R_n = \tfrac{1}{2}, \tag{5.28}$$

which, in conjunction with the condition $R_0 = 0$, reproduces the classical fact, that for the Gaussian weight (5.24), the recurrent coefficients R_n are given by the equation,

$$R_n = \frac{n}{2}, \quad n = 0, 1, 2, \ldots \tag{5.29}$$

It is also easy to check that the second equation (there is no third equation!) of system (5.14) for the Gaussian weight (5.24) yields the parabolic cylinder equation,

$$\frac{d^2y}{dz^2} + (1 + 2n - z^2)y = 0,$$ (5.30)

for

$$y(z) := \Psi_{11}(z).$$

This is again a classical fact: the polynomials $P_n(z)$ we are talking about in this subsection are, in fact, Hermite polynomials,

$$P_n(z) = \frac{1}{2^n}H_n(z).$$

We conclude the consideration of this simple example by noticing that using trivial scaling transformations of the spectral variable z, any quadratic weight can be reduced to the normalized form (5.24).

5.2.2 d-Painlevé I

A natural next to the Gaussian case is the even quartic potential,

$$V(z) = \frac{a}{2}z^2 + \frac{t}{4}z^4, \qquad t > 0.$$ (5.31)

Again, the evenness of the potential implies that

$$Q_n = 0.$$

The matrix $A(z)$ is now a polynomial of the third degree, and its representation in terms of R_n and h_n is given by the equation (see [3, 20]),

$$A(z) = N\begin{pmatrix} -\frac{t}{2}z^3 - \left(\frac{a}{2} + tR_n\right)z & \frac{i}{2\pi}h_n(tz^2 + a + t(R_n + R_{n+1})) \\ \frac{2\pi i}{h_{n-1}}(tz^2 + a + t(R_{n-1} + R_n)) & \frac{t}{2}z^3 + \left(\frac{a}{2} + tR_n\right)z \end{pmatrix}$$ (5.32)

Substitution of (5.32) and (5.19) into the Lax equation (5.21), leads this time to the nonlinear difference equation on the recurrence coefficients R_n,

$$\frac{n}{N} = R_na + tR_n(R_{n-1} + R_n + R_{n+1}).$$ (5.33)

This is the *discrete Painlevé I* equation. This equation has been extensively investigated since its appearance in the planar Feynman diagram expansions of Hermitian matrix model, which was introduced and studied in the classical works [2, 7, 29]. We refer the reader to [39] for more on the history of this equation, and especially on the history of its orthogonal polynomial connections. The Lax pair representation (5.21) of equation (5.33) was pointed out in

[20] which allowed to apply to the asymptotic analysis of (5.33) the Riemann–Hilbert method of the theory of integrable systems.

Discrete Painlevé I equation (5.33) admits two distinguished continuous limits. The first one maps (5.33) to the continuous Painlevé I equation, which explains the name of equation (5.33). The second continuous limit is to the continuous Painlevé II equation.

Continuous limit to Painlevé I

Put $a = 1$ and, assuming that

$$n, N \to \infty, \qquad \frac{n}{N} = -\frac{1}{12t} + c_1 N^{-4/5} x,$$

$$t < 0, \qquad c_1 = \frac{1}{|t|^{1/8}} 3^{-3/8} 2^{-1/4}, \tag{5.34}$$

consider the following asymptotic ansatz,

$$R_n \cong -\frac{1}{6t} - c_2 N^{-2/5} u(x), \qquad c_2 = \frac{1}{|t|^{3/8}} 3^{-1/8} 2^{1/4}. \tag{5.35}$$

Noticing that the shift $n \to n \pm 1$ corresponds to the shift $x \to x \pm h$, with $h = 1/c_1 N^{-1/5}$, we derive that

$$R_{n\pm1} \cong -\frac{1}{6t} - c_2 N^{-2/5} \left(u(x) \pm h u'(x) + \frac{h^2}{2} u''(x) \right). \tag{5.36}$$

Formal substitution of (5.34)–(5.36) into equation (5.33) maps the latter to the first Painlevé equation for the function $u(x)$,

$$u'' = 6u^2 + x. \tag{5.37}$$

This fact was discovered in the well-known works [6, 15, 25] and used there for a nonperturbative formulation of the 2D quantum gravity. Moreover, in these works it was indicated that the solution $u(x)$ belongs to the one-parameter family of *tronquée* solutions, i.e., the solutions each satisfying the same asymptotic condition,

$$u(x) \cong \sqrt{-\frac{x}{6}} \left[1 + \sum_{k=1}^{\infty} a_k (-x)^{-5k/2} \right], \qquad x \to -\infty, \tag{5.38}$$

were the coefficients a_k determined uniquely by the formal substitution of series (5.38) into equation (5.37). The exact rigorous statement and the completion of characterization of the solution $u(x)$ were obtained in [17, 20, 22, 28] with the help of the Riemann–Hilbert approach, and they read as follows.

Let us replace, in the definition of the orthogonal polynomials (5.1), the integration along the real line by the integration along the rays $\arg z = \pm\pi/4$,

$\pm 3\pi/4$,

$$\int_{-\infty}^{\infty} \to \alpha \int_{\infty e^{3i\pi/4}}^{\infty e^{i\pi/4}} + (1 - \alpha) \int_{\infty e^{3i\pi/4}}^{\infty e^{-i\pi/4}} \equiv \int_{\alpha}. \tag{5.39}$$

Here, α is an arbitrary number. That is, we take $t < 0$ and consider the following initial data for equation (5.33),

$$R_0 = R_{-1} = 0, \qquad R_1 = \int_{\alpha} z^2 e^{-N(z^2/2 + tz^4/4)} \, dz \left(\int_{\alpha} e^{-N(z^2/2 + tz^4/4)} \, dz \right)^{-1}. \tag{5.40}$$

Then, under the asymptotic assumption (5.34), the large n behavior of the solution R_n is described by the relations,

$$R_n = -\frac{1}{6t} - c_2 N^{-2/5} u_\alpha(x) + O(N^{-3/5}), \tag{5.41}$$

where $u_\alpha(x)$ is the solution of (5.37) *uniquely* characterized by the asymptotic relation,

$$u_\alpha(x) = u_{\text{tritronq}}(x) - \frac{i(\alpha - 1)}{\sqrt{\pi} 2^{11/8} 3^{1/8}} (-x)^{-1/8} e^{-2^{11/4} 3^{1/4} (-x)^{5/4}/5} (1 + O(x^{-3/8})),$$

$$x \to -\infty. \tag{5.42}$$

Here, $u_{\text{tritronq}}(x)$ denote Boutroux's *tri-tronquée* solution of the Painlevé equation (5.37), i.e., the *unique* among all *tronquée* solutions of (5.37) which has asymptotic representation (5.38) in the whole sector,

$$-\frac{\pi}{5} < \arg x < \frac{7\pi}{5}.$$

(For more on the asymptotic properties of the *tronquée* solution see [33].)

It should be noticed, that

$$u_1(x) = u_{\text{tritronq}}(x),$$

and that the choice $\alpha = 1$ corresponds to the 2D quantum gravity model considered in the above mentioned physical works. Using physical arguments, the characterization of the solution $u(x)$ in the ansatz (5.35) as $u_{\text{tritronq}}(x)$ was also established in [12].

Asymptotics (5.41) is a *critical asymptotics* of the discrete Painlevé function R_n. Away from the critical ratio, $n/N = -1/12t$, the asymptotics is given by elementary functions [17] (see also the above mentioned physical works for a heuristic derivation),

$$R_n \cong \frac{\sqrt{1 + 12t\gamma} - 1}{6t} + O\left(\frac{1}{N}\right), \qquad n, N \to \infty, \frac{n}{N} \equiv \gamma > -\frac{1}{12t}. \tag{5.43}$$

The indicated error term was proven in [17], together with the generalization of the whole result for a two-parameter combination of the integrals along the rays is also obtained.

The behavior of R_n in the case when

$$\frac{n}{N} \equiv \gamma < -\frac{1}{12t},$$

is an interesting open problem.

Continuous limit to Painlevé II

Suppose, that in (5.31),

$$a < 0, \qquad \text{and} \qquad t > 0,$$

and assume that

$$n, N \to \infty, \qquad \frac{n}{N} = \frac{a^2}{4t} + d_1 N^{-2/3} x, \qquad d_1 = \left(\frac{a^2}{2t}\right)^{1/3}. \tag{5.44}$$

Using the Riemann–Hilbert method, it was proven in [4] that, under this asymptotic assumption, the behavior of the recurrence coefficients R_n is given by the equation,

$$R_n = -\frac{a}{2t} + d_2(-1)^{n+1} N^{-1/3} u(x) + d_3 N^{-2/3}(x + 2u^2) + O(N^{-1}), \tag{5.45}$$

where

$$d_2 = \left(-\frac{2a}{t^2}\right)^{1/3}, \qquad d_3 = \frac{1}{2}\left(-\frac{1}{2at}\right)^{1/3},$$

and $u(x)$ is the solution of the (continuous) second Painlevé equation,

$$u'' = xu + 2u^3, \tag{5.46}$$

uniquely characterized by the asymptotic condition,

$$u(x) \cong \frac{1}{2\sqrt{\pi}} x^{-1/4} e^{-2x^{3/2}/3}, \qquad x \to +\infty. \tag{5.47}$$

This solution of (5.46) was first singled out by Hastings and McLeod in [26] where they also showed that

$$u(x) \cong \sqrt{-\frac{x}{2}}, \qquad x \to -\infty. \tag{5.48}$$

The asymptotics (5.45) was first suggested in physical papers [11, 16, 37].

The noncritical asymptotics of R_n are the following [3–5]:

$$R_n \cong \frac{\sqrt{a^2 + 12t\gamma} - a}{6t} + O\left(\frac{1}{N}\right), \qquad n, N \to \infty, \tag{5.49}$$

if

$$\frac{n}{N} \equiv \gamma > \frac{a^2}{4t},$$

and

$$R_n \cong \frac{(-1)^{n+1} \sqrt{a^2 - 4t\gamma} - a}{2t} + O\left(\frac{1}{N}\right), \qquad n, N \to \infty, \qquad (5.50)$$

if

$$\frac{n}{N} \equiv \gamma < \frac{a^2}{4t},$$

The physical derivation of (5.49)) was also obtained in [6, 15, 25]. It is worth noticing, that due to the asymptotic formulae (5.47) and (5.48), the critical asymptotic behavior (5.45) matches the noncritical asymptotics.

Continuous Painlevé before the large N limit

For the purposes of this subsection, it is convenient to choose the following normalization of the potential $V(z)$,

$$N = 1, \qquad t = 1, \qquad a = 2x,$$

so that the only t_j-parameter is quantity x. With these normalizations, the third equation of the Lax pair (5.14) can be written in the form,

$$\frac{\partial \Psi}{\partial x} = \begin{pmatrix} -\frac{1}{2}z^2 + yv & izv \\ izy & \frac{1}{2}z^2 - yv \end{pmatrix} \Psi, \qquad (5.51)$$

where

$$v := \frac{h_n}{2\pi}, \qquad \text{and} \qquad y := \frac{2\pi}{h_{n-1}}.$$

Note that

$$R_n = yv. \qquad (5.52)$$

Simultaneously, with these new notations, the second equation of the Lax pair (5.14) becomes (cf. (5.32)) the equation,

$$\frac{\partial \Psi}{\partial z} = \begin{pmatrix} -\frac{1}{2}z^3 - zx - zyv & iz^2v + 2ixv + iv(R_n + R_{n+1}) \\ iz^2y + 2ixy + iy(R_{n-1} + R_n) & \frac{1}{2}z^3 - zx - zyv \end{pmatrix} \Psi. \qquad (5.53)$$

Placing the pair (5.51), (5.53) against the Jimbo–Miwa list of the Painlevé Lax pairs [30], and taking into account (5.52) we immediately conclude that the recurrence coefficient R_n, taken for fixed n and considered as a function of x,

$$R_n \equiv u(x),$$

satisfies the fourth Painlevé equation,

$$u'' = \frac{1}{2u}(u')^2 + \frac{3}{2}u^3 + 4xu^2 + 2\left(x^2 + \frac{n}{2}\right) - \frac{n^2}{2u}. \tag{5.54}$$

This fact was first observed by A. Kitaev in 1990.

5.2.3 d-Painlevé XXXIV

Another possible "next to Gaussian" case can be obtained by introducing a jump discontinuity to the Gaussian weight e^{-z^2}. That is, we are suggesting to consider the orthogonal polynomials on the line with respect to the weight,

$$w(z) = e^{-z^2} \omega(z), \tag{5.55}$$

where

$$\omega(z) = \begin{cases} e^{i\beta\pi}, & z < \mu \\ e^{-i\beta\pi}, & z > \mu \end{cases}, \tag{5.56}$$

μ is an arbitrary real parameter, and β is an arbitrary complex parameter. The setting of the Riemann–Hilbert problem needs to be modified in order to account for the discontinuity of the weight at point $z = \mu$. We shall give the precise formulation of the Ψ-version of the Riemann–Hilbert problem, i.e., for the function,

$$\Psi(z) = Y(z)e^{-(z^2/2)\sigma_3}.$$

The problem reads as follows.

- $\Psi(z) \in H(\mathbb{C} \setminus \mathbb{R})$
- $\Psi_+(z) = \Psi_-(z)\left(\begin{smallmatrix} 1 & \omega(z) \\ 0 & 1 \end{smallmatrix}\right), z \in \mathbb{R}, z \neq \mu$
- $\Psi(z) = (I + O(1/z))e^{-z^2/2\sigma_3 + n \ln z\sigma_3}, z \to \infty$
- In the neighborhood of the point $z = \mu$, the function $\Psi(z)$ admits the representation,

$$\Psi(z) = \Psi_0(z)\left(I + \frac{1}{\pi}\sin\pi\beta\ln(z - \mu)\begin{pmatrix} 0 & 1 \\ 0 & 0 \end{pmatrix}\right)C,$$

where $\Psi_0(z)$ is a function holomorphic at $z = \mu$, the branch of the logarithmic function is fixed by the condition $-\pi/2 < \arg(z - \mu) < 3\pi/2$, and C is a piecewise constant function defined by the formula,

$$
C = \begin{cases} I, & 0 < \arg(z - \mu) < \pi \\[2mm] \begin{pmatrix} 1 & e^{-i\pi\beta} \\ 0 & 1 \end{pmatrix}, & -\dfrac{\pi}{2} < \arg(z - \mu) < 0 \\[4mm] \begin{pmatrix} 1 & -e^{i\pi\beta} \\ 0 & 1 \end{pmatrix}, & \pi < \arg(z - \mu) < \dfrac{3\pi}{2}. \end{cases}
$$

The parameter μ plays the role of the t_j-parameter. The presence of the fourth condition in the Riemann–Hilbert setting implies that the matrix valued functions $A(z)$ and $V_j(z) \equiv V_\mu(z)$ are now the rational functions – they both have a simple pole at $z = \mu$ (cf. [31]). Indeed, the basic linear system (5.14) associated with the Riemann–Hilbert problem for the weight (5.55) is the system,

$$
\begin{cases} \Psi^{(n+1)}(z) = U(z)\Psi^{(n)}(z) \\[2mm] \dfrac{\partial \Psi(z)}{\partial z} = \left(-z\sigma_3 + A + \dfrac{B}{z - \mu} \right) \Psi(z) \\[4mm] \dfrac{\partial \Psi(z)}{\partial \mu} = -\dfrac{B}{z - \mu} \Psi(z). \end{cases} \tag{5.57}
$$

Here, the expression for $U(z)$ is the same as in (5.19), while the formulae for the coefficient matrices A and B are given by the equations (cf. (5.26)),

$$
A = \begin{pmatrix} 0 & -\dfrac{h_n}{\pi i} \\[3mm] \dfrac{4\pi i}{h_{n-1}} & 0 \end{pmatrix}, \tag{5.58}
$$

and

$$
B = \frac{e^{-\mu^2}}{\pi} \sin \pi\beta \begin{pmatrix} -pq & p^2 \\ -q^2 & pq \end{pmatrix}, \qquad p := P_n(\mu), q := -\frac{2\pi i}{h_{n\,1}} P_{n-1}(\mu). \tag{5.59}
$$

Discrete string equation (5.21) is equivalent to the mapping,

$$
(Q_n, R_n) \to (Q_{n+1}, R_{n+1}),
$$

which is described by the relations,

$$
\begin{cases} R_{n+1} = n + \dfrac{1}{2} - R_n + \mu Q_n - Q_n^2 \\[4mm] Q_{n+1} Q_n R_{n+1} = \left(R_{n+1} - \dfrac{n+1}{2} \right)^2 \end{cases} \tag{5.60}
$$

System (5.21) also yields the equations,

$$\begin{cases} R_n - \dfrac{n}{2} = \dfrac{\sin \pi \beta}{2\pi} p_n q_n \\ Q_n h_n = -i \sin \pi \beta p_n^2 \\ Q_n h_n^{-1} = \dfrac{i}{4\pi^2} \sin \pi \beta q_{n+1}^2, \end{cases} \tag{5.61}$$

which allow to express p_n and q_n in terms of R_n and Q_n.

System (5.60) is an integrable 2×2 difference system of the first order, and hence it should be identified with one of the discrete Painlevé equations from the Grammaticos–Ramani–Nijhoff list. We, however, have not yet succeeded with this identification, and have temporarily attached to system (5.60) the name of "discrete thirty fourth Painlevé equation" – d-Painlevé XXXIV equation. The motivation is provided in the next subsection.

Continuous limit to Painlevé XXXIV

Assume that

$$n \to \infty, \quad \mu = \sqrt{2n}(1 + \tfrac{1}{2}n^{-2/3}x). \tag{5.62}$$

Then the asymptotic behavior of the recurrence coefficients Q_n and R_n is given by the equations [10],

$$Q_n = -\frac{1}{\sqrt{2}} n^{-1/6} v(x) + o(n^{-1/6}), \tag{5.63}$$

and

$$R_n = \frac{n}{2} - \frac{1}{2} n^{1/3} v(x) + o(n^{1/3}), \tag{5.64}$$

respectively. The function $v(x)$ is the solution of a particular case of thirty fourth equation from the original Painlevé–Gambier list,

$$v'' = 4v^2 + 2xv + \frac{(v')^2}{2v}, \tag{5.65}$$

which, in fact, can be reduced to the second Painlevé equation (5.46) by the substitution,

$$v = u^2. \tag{5.66}$$

This time, however, the second Painlevé function $u(x)$ is not the Hastings-McLeod solution. To avoid too involved formulae, we shall present the asymptotic characterization of the function $u(x)$ for the particular but important case of a pure imaginary β,

$$\beta = i\kappa, \quad \kappa > 0.$$

With this assumption, the solution $u(x)$ is pure imaginary, and it is uniquely determined either by its asymptotics at $x = +\infty$,

$$u(x) \cong \frac{i\rho}{2\sqrt{\pi}} x^{-1/4} e^{-2x^{3/2}/3}, \qquad x \to +\infty, \qquad (5.67)$$

where

$$\rho = \sqrt{e^{2\kappa\pi} - 1},$$

or by its asymptotics at $x = -\infty$,

$$u(x) = i(-x)^{-1/4} \sqrt{2\kappa} \sin\left(\frac{2}{3}(-x)^{3/2} + \frac{3\kappa}{2}\ln(-x) + \phi\right), \qquad x \to -\infty, \quad (5.68)$$

where

$$\phi = 3\kappa \ln 2 - \frac{\pi}{4} - \arg \Gamma(i\kappa).$$

This one parameter family of the second Painlevé transcendents was first singled out by Ablowitz and Segur [1]. (For more on the global asymptotic analysis of the Painlevé equations see monograph [23].)

Similar to the equations (5.41) and (5.45), equations (5.62) and (5.65) represent the critical large n behavior of the recurrence coefficients R_n and Q_n corresponding to weight (5.55). The non-critical asymptotics of R_n and Q_n arise under the assumption that,

$$\mu = \sqrt{2n}\lambda, \qquad \lambda \neq 1. \qquad (5.69)$$

In the regime,

$$|\lambda| < 1,$$

these asymptotics are given by the equations [27],

$$R_n = \frac{n}{2} + \frac{\kappa\lambda}{2\sqrt{1-\lambda^2}} + \frac{\kappa}{2(1-\lambda^2)} \cos(\arcsin\lambda + \lambda\sqrt{1-\lambda^2})$$

$$\times \cos\left((2n+1)\left(\arcsin\lambda + \lambda\sqrt{1-\lambda^2} - \frac{\pi}{2}\right) - 2\kappa\ln 8n + \phi_1\right)$$

$$+ O\left(\frac{1}{n}\right), \qquad n \to \infty, \quad (5.70)$$

and

$$Q_n = \frac{2\kappa}{\sqrt{2n(1-\lambda^2)}} \sin^2\left(n\left(\arcsin\lambda + \lambda\sqrt{1-\lambda^2} - \frac{\pi}{2}\right) - 2\kappa\ln 8n + \phi_2\right)$$

$$\times \left(1 + O\left(\frac{1}{n}\right)\right), \qquad n \to \infty, \quad (5.71)$$

where the formulae for the phases $\phi_{1,2}$ read,

$$\phi_1 = 2 \arg \Gamma(i\kappa) - 3\kappa \ln(1 - \lambda^2) - \arcsin \lambda, \qquad (5.72)$$

and

$$\phi_2 = \arg \Gamma(i\kappa) - \frac{3\kappa}{2} \ln(1 - \lambda^2) + \arctan \frac{\lambda}{1 + \sqrt{1 - \lambda^2}}. \qquad (5.73)$$

For the case $\lambda = 0$ and without the formulae for phases $\phi_{1,2}$ the asymptotics (5.71), (5.70) where first suggested in [9].

In the regime,

$$|\lambda| > 1,$$

the asymptotics of R_n and Q_n should be expected to reproduce the recurrence coefficients of the Hermit polynomials, i.e.,

$$R_n = \frac{n}{2} + O(e^{-cn}) \quad \text{and} \quad Q_n = O(e^{-cn}), \qquad n \to \infty. \qquad (5.74)$$

These formulae are proven in [32]. As in the previous section, the critical asymptotics (5.63), (5.64) match the noncritical regimes (5.71), (5.70) and (5.74).

The Hankel determinant,

$$D_n(\beta) = \det\left\{ \int_{-\infty}^{\infty} w(z) z^{j+k} \, dz \right\}, \quad j,k = 0, \ldots, n-1,$$

plays the role of the τ-function for the discrete system (5.60). The asymptotics of $D_n(\beta)$ can be also evaluate in closed form. In the regime, $|\lambda| < 1$, it is given by the equation [27],

$$\frac{D_n(\beta)}{D_n(0)} = G(1 + \beta)G(1 - \beta)(1 - \lambda^2)^{-3\beta^2/2}(8n)^{-\beta^2}$$

$$\times \exp\left\{ 2in\beta\left(\arcsin \lambda + \lambda \sqrt{1 - \lambda^2} \right) \right\}$$

$$\times \left(1 + O\left(\frac{\ln n}{n^{1 - 2|\Re\beta|}} \right) \right), \qquad n \to \infty, \quad (5.75)$$

where $G(s)$ is Barnes' G-function.

Continuous Painlevé before the large n limit

Comparing the last two equations of the system (5.57) with the Painlevé Lax pairs from the Jimbo–Miwa list, one can, after some obvious simple gauge transformations, recognize these two equations as the alternative to (5.53)–(5.51) Lax pair for the fourth Painlevé equation. Indeed, if we put,

$$x := t, \qquad u(x) := -2Q_{n-1}, \qquad w(x) := n - 2R_n, \qquad (5.76)$$

then the function $u(x)$ would satisfy the (another) particular case of the fourth Painlevé equation,

$$u'' = \frac{1}{2u}(u')^2 + \frac{3}{2}u^3 + 4xu^2 + 2(x^2 + 1 - 2n)u, \qquad (5.77)$$

and the function $w(x)$ will be given in terms of $u(x)$,

$$4w = -u' + u^2 + 2xu.$$

Equation (5.77) was first derived, using a more direct approach, in [9].

References

[1] Ablowitz, M. J., and Segur, H. 1976/77. Asymptotic solutions of the Korteweg–de Vries equation. *Studies in Appl. Math.*, **57**(1), 13–44.

[2] Bessis, D., Itzykson, C., and Zuber, J.-B. 1980. Quantum field theory techniques in graphical enumeration. *Adv. in Appl. Math.*, **1**(2), 109–157.

[3] Bleher, P., and Its, A. 1999. Semiclassical asymptotics of orthogonal polynomials, Riemann–Hilbert problem, and universality in the matrix model. *Ann. of Math. (2)*, **150**(1), 185–266.

[4] Bleher, P., and Its, A. 2003. Double scaling limit in the random matrix model. the Riemann–Hilbert approach. *Comm. Pure Appl. Math.*, **56**(4), 433–516.

[5] Bleher, P. M., and Its, A. R. 2005. Asymptotics of the partition function of a random matrix model. *Ann. Inst. Fourier (Grenoble)*, **55**(6), 1943–2000.

[6] Brézin, É., and Kazakov, V. A. 1990. Exactly solvable field theories of closed strings. *Phys. Lett. B*, **236**(2), 144–150.

[7] Brézin, É., Itzykson, C., Parisi, G., and Zuber, J.-B. 1978. Planar diagrams. *Comm. Math. Phys.*, **59**(1), 35–51.

[8] Chen, Y., and Ismail, M. E. H. 1997. Ladder operators and differential equations for orthogonal polynomials. *J. Phys. A*, **30**(22), 7817–7829.

[9] Chen, Y., and Pruessner, G. 2005. Orthogonal polynomials with discontinuous weights. *J. Phys. A*, **38**(12), L191–L198.

[10] Claeys, T., Its, A., and Krasovsky, I. *Hankel determinant and orthogonal polynomials for the Gaussian weight with a jump II*. In preparation.

[11] Crnković, Č., and Moore, G. 1991. Multicritical multi-cut matrix models. *Phys. Lett. B*, **257**(3-4), 322–328.

[12] David, F. 1990. Loop equations and nonperturbative effects in two-dimensional quantum gravity. *Modern Phys. Lett. A*, **5**(13), 1019–1029.

[13] Deift, P. A. 1999. *Orthogonal Polynomials and Random Matrices: A Riemann–Hilbert Approach*. Courant Lect. Notes Math., vol. 3. New York: New York Univ., Courant Inst. Math. Sci.

[14] Deift, P. A., Its, A. R., and Zhou, X. 1993. Long-time asymptotics for integrable nonlinear wave equations. Pages 181–204 of: *Important Developments in Soliton Theory*. Springer Ser. Nonlinear Dynam. Berlin: Springer.

[15] Douglas, M. R., and Shenker, S. H. 1990. Strings in less than one dimension. *Nuclear Phys. B*, **335**(3), 635–654.

[16] Douglas, M. R., Seiberg, N., and Shenker, S. H. 1990. Flow and instability in quantum gravity. *Phys. Lett. B*, **244**(3-4), 381–386.

[17] Duits, M., and Kuijlaars, A. B. J. 2006. Painlevé I asymptotics for orthogonal polynomials with respect to a varying quartic weight. *Nonlinearity*, **19**(10), 2211–2245.

[18] Eynard, B. 2001. *A concise expression for the ODEs of orthogonal polynomials.* arXiv:math-ph/0109018.

[19] Faddeev, L. D., and Takhtajan, L. A. 1987. *Hamiltonian methods in the theory of solitons.* Springer Ser. in Soviet Math. Berlin: Springer.

[20] Fokas, A. S., Its, A. R., and Kitaev, A. V. 1991. Discrete Painlevé equations and their appearance in quantum gravity. *Comm. Math. Phys.*, **142**(2), 313–344.

[21] Fokas, A. S., Its, A. R., and Zhou, X. 1992a. Continuous and discrete Painlevé equations. Pages 33–47 of: *Painlevé Transcendents*. NATO Adv. Sci. Inst. Ser. B Phys., vol. 278. New York: Plenum.

[22] Fokas, A. S., Its, A. R., and Kitaev, A. V. 1992b. The isomonodromy approach to matrix models in 2D quantum gravity. *Comm. Math. Phys.*, **147**(2), 395–430.

[23] Fokas, A. S., Its, A. R., Kapaev, A. A., and Novokshenov, V. Yu. 2006. *Painlevé Transcendents. The Riemann–Hilbert Approach.* Math. Surveys Monogr., vol. 128. Providence, RI: Amer. Math. Soc.

[24] Grammaticos, B., Nijhoff, F. W., and Ramani, A. 1999. Discrete Painlevé Equations. Pages 413–516 of: *The Painlevé Property*. CRM Ser. Math. Phys. New York: Springer.

[25] Gross, D. J., and Migdal, A. A. 1990. Nonperturbative two-dimensional quantum gravity. *Phys. Rev. Lett.*, **64**(2), 127–130.

[26] Hastings, S. P., and McLeod, J. B. 1980. A boundary value problem associated with the second Painlevé transcendent and the Korteweg–de Vries equation. *Arch. Rational Mech. Anal.*, **73**(1), 31–51.

[27] Its, A., and Krasovsky, I. 2008. Hankel determinant and orthogonal polynomials for the Gaussian weight with a jump. Pages 215–247 of: *Integrable Systems and Random Matrices*. Contemp. Math., vol. 458. Providence, RI: Amer. Math. Soc.

[28] Its, A. R., Kitaev, A. V., and Fokas, A. S. 1990. An isomonodromy approach to the theory of two-dimensional quantum gravity. *Uspekhi Mat. Nauk*, **45**(6(276)), 135–136. in Russian.

[29] Itzykson, C., and Zuber, J.-B. 1980. The planar approximation. II. *J. Math. Phys.*, **21**(3), 411–421.

[30] Jimbo, M., and Miwa, T. 1981. Monodromy preserving deformation of linear ordinary differential equations with rational coefficients. II. *Phys. D*, **2**(3), 407–448.

[31] Jimbo, M., Miwa, T., and Ueno, K. 1981. Monodromy preserving deformation of linear ordinary differential equations with rational coefficients. I. General theory and τ-function. *Phys. D*, **2**(2), 306–352.

[32] Johansson, K. 1998. On fluctuations of eigenvalues of random Hermitian matrices. *Duke Math. J.*, **91**(1), 151–204.

[33] Kapaev, A. A. 2004. Quasi-linear Stokes phenomenon for the Painlevé first equation. *J. Phys. A*, **37**(46), 11149–11167.

[34] Magnus, A. P. 1995. Painlevé-type differential equations for the recurrence coefficients of semi-classical orthogonal polynomials. *J. Comput. Appl. Math.*, **57**(1-2), 215–237.

[35] Magnus, A. P. 1999. Freud's equations for orthogonal polynomials as discrete Painlevé equations. Pages 228–243 of: *Symmetries and Integrability of Difference Equations*. London Math. Soc. Lecture Note Ser., vol. 255. Cambridge: Cambridge Univ. Press.

[36] Mehta, M. L. 1991. *Random Matrices*. 2nd edn. Boston, MA: Academic Press.

[37] Periwal, V., and Shevitz, D. 1990. Exactly solvable unitary matrix models: multicritical potentials and correlations. *Nuclear Phys. B*, **344**(3), 731–746.

[38] Tracy, C. A., and Widom, H. 1994. Fredholm determinants, differential equations and matrix models. *Comm. Math. Phys.*, **163**(1), 33–72.

[39] Van Assche, W. 2007. Discrete Painlevé equations for recurrence coefficients of orthogonal polynomials. Pages 687–725 of: *Difference Equations, Special Functions and Orthogonal Polynomials*. Hackensack, NJ: World Sci. Publ.

[40] van Moerbeke, P. 2002. Random matrices and permutations, matrix integrals and integrable systems. *Astérisque*, 411–433.

6

Generalized Lie Symmetries for
Difference Equations

Decio Levi and Ravil I. Yamilov

6.1 Introduction

In this chapter we discuss the application of generalized symmetries to the investigation of difference and differential-difference equations. This is a sequel to the presentation of P. Winternitz where Lie point symmetries for difference equations have been introduced and studied in detail. In particular it has been shown there that for a given discrete equation, unless we allow for variable lattices, i.e., we consider a difference scheme, very few symmetries are present. So, if we want to get symmetries for difference equations, either we consider the point symmetries of a difference scheme or we extend the class of symmetries to the case of the generalized symmetries. In the following we will proceed in this second direction and analyze the structure of the generalized symmetries for a difference equation. We will limit ourselves to consider just partial difference equations (with two independent variables) where the lattice is fixed and non-transformable and either all independent variables are discrete (n, m) or one is discrete n and one is continuous t. We will limit our discussion to the case of scalar equations of a low order, i.e., when the dependent variable is a scalar and the differential difference equations involve at most derivatives of the second order of the fields and nearest neighboring interactions.

So, here we have either $u_n(t)$ or $u_{n,m}$, and the equations will have the form

$$\mathcal{E}\left(\ddot{u}_{n,m}, \dot{u}_{n,m}, \{u_{n+i,m+j}\}_{(i,j)=(i_a,j_a)}^{(i,j)=(i_b,j_b)}\right) = 0; \qquad i_b \geq i_a, \ j_b \geq j_a.$$

The proper formulation of the Lie point symmetry generator, valid when all independent variables are discrete or some are continuous and some are discrete, is given always by the evolutionary form, i.e., when the Lie point symmetry generator takes the form

$$\widehat{X}_e = Q_{n,m}(t, u_{n,m}, \dot{u}_{n,m})\partial_{u_{n,m}} \tag{6.1}$$

with $Q_{n,m}$ linear in $\dot{u}_{n,m}$. A Lie group transformation is then obtained by solving the differential equation

$$\frac{du_{n,m}}{d\epsilon} = Q_{n,m}(t, u_{n,m}, \dot{u}_{n,m}). \tag{6.2}$$

Emma Noether was the first to notice in 1918 [20] that one can generalize the symmetries of a differential equation by including higher derivatives of the dependent variables in the transformation. On the lattice this implies, apart from the possible terms containing higher derivatives with respect to continuous variables, that $Q_{n,m}$ will depend on a set of lattice points $\{u_{n+i,m+j}\}$, with i and j running on a finite number of positive and negative integers. The order of the generalized symmetry depends from the number of points involved.

A generalized symmetry *will not* give rise to a group transformation, as, in general, we cannot get the solution of (6.2) for any initial data. Symmetry reduction, however, can still provide particular solutions.

6.1.1 Direct construction of generalized symmetries: an example

Let us consider the class of nonlinear partial difference equations defined on a square:

$$\mathcal{E}_1(u_{n,m}, u_{n+1,m}, u_{n,m+1}, u_{n+1,m+1}) = 0, \tag{6.3}$$

where the function $\mathcal{E}_1$ is solvable with respect to all its variables.

Let us consider a generalized symmetry generator depending on 9 points defined on a square of vertices $u_{n-1,m-1}$, $u_{n-1,m+1}$, $u_{n+1,m+1}$ and $u_{n+1,m-1}$. By taking into account the difference equation (6.3), we can express the extremal points $u_{n-1,m-1}$, $u_{n-1,m+1}$, $u_{n+1,m+1}$ and $u_{n+1,m-1}$ in terms of the remaining five points $u_{n-1,m}$, $u_{n+1,m}$, $u_{n,m}$, $u_{n,m-1}$ and $u_{n,m+1}$. In this way the most general n, m–independent 9 points generator is represented by the infinitesimal symmetry generator

$$\widehat{X} = Q(u_{n-1,m}, u_{n+1,m}, u_{n,m}, u_{n,m-1}, u_{n,m+1})\partial_{u_{n,m}},$$

and the prolongation necessary to construct the determining equation is given

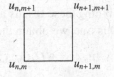

Figure 6.1 A square lattice

by

$$\mathrm{pr}\,\widehat{X} = \sum_{i=0,1}\sum_{j=0,1} Q(u_{n-1+i,m+j}, u_{n+1+i,m+j}, u_{n+i,m+j}, u_{n+i,m-1+j}, u_{n+i,m+1+j})\partial_{u_{n+i,m+j}}.$$

Applying the prolonged vector field onto (6.3), we get:

$$Q\frac{\partial \mathcal{E}_1}{\partial u_{n,m}} + [T_1 Q]\frac{\partial \mathcal{E}_1}{\partial u_{n+1,m}} + [T_2 Q]\frac{\partial \mathcal{E}_1}{\partial u_{n,m+1}} + [T_1 T_2 Q]\frac{\partial \mathcal{E}_1}{\partial u_{n+1,m+1}} = 0, \quad (6.4)$$

where $T_1 f_{n,m} = f_{n+1,m}$ and $T_2 f_{n,m} = f_{n,m+1}$. Equation (6.4) contains $u_{n+i,m+j}$ with $i = -1, 0, 1, 2$, $j = -1, 0, 1, 2$.

The invariance condition requires that (6.4) be satisfied on the solutions of (6.3). To be able to do so, as (6.3) is a relation between points on a square, we need to choose a set of independent variables (not on a square) for which (6.4) must be identically satisfied. A natural choice is to take as independent variables the values on the two axis, i.e., $u_{n,m+2}, u_{n,m+1}, u_{n,m-1}, u_{n,m}, u_{n+2,m}, u_{n+1,m}$ and $u_{n-1,m}$. By substituting

- $u_{n+2,m+1} \rightarrow (u_{n+2,m}, u_{n+1,m+1}, u_{n+1,m})$,
- $u_{n+1,m+1} \rightarrow (u_{n+1,m}, u_{n,m+1}, u_{n,m})$,
- $u_{n-1,m+1} \rightarrow (u_{n-1,m}, u_{n,m+1}, u_{n,m})$,

etc., we reduce the determining equation to an equation, written just in terms of independent variables, which thus must be identically satisfied. Differentiating it with respect to $u_{n,m+2}$ and to $u_{n+2,m}$, we get

$$\frac{\partial^2 T_1 T_2 Q}{\partial u_{n+2,m+1}\partial u_{n+1,m+2}} = T_1 T_2 \frac{\partial^2 Q}{\partial u_{n+1,m}\partial u_{n,m+1}} = 0. \quad (6.5)$$

Consequently the symmetry generator coefficient Q is the sum of *two simpler functions*,

$$Q = Q_0(u_{n-1,m}, u_{n+1,m}, u_{n,m}, u_{n,m-1}) + Q_1(u_{n-1,m}, u_{n,m}, u_{n,m-1}, u_{n,m+1}).$$

Introducing this result into the determining equation (6.4) and differentiating it with respect to $u_{n,m+2}$ and to $u_{n-1,m}$, we have that Q_1 reduces to $Q_1 = Q_{10}(u_{n-1,m}, u_{n,m}, u_{n,m-1}) + Q_{11}(u_{n,m}, u_{n,m-1}, u_{n,m+1})$. In a similar way, if we differentiate the resulting determining equation with respect to $u_{n+2,m}$ and to $u_{n,m-1}$, we have that Q_0 reduces to $Q_0 = Q_{00}(u_{n-1,m}, u_{n,m}, u_{n,m-1}) + Q_{01}(u_{n-1,m}, u_{n+1,m}, u_{n,m})$. Combining these results and taking into account that also $\partial^2 Q/(\partial u_{n-1,m}\partial u_{n,m-1}) = 0$ is true, we obtain the following form for Q:

$$Q = Q_0(u_{n,m-1}, u_{n,m}, u_{n,m+1}) + Q_1(u_{n-1,m}, u_{n,m}, u_{n+1,m}).$$

So, *the infinitesimal symmetry coefficient Q is the sum of functions that either involve shifts only in n with m fixed or only in m with n fixed* [12–14, 22].

Let us consider the case when the symmetry generator is given by

$$\frac{\mathrm{d}u_{n,m}}{\mathrm{d}\epsilon} = Q_1(u_{n-1,m}, u_{n,m}, u_{n+1,m}). \tag{6.6}$$

This is a differential difference equation depending parametrically on m. Setting $u_{n,m} = u_n$ and $u_{n,m+1} = \tilde{u}_n$, a different solution, the compatible partial difference equation (6.3) turns out to be an ordinary difference equation for a new solution $\tilde{u}_n$ of (6.6),

$$\mathcal{E}_1(u_n, u_{n+1}, \tilde{u}_n, \tilde{u}_{n+1}) = 0,$$

i.e., a Bäcklund transformation [15]. A similar result is obtained in case of Q_0.

The same splitting will also appear for higher-order symmetries of this class of equations.

To find the specific form of Q_1 we have to differentiate the determining equation (6.4) with respect to the independent variables and get some further necessary conditions for its form. If, for example, we differentiate with respect to $u_{n+2,m}$,

$$\frac{\partial \mathcal{E}_1}{\partial u_{n+1,m}} \frac{\partial T_1 Q_1}{\partial u_{n+2,m}} + \frac{\partial \mathcal{E}_1}{\partial u_{n+1,m+1}} \frac{\partial T_1 T_2 Q_1}{\partial u_{n+2,m}} = 0.$$

Dividing by $\partial \mathcal{E}_1 / \partial u_{n+1,m}$ and differentiating the resulting expression with respect to $u_{n,m+1}$, we get

$$\frac{\partial}{\partial u_{n,m+1}} \Big[\frac{\partial \mathcal{E}_1 / \partial u_{n+1,m+1}}{\partial \mathcal{E}_1 / \partial u_{n+1,m}} \frac{\partial T_1 T_2 Q_1}{\partial u_{n+2,m}} \Big] = 0.$$

This is a partial differential equation for Q_1 which constrains its form, and by solving it it will give an expression in terms of functions of lower number of independent variables and possibly some integration constants. Proceeding further in the general study of the class of equations possessing generalized symmetries is extremely hard. So we go over to the construction of generalized symmetries for given equations. In the specific case when

$$\mathcal{E}_1 = \alpha(u_{n,m}u_{n+1,m} + u_{n+1,m+1}u_{n,m+1}) - \beta(u_{n,m}u_{n,m+1} + u_{n+1,m+1}u_{n+1,m}) + \delta(\alpha^2 - \beta^2),$$

where α, β and δ are constants, we get

$$Q_1 = \frac{u_{n,m}(u_{n+1,m} + u_{n-1,m}) + 2\delta\alpha}{u_{n+1,m} - u_{n-1,m}}.$$

This generalized symmetry was quite complicated to derive. It would be extremely complicated to derive in this way generalized symmetries involving higher number of points. So we need a different procedure to get them in compact form. They can be derived, using the *Recursion Operator*, which can be derived from the linear problem associated to the nonlinear discrete equation.

6.2 Generalized symmetries from the integrability properties

Equations with generalized symmetries are rare. Here we present results on the Toda Lattice and discrete-time Toda Lattice, as for these equations their integrability is well known and studied [5, 16, 23].

6.2.1 Toda Lattice

The Toda Lattice equation reads

$$\ddot{v}_n(t) = e^{v_{n-1}(t)-v_n(t)} - e^{v_n(t)-v_{n+1}(t)}, \tag{6.7}$$

equivalently written as

$$\dot{a}_n(t) = a_n(t)(b_n(t) - b_{n+1}(t)), \qquad \dot{b}_n(t) = a_{n-1}(t) - a_n(t), \tag{6.8}$$

where

$$b_n = \dot{v}_n, \qquad a_n = e^{v_n - v_{n+1}}. \tag{6.9}$$

The Toda Lattice can be associated to the isospectral deformation ($\lambda_t = 0$) of the discrete Schrödinger spectral problem

$$L(a_n, b_n)\psi(n, t; \lambda) = \psi(n-1, t; \lambda) + b_n\psi(n, t; \lambda) + a_n\psi(n+1, t; \lambda)$$
$$= \lambda\psi(n, t; \lambda). \tag{6.10}$$

The t–evolution of the function $\psi(n, t; \lambda)$ corresponding to the Toda Lattice is given by

$$\psi_t(n, t; \lambda) = -M(a_n, b_n)\psi(n, t; \lambda) = -a_n\psi(n+1, t; \lambda). \tag{6.11}$$

To carry out the compatibility of the Lax pair (6.10), (6.11), we just differentiate (6.10) with respect to time and substitute the time evolution of the function ψ provided by (6.11). In such a way, in the isospectral case ($\lambda_t = 0$), we get the Lax equation

$$L_t(a_n, b_n) = [L(a_n, b_n), M(a_n, b_n)]. \tag{6.12}$$

If $\lambda_t = f(\lambda, t)$, i.e., in the non-isospectral case, where $f(z, t)$ is an entire function of its first argument, the Lax equation reads

$$L_t(a_n, b_n) = [L(a_n, b_n), M(a_n, b_n)] + f(L(a_n, b_n), t). \tag{6.13}$$

When the following boundary conditions

$$\lim_{|n|\to\infty} a_n - 1 = \lim_{|n|\to\infty} b_n = 0 \tag{6.14}$$

are imposed on the fields a_n and b_n, the solution of the discrete Schrödinger spectral problem (6.10), for $\lambda = z + z^{-1}$ and $|z| = 1$, has the following asymptotic behavior:

$$\psi(n, z, t) \to z^{-n} + R(z, t)z^n, \qquad (n \to +\infty), \tag{6.15}$$

$$\psi(n, z, t) \to T(z, t)z^{-n}, \qquad (n \to -\infty), \tag{6.16}$$

where $T(z, t)$ and $R(z, t)$ are respectively the *transmission* and *reflection coefficients*.

When $v(n, t)$ evolves according to the Toda Lattice, $\dot{T}(z, t) = 0$ and

$$\dot{R}(z, t) = \mu R(z, t), \qquad \mu = \frac{1}{z} - z. \tag{6.17}$$

For $z_j, \ j = 1, 2, \dots N$, belonging to the real interval $(-1, 1)$, the function $\psi(n, z_j) = f_j(n)$ is bounded at infinity, and we can define its normalization coefficient, ρ_j:

$$\rho_j = \left[\sum_{n=-\infty}^{\infty} |f_j(n)|^2 \right]^{-1/2}, \qquad j = 1, 2, \dots N. \tag{6.18}$$

Summing up, for any solution of the Toda Lattice satisfying the condition (6.14), we can associate to the discrete Schrödinger spectral problem (6.10) its spectrum $S[a_n, b_n]$ defined in the complex plane of the variable z:

$$S[a_n, b_n] = \{R(z, t), |z| = 1; z_j \in \mathbb{R}, \rho_j(t), |z_j| < 1, j = 1, 2, \dots, N\}. \tag{6.19}$$

When a_n, b_n satisfy the boundary conditions (6.14), the potentials define the spectral data in a unique way. Moreover, by solving the inverse problem one can show that, when the boundary conditions (6.14) are satisfied, the spectral data defines the potentials in a unique way. Thus, there is a *one-to-one correspondence between the evolution of the potentials $a_n(t), b_n(t)$ of the discrete Schrödinger spectral problem and that of the reflection coefficient $R(z, t)$.*

To the spectral problem (6.10) we can associate an infinite number of nonlinear differential difference equations (*the Toda Lattice hierarchy*):

$$\begin{pmatrix} \dot{a}_n \\ \dot{b}_n \end{pmatrix} = f_1(\mathcal{L}, t) \begin{pmatrix} a_n(b_n - b_{n+1}) \\ a_{n-1} - a_n \end{pmatrix}, \tag{6.20}$$

where $f_1(\mathcal{L}, t)$ is an entire function of its first argument, and the recursion operator $\mathcal{L}$ is given by

$$\mathcal{L} \begin{pmatrix} p_n \\ q_n \end{pmatrix} = \begin{pmatrix} p_n b_{n+1} + a_n(q_n + q_{n+1}) + (b_n - b_{n+1})s_n \\ b_n q_n + p_n + s_{n-1} - s_n \end{pmatrix}, \tag{6.21}$$

s_n being a bounded solution of the nonhomogeneous first-order difference equation

$$s_{n+1} = \frac{a_{n+1}}{a_n}(s_n - p_n). \tag{6.22}$$

There are various techniques one can use to get the nonlinear equations starting from the spectral problem. We will consider here, because of its algebraic simplicity and large applicability, the Lax Technique, a constructive procedure based on the request that the Lax equation must be satisfied. Let us start from the Lax equation (6.12) with L written in operator form in terms of the shift operator T as $L = T^{-1} + b_n + a_n T$. We assume that:

1. We can associate a sequence of nonlinear equations to the same Spectral Problem.
2. The equations are related among themselves so that a recursion operator will exist.

From these assumptions it follows that we can write, due to the assumed form of L,

$$[L, M] = u_n + w_n T, \quad [L, \widetilde{M}] = \tilde{u}_n + \widetilde{w}_n T, \quad \widetilde{M} = LM + F_n T + G_n,$$

where u_n, w_n, $\tilde{u}_n$ and $\widetilde{w}_n$ are scalar functions, and the relation between the operators M and $\widetilde{M}$ is such that $\tilde{u}_n$ and $\widetilde{w}_n$ depend explicitly on u_n and w_n. Introducing the postulated expression of $\widetilde{M}$ into the commutation relation and taking into account the result of the commutation relation between L and M, we get that the coefficient of T^2 and T^{-1} must be zero, i.e., F_n and G_n must satisfy the following first-order difference equations:

$$G_n = G_{n-1} + u_{n-1}, \quad a_{n+1}F_n = a_n(F_{n+1} + w_{n+1}).$$

The solution of the discrete equations for F_n and G_n provide us with an explicit expression of F_n and G_n as functions of u_n and w_n and of some integration constants. Taking into account that $\tilde{u}_n$ must equal the coefficient of T^0 and $\widetilde{w}_n$ that of T, we get both the recursion operator (6.21) and the starting equations (6.20). The starting equations are obtained as coefficients of the constants appearing in the solutions of the equations for F_n and G_n.

For any equation of the hierarchy we can write down an explicit evolution equation for the function $\psi(n, t; \lambda)$. This is possible if

$$\lim_{|n|\to\infty} a_n - 1 = \lim_{|n|\to\infty} b_n = \lim_{|n|\to\infty} s_n = 0, \tag{6.23}$$

as

$$\frac{dR(z, t)}{dt} = \mu f_1(\lambda, t)R(z, t). \tag{6.24}$$

A class of independent symmetries for the Toda Lattice are given by the following equations:

$$\begin{pmatrix} a_{n,\epsilon_k} \\ b_{n,\epsilon_k} \end{pmatrix} = \mathcal{L}^k \begin{pmatrix} a_n(b_n - b_{n+1}) \\ a_{n-1} - a_n \end{pmatrix}, \qquad k = 0, 1, \dots. \tag{6.25}$$

Equations (6.25) commute with the Toda Lattice (see the proof in Section 6.2.2). Here k is any positive integer and ϵ_k is a variable. In correspondence with (6.25), when the functions a_n and b_n satisfy the boundary conditions (6.23), the spectrum evolves according to the equation

$$\frac{\mathrm{d}R}{\mathrm{d}\epsilon_k} = \mu \lambda^k R. \tag{6.26}$$

Let us show here the correspondence between symmetries, defined as *point transformations of the independent and dependent variables which leave the equation invariant*, and commuting flows, i.e., *equations providing an evolution in the group parameter which is compatible with our equation*. To do so in the whole generality, let us consider a generic differential difference equation defined on three neighboring points

$$\mathcal{F}_n(u_n, u_{n+1}, u_{n-1}, \dot{u}_n) = 0.$$

By the First Lie Theorem a Lie point group transformation is equivalent on one side to the exponential of the infinitesimal generator and on the other to the solution of a differential equation with arbitrary initial condition, which for a generator $\widehat{X}_e = Q_n(t, u_n, \dot{u}_n)\partial_{u_n}$ written in evolutionary form reads:

$$\frac{\partial u_n}{\partial \epsilon} = Q_n(t, u_n, \dot{u}_n). \tag{6.27}$$

The necessary prolongation of this generator is given by

$$\mathrm{pr}\,\widehat{X} = Q_n \partial_{u_n} + D_t Q_n \partial_{\dot{u}_n} + Q_{n+1} \partial_{u_{n+1}} + Q_{n-1} \partial_{u_{n-1}}.$$

From the point of view of Lie theory the invariance condition $\mathrm{pr}\,\widehat{X}\mathcal{F}_n|_{\mathcal{F}=0} = 0$ implies

$$Q_n \mathcal{F}_{n,u_n} + Q_{n+1}\mathcal{F}_{n,u_{n+1}} + Q_{n-1}\mathcal{F}_{n,u_{n-1}} + D_t Q_n \mathcal{F}_{n,\dot{u}_n} = 0 \tag{6.28}$$

when $\mathcal{F}_n = 0$. On the other hand, the compatibility of the equation $\mathcal{F}_n = 0$ with the group flow (6.27) requires that the total derivative with respect to ϵ of the equation $\mathcal{F}_n = 0$ be identically satisfied. This gives

$$\frac{\mathrm{d}\mathcal{F}_n}{\mathrm{d}\epsilon} = u_{n,\epsilon}\mathcal{F}_{n,u_n} + u_{n+1,\epsilon}\mathcal{F}_{n,u_{n+1}} + u_{n-1,\epsilon}\mathcal{F}_{n,u_{n-1}} + \dot{u}_{n,\epsilon}\mathcal{F}_{n,\dot{u}_n} = 0. \tag{6.29}$$

As from (6.27) $u_{n+1,\epsilon} = Q_{n+1}$, $u_{n-1,\epsilon} = Q_{n-1}$ and $\dot{u}_{n,\epsilon} = D_t Q_n$, it is immediate to see that (6.28) and (6.29) are equal.

We can add to the class of symmetries introduced before in (6.25) the non-isospectral deformations satisfying the Lax equation (6.13). They are

$$\begin{pmatrix} a_{n,\epsilon_k} \\ b_{n,\epsilon_k} \end{pmatrix} = \mathcal{L}^{k+1} t \begin{pmatrix} a_n(b_n - b_{n+1}) \\ a_{n-1} - a_n \end{pmatrix}$$

$$+ \mathcal{L}^k \begin{pmatrix} a_n[(2n + 3)b_{n+1} - (2n - 1)b_n] \\ b_n^2 - 4 + 2[(n + 1)a_n - (n - 1)a_{n-1}] \end{pmatrix}, \quad (6.30)$$

where

$$\frac{dR}{d\epsilon_k} = \mu \lambda^{k+1} t R, \qquad \lambda_{\epsilon_k} = \mu^2 \lambda^k. \quad (6.31)$$

In addition to the above two hierarchies of symmetries, there are two further flows commuting with the Toda Lattice (6.8), which do not satisfy the asymptotic boundary conditions (6.23):

$$\begin{pmatrix} a_{n,\epsilon} \\ b_{n,\epsilon} \end{pmatrix} = \begin{pmatrix} 0 \\ 1 \end{pmatrix}, \qquad \begin{pmatrix} a_{n,\epsilon} \\ b_{n,\epsilon} \end{pmatrix} = t \begin{pmatrix} a_n(b_n - b_{n+1}) \\ a_{n-1} - a_n \end{pmatrix} + \begin{pmatrix} 2a_n \\ b_n \end{pmatrix}. \quad (6.32)$$

These two exceptional cases correspond to the point symmetries of the Toda Lattice. In fact, taking into account the definitions (6.9), the first system is equivalent to $\dot{v}_{n,\epsilon} = 1$ and the second one to $v_{n,\epsilon} - v_{n+1,\epsilon} = t(\dot{v}_n - \dot{v}_{n+1}) + 2$. The first partial differential-difference equation is equivalent to the evolutionary equation $v_{n,\epsilon} = t + c$, where c is an arbitrary constant. So we get the two symmetries $\widehat{W} = t\partial_{v_n}$ and $\widehat{U} = \partial_{v_n}$. Similarly, integrating the second equation, we get the remaining two point symmetries of the Toda Lattice: $\widehat{D} = t\partial_t + 2n\partial_{v_n}$ and $\widehat{T} = \partial_t$.

6.2.2 The symmetry algebra for the Toda Lattice

The structure of the symmetry algebra for the Toda Lattice, is obtained by computing the commutation relations between the symmetries.

If we define

$$\mathcal{L}^k = \begin{pmatrix} \mathcal{L}_{11}^{(k)} & \mathcal{L}_{12}^{(k)} \\ \mathcal{L}_{21}^{(k)} & \mathcal{L}_{22}^{(k)} \end{pmatrix}, \quad (6.33)$$

then the infinitesimal generator of the corresponding symmetry is given by

$$\widehat{X}_k = \{\mathcal{L}_{11}^{(k)}[a_n(b_n - b_{n+1})] + \mathcal{L}_{12}^{(k)}(a_{n-1} - a_n)\}\partial_{a_n}$$

$$+ \{\mathcal{L}_{21}^{(k)}[a_n(b_n - b_{n+1})] + \mathcal{L}_{22}^{(k)}(a_{n-1} - a_n)\}\partial_{b_n}. \quad (6.34)$$

We can also write an infinitesimal generator for the equation satisfied by the reflection coefficient:

$$\widehat{X}_k = \mu \lambda^k R \partial_R. \quad (6.35)$$

Using the one-to-one correspondence between a_n, b_n and $R(z)$, it is trivial to show the commutativity of the infinitesimal generators (6.34) without calculating it explicitly. In fact, for all values of k and m, we have

$$[\widehat{X}_k, \widehat{X}_m] = [\mu\lambda^k R\partial_R, \mu\lambda^m R\partial_R] = 0 \rightarrow [\widehat{X}_k, \widehat{X}_m] = 0. \tag{6.36}$$

So all isospectral flows commute.

We can introduce the generators of the non-isospectral symmetries for the Toda Lattice in the same way. They are:

$$\begin{aligned}
\widehat{Y}_k = & \{t[\mathcal{L}_{11}^{(k+1)}[a_n(b_n - b_{n+1})] + \mathcal{L}_{12}^{(k+1)}(a_{n-1} - a_n)] \\
& + \mathcal{L}_{11}^{(k)}[a_n((2n+3)b_{n+1} - (2n-1)b_n)] \\
& + \mathcal{L}_{12}^{(k)}[b_n^2 - 4 + 2(n+1)a_n - 2(n-1)a_{n-1}]\}\partial_{a_n} \\
& + \{t[\mathcal{L}_{21}^{(k+1)}[a_n(b_n - b_{n+1})] + \mathcal{L}_{22}^{(k+1)}(a_{n-1} - a_n)] \\
& + \mathcal{L}_{21}^{(k)}[a_n((2n+3)b_{n+1} - (2n-1)b_n)] \\
& + \mathcal{L}_{22}^{(k)}[b_n^2 - 4 + 2(n+1)a_n - 2(n-1)a_{n-1}]\}\partial_{b_n}, \tag{6.37}
\end{aligned}$$

and from the corresponding equations satisfied by the reflection coefficients we get

$$\widehat{\mathcal{Y}}_k = \mu\lambda^{k+1} t R\partial_R + \mu^2\lambda^k\partial_\lambda. \tag{6.38}$$

Commuting $\widehat{\mathcal{Y}}_k$ with $\widehat{\mathcal{Y}}_m$ we get:

$$[\widehat{\mathcal{Y}}_k, \widehat{\mathcal{Y}}_m] = (m-k)[\widehat{\mathcal{Y}}_{k+m+1} - 4\widehat{\mathcal{Y}}_{k+m-1}], \tag{6.39}$$

while commuting $\widehat{X}_k$ with $\widehat{\mathcal{Y}}_m$ we get:

$$[\widehat{X}_k, \widehat{\mathcal{Y}}_m] = -(1+k)\widehat{X}_{k+m+1} + 4k\widehat{X}_{k+m-1}. \tag{6.40}$$

So the non-isospectral flows are not commuting among themselves and form a Kac–Moody algebra. Moreover the commutation of the non-isospectral flows with the isospectral ones give higher-order isospectral flows, showing that they behave like master symmetries [6].

Let us now consider the commutation relations involving the exceptional symmetries:

$$\widehat{Z}_0 = \partial_{b_n}, \tag{6.41}$$

$$\widehat{Z}_1 = (2a_n + t\dot{a}_n)\partial_{a_n} + (b_n + t\dot{b}_n)\partial_{b_n}. \tag{6.42}$$

We calculate explicitly their commutation relations with the lowest-order symmetries $\widehat{X}_0, \widehat{X}_1$ and $\widehat{Y}_0$, as we cannot associate any reflection coefficient to the exceptional cases. The non-zero commutation relations are:

$$[\widehat{X}_0, \widehat{Z}_1] = -\widehat{X}_0, \qquad [\widehat{Z}_0, \widehat{Z}_1] = \widehat{Z}_0,$$

$$[\widehat{Y}_0, \widehat{Z}_0] = -2\widehat{Z}_1, \qquad [\widehat{Y}_0, \widehat{Z}_1] = -\widehat{Y}_0 - 8\widehat{Z}_0, \qquad [\widehat{X}_1, \widehat{Z}_0] = -2\widehat{X}_0, \qquad (6.43)$$

$$[\widehat{X}_1, \widehat{Z}_1] = -2\widehat{X}_1, \qquad [\widehat{X}_0, \widehat{Y}_0] = -\widehat{X}_1, \qquad [\widehat{X}_1, \widehat{Y}_0] = -2\widehat{X}_2 + 4\widehat{X}_0.$$

So, the structure of the symmetry Lie algebra for the Toda system is the semi-direct sum of the two infinite algebras L_0 and L_1, i.e.

$$L = L_0 \ni L_1, \qquad L_0 = \{\hat{h}, \hat{e}, \hat{f}, \widehat{Y}_1, \widehat{Y}_2, \ldots\}, \; L_1 = \{\widehat{X}_0, \widehat{X}_1, \ldots\}, \qquad (6.44)$$

where $\{\hat{h} = \widehat{Z}_1, \hat{e} = \widehat{Z}_0, \hat{f} = \widehat{Y}_0 + 4\widehat{Z}_0\}$ denotes a $sl(2, \mathbb{R})$ subalgebra with $[\hat{h}, \hat{e}] = \hat{e}, [\hat{h}, \hat{f}] = -\hat{f}, [\hat{e}, \hat{f}] = 2\hat{h}$. The algebra L_0 is perfect, i.e., we have $[L_0, L_0] = L_0$.

Note that $\widehat{Z}_0, \widehat{Z}_1$ and $\widehat{X}_0$ are point symmetries while all the others are generalized symmetries.

We write here in the following as an example the simplest symmetries, written in evolutionary form:

- $a_{n,\mu_0} = 0, b_{n,\mu_0} = 1$;
- $a_{n,\mu_1} = 2a_n + t\dot{a}_n, b_{n,\mu_1} = b_n + t\dot{b}_n$;
- $a_{n,\epsilon_0} = \dot{a}_n, b_{n,\epsilon_0} = \dot{b}_n$;
- $a_{n,\epsilon_1} = a_n(b_n^2 - b_{n+1}^2 + a_{n-1} - a_{n+1}), b_{n,\epsilon_1} = a_{n-1}(b_n + b_{n-1}) - a_n(b_{n+1} + b_n)$;
- $a_{n,\nu_0} = ta_{n,\epsilon_1} + a_n[(2n+3)b_{n+1} - (2n-1)b_n], b_{n,\nu_0} = tb_{n,\epsilon_1} + b_n^2 - 4 + 2[(n+1)a_n - (n-1)a_{n-1}]$.

6.2.3 The continuous limit of the Toda symmetry algebras

We have seen above that the Toda Lattice has a four-dimensional algebra of point symmetries, and it is well known that the Korteweg–de Vries (KdV) equation is its continuous limit. Also the KdV [21] has a four-dimensional algebra of point symmetries. However the symmetry algebra of the KdV contains a dilation symmetry which is absent in the case of the Toda Lattice. So it is interesting to consider the continuous limit of both the Toda Lattice and its symmetry algebra to elucidate the origin of the dilation symmetry of the KdV in the symmetries of the Toda Lattice.

Let us consider the continuous limit which carries the Toda Lattice into the KdV equation. Defining

$$v_n(t) = -\frac{h}{2}u(x, \tau), \qquad x = (n - t)h, \qquad \tau = -\frac{h^3}{24}t \qquad (6.45)$$

in the limit when h goes to zero, we can write the Toda Lattice equation as

$$(u_\tau - u_{xxx} - 3u_x^2)_x = O(h^2),\qquad(6.46)$$

i.e., the once differentiated potential Korteweg–de Vries equation. With the same level of approximation in h, the simplest infinitesimal symmetry generators of the infinite-dimensional symmetry algebra of the Toda Lattice read:

$$\widehat{X}_0 = \left\{-u_x(x,\tau)h - \frac{1}{24}u_\tau(x,\tau)h^3\right\}\partial_u,\qquad(6.47)$$

$$\widehat{X}_1 = \left\{-2u_x(x,\tau)h - \frac{1}{3}u_\tau(x,\tau)h^3 + O(h^5)\right\}\partial_u,\qquad(6.48)$$

$$\widehat{X}_2 = \left\{-4u_x(x,\tau)h - \frac{7}{6}u_\tau(x,\tau)h^3 + O(h^5)\right\}\partial_u,\qquad(6.49)$$

$$\widehat{Y}_0 = \{2[u(x,\tau) + xu_x(x,\tau) + 3\tau u_\tau(x,\tau)] + O(h)\}\partial_u,\qquad(6.50)$$

$$\widehat{Z}_{-1} = -\frac{2}{h}\partial_u,\qquad \widehat{Z}_0 = \frac{48}{h^4}\tau\partial_u,\qquad(6.51)$$

$$\widehat{Z}_1 = \left\{-\frac{96}{h^4}\tau + \frac{4}{h^2}[x + 6\tau u_x(x,\tau)] + O(1)\right\}\partial_u.\qquad(6.52)$$

Defining

$$\widetilde{P}_0 = \frac{4}{h^3}(2\widehat{X}_0 - \widehat{X}_1), \qquad \widetilde{P}_1 = -\frac{1}{h}\widehat{X}_0, \qquad \widetilde{D} = \frac{1}{2}\widehat{Y}_0,\qquad(6.53)$$

$$\widetilde{B} = \frac{h^2}{4}(2\widehat{Z}_0 + \widehat{Z}_1), \qquad \widetilde{\Gamma} = -\frac{h}{2}\widehat{Z}_{-1}\qquad(6.54)$$

and taking into account their commutation relations, we get.

	$\widetilde{P}_0$	$\widetilde{P}_1$	$\widetilde{B}$	$\widetilde{D}$	$\widetilde{\Gamma}$
$\widetilde{P}_0$	0	0	$-6\widetilde{P}_1 + O(h^2)$	$-3\widetilde{P}_0 + O(h^2)$	0
$\widetilde{P}_1$		0	$-\widetilde{\Gamma} + O(h^2)$	$-\widetilde{P}_1 + O(h^2)$	0
$\widetilde{B}$			0	$2\widetilde{B} + O(h^2)$	0
$\widetilde{D}$				0	$-\widetilde{\Gamma}$
$\widetilde{\Gamma}$					0

$$(6.55)$$

This table corresponds to the commutation rules of the point symmetry generators for the potential KdV equation $u_\tau = u_{xxx} + 3u_x^2$:

$$\widehat{P}_0 = \partial_\tau, \qquad \widehat{P}_1 = \partial_x, \qquad \widehat{B} = x\partial_u - 6\tau\partial_x,\qquad(6.56)$$

$$\widehat{D} = u\partial_u - x\partial_x - 3\tau\partial_\tau, \qquad \widehat{\Gamma} = \partial_u.\qquad(6.57)$$

This simple calculation has shown that in the continuous limit the infinite symmetry group of the Toda Lattice has a finite subgroup which contains more point symmetries than the original one. This is an example of symmetry contraction [8]. The dilation symmetry of the KdV equation appears as the lowest-order non-isospectral generalized symmetry $\widehat{Y}_0$ of the Toda lattice.

6.2.4 Bäcklund transformations for the Toda equation

Among the important properties that one can associate to an integrable non-linear system, certainly an important role is played by the Bäcklund transformations. They can be obtained by a straightforward modification of the Lax technique we have used to get the hierarchy of equations (and thus the symmetries).

Let us assume that we are given a spectral problem $L\psi(n, t; \lambda) = \lambda\psi(n, t; \lambda)$, where L depends on a field $v(n, t)$. The Bäcklund transformations are a relation between two different solutions of the nonlinear system. If they exist, then we must be able to find two different solutions of the spectral problem, $\psi(n, t; \lambda)$ and $\tilde{\psi}(n, t; \lambda)$. These two solutions will be associated to two different solutions of the Lax equation, $v(n, t)$ and $\tilde{v}(n, t)$, and consequently two different Lax pairs, $(L(v), M(v))$ and $(L(\tilde{v}), M(\tilde{v}))$. So, if a transformation exists between v and $\tilde{v}$, then there must exist a transformation between $\psi(n, t; \lambda)$ and $\tilde{\psi}(n, t; \lambda)$ and between $(L(v), M(v))$ and $(L(\tilde{v}), M(\tilde{v}))$. Introducing an operator D which transforms $\psi(n, t; \lambda)$ into $\tilde{\psi}(n, t; \lambda)$:

$$\tilde{\psi}(x, t; \lambda) = D(v, \tilde{v})\psi(x, t; \lambda), \tag{6.58}$$

we get that the Lax equations for the Bäcklund transformations is given by

$$L(\tilde{v})D(v, \tilde{v}) = D(v, \tilde{v})L(v), \tag{6.59}$$

$$D_t(v, \tilde{v}) = D(v, \tilde{v})M(v) - M(\tilde{v})D(v, \tilde{v}). \tag{6.60}$$

Using the same ideas which were at the base of the Lax technique, we can obtain the class of Bäcklund transformations. The class of Bäcklund transformations associated to the Toda system is given by

$$\gamma(\Lambda)\begin{pmatrix} \tilde{a}_n - a_n \\ \tilde{b}_n - b_n \end{pmatrix} = \delta(\Lambda)\begin{pmatrix} \widetilde{\Pi}(n)\Pi^{-1}(n+1)(\tilde{b}_n - b_{n+1}) \\ \widetilde{\Pi}(n-1)\Pi^{-1}(n) - \tilde{\Pi}(n)\Pi^{-1}(n+1) \end{pmatrix}, \tag{6.61}$$

where $\gamma(z)$ and $\delta(z)$ are entire functions of their argument and

$$\Pi(n) = \prod_{j=n}^{\infty} a_j, \qquad \widetilde{\Pi}(n) = \prod_{j=n}^{\infty} \tilde{a}_j.$$

Λ is a recursion operator:

$$\Lambda\begin{bmatrix}p(n)\\q(n)\end{bmatrix}$$
$$=\begin{bmatrix}p(n)b_{n+1}+\tilde{a}_n[q(n)+q(n+1)]+\Sigma(n)[\tilde{b}_n-b_{n+1}]+[a_n-\tilde{a}_n]\sum_{j=n}^{\infty}p(j)\\p(n)+\tilde{b}_nq(n)-\Sigma(n)+\Sigma(n-1)+[b_n-\tilde{b}_n]\sum_{j=n}^{\infty}q(j)\end{bmatrix},$$

where

$$\Sigma(n)=\tilde{\Pi}(n)\left[\sum_{j=n}^{\infty}\tilde{\Pi}(j)^{-1}p(j)\Pi(j+1)\right]\Pi^{-1}(n+1).$$

Whenever (a_n,b_n) and $(\tilde{a}_n,\tilde{b}_n)$ satisfy the asymptotic conditions (6.14), the corresponding reflection coefficients, $R(\lambda)$ and $\widetilde{R}(\lambda)$ will satisfy the transformation

$$\widetilde{R}(\lambda)=\frac{\gamma(\lambda)-\delta(\lambda)z}{\gamma(\lambda)-\delta(\lambda)/z}R(\lambda),\tag{6.62}$$

corresponding to the Bäcklund transformations (6.61). When $\gamma(\lambda)$ and $\delta(\lambda)$ are constant, we get

$$\dot{v}(n)-\dot{v}(n+1)=\frac{1}{\beta}\left[e^{v(n+1)-\tilde{v}(n+1)}-e^{v(n)-\tilde{v}(n)}\right]-\beta\left[e^{\tilde{v}(n-1)-v(n)}-e^{\tilde{v}(n)-v(n+1)}\right],\tag{6.63}$$

the simplest *adding one soliton* Bäcklund transformation, where $\beta=\delta/\gamma$.

6.2.5 Bäcklund transformations vs. generalized symmetries

We have seen that an integrable nonlinear differential-difference equation associated to a linear Spectral Problem, as the Toda Lattice, has an infinite algebra of symmetries and an infinite number of Bäcklund transformations. Are these two properties correlated? In this subsection we will provide a positive answer to this question which will also support the idea that the existence of a generalized symmetry and of a Bäcklund transformation is strictly related to the integrability of a system.

A general isospectral higher symmetry of the Toda equation is given by

$$\begin{pmatrix}a_{n,\epsilon}\\b_{n,\epsilon}\end{pmatrix}=\phi(\mathcal{L})\begin{pmatrix}a_n(b_n-b_{n+1})\\a_{n-1}-a_n\end{pmatrix},\tag{6.64}$$

where the function ϕ is an entire function of its argument. To this symmetry there will correspond the following evolution of the spectrum

$$\frac{dR(\lambda,\epsilon)}{d\epsilon}=\mu\phi(\lambda)R(\lambda,\epsilon).\tag{6.65}$$

Equation (6.65) can be solved and gives

$$R(\lambda, \epsilon) = e^{\mu \epsilon \phi(\lambda)} R(\lambda, 0).$$ (6.66)

Taking into account the definitions of λ and μ in terms of z:

$$\lambda = \frac{1}{z} + z, \qquad \mu = \frac{1}{z} - z, \qquad \mu^2 = \lambda^2 - 4,$$ (6.67)

$$z = \frac{\lambda - \mu}{2}, \qquad \frac{1}{z} = \frac{\lambda + \mu}{2},$$ (6.68)

we can rewrite the relation between the reflection coefficients (6.62) associated to the general Bäcklund transformation (6.61) as

$$\widetilde{R}(\lambda) = \frac{2 - (\lambda - \mu)\beta(\lambda)}{2 - (\lambda + \mu)\beta(\lambda)} R(\lambda), \qquad \beta(\lambda) = \frac{\delta(\lambda)}{\gamma(\lambda)}.$$

To identify a general symmetry transformation with a Bäcklund transformation, we define $R(\lambda, 0) \equiv R(\lambda)$, $R(\lambda, \epsilon) \equiv \widetilde{R}(\lambda)$ and consequently we have

$$e^{\mu \epsilon \phi(\lambda)} = \frac{2 - (\lambda - \mu)\beta(\lambda)}{2 - (\lambda + \mu)\beta(\lambda)},$$ (6.69)

i.e.,

$$\phi(\lambda) = \frac{1}{\epsilon \mu} \ln\left[\frac{2 - (\lambda - \mu)\beta(\lambda)}{2 - (\lambda + \mu)\beta(\lambda)}\right].$$ (6.70)

Taking into account the relation (6.67) between μ and λ we can rewrite the left-hand side of (6.69) as

$$e^{\mu \epsilon \phi(\lambda)} = \cosh[\mu \epsilon \phi(\lambda)] + \mu \frac{\sinh[\mu \epsilon \phi(\lambda)]}{\mu} = E_0(\lambda) + \mu E_1(\lambda),$$ (6.71)

where $E_0(\lambda)$ and $E_1(\lambda)$ are just functions of λ. Introducing (6.71) into (6.69) and identifying the powers (0th and 1st) of μ we get the following system of two compatible equations:

$$-(2 - \lambda\beta)E_0 + (\lambda^2 - 4)\beta E_1 = -(2 - \lambda\beta),$$ (6.72)

$$-\beta E_0 + (2 - \lambda\beta)E_1 = \beta.$$ (6.73)

From (6.72), (6.73) we derive an expression for β in terms of ϕ:

$$\beta(\lambda) = \frac{\delta(\lambda)}{\gamma(\lambda)} = \frac{2E_1}{E_0 + \lambda E_1 + 1}$$

$$= \frac{2\sinh[\mu \epsilon \phi(\lambda)]/\mu}{\cosh[\mu \epsilon \phi(\lambda)] + \lambda \sinh[\mu \epsilon \phi(\lambda)]/\mu + 1}.$$ (6.74)

From (6.74) we see that whatsoever be the symmetry, we find a Bäcklund transformation, i.e., for an arbitrary function ϕ we obtain two entire functions γ

and δ. Vice versa, given a general Bäcklund transformation, we can find the corresponding generalized symmetry

$$\phi(\lambda) = \frac{1}{\epsilon\mu}\sinh^{-1}\left[-\mu\frac{(\lambda\beta - 2)\beta}{2(\beta^2 - \lambda\beta + 1)}\right].$$

The relation between Bäcklund transformations and symmetries is given by an entire transcendent function. So the simplest Bäcklund transformation, obtained by choosing β constant, will correspond to the infinite sum of symmetries, and viceversa the simplest generalized symmetry, obtained by choosing ϕ constant, will correspond to the infinite sum of Bäcklund transformations.

6.2.6 Generalized symmetries for PΔE's

We apply the same approach we used for the Toda Lattice to the case of the time-discrete Toda Lattice. The spectral problem is the same, but the time-discrete evolution is different. We have

$$\psi(n - 1, m; \lambda) + b_{n,m}\psi(n, m; \lambda) + a_{n,m}\psi(n + 1, m; \lambda) = \lambda\psi(n, m; \lambda),$$

$$\psi(n, m + 1; \lambda) = -M(a_{n,m}, b_{n,m})\psi(n, m; \lambda).$$

Consequently also the Lax equations are different. Applying the Lax Technique in this case, we get

$$\begin{pmatrix} a_{n,m+1} - a_{n,m} \\ b_{n,m+1} - b_{n,m} \end{pmatrix} = f_m^1(\widetilde{\mathcal{L}}_{n,m})\begin{pmatrix} (b_{n,m+1} - b_{n+1,m})\pi_{n,m+1}/\pi_{n+1,m} \\ \pi_{n-1,m+1}/\pi_{n,m} - \pi_{n,m+1}/\pi_{n+1,m} \end{pmatrix}$$

$$+ f_m^2(\widetilde{\mathcal{L}}_{n,m})\begin{pmatrix} a_{n,m+1} - a_{n,m} \\ b_{n,m+1} - b_{n,m} \end{pmatrix}. \quad (6.75)$$

Here f_m^1 and f_m^2 are entire functions of their argument, and $\widetilde{\mathcal{L}}$ is the recursion operator of the hierarchy:

$$\widetilde{\mathcal{L}}_{n,m}\begin{pmatrix} p_{n,m} \\ q_{n,m} \end{pmatrix} = \begin{pmatrix} a_{n,m+1}S_{n+2,m} - a_{n,m}S_{n,m} \\ p_{n-1,m} + \Sigma_{n-1,m}\pi_{n-1,m+1}/\pi_{n,m} - \Sigma_{n,m}\pi_{n,m+1}/\pi_{n+1,m} \end{pmatrix}$$

$$+ \begin{pmatrix} b_{n,m+1}p_{n,m} + (b_{n,m+1} - b_{n+1,m})\Sigma_{n,m}\pi_{n,m+1}/\pi_{n+1,m} \\ b_{n,m+1}q_{n,m} + (b_{n,m+1} - b_{n,m})S_{n,m} \end{pmatrix}. \quad (6.76)$$

The function $\pi_{n,m}$ is given by

$$\pi_{n,m} = \Pi_{j=n}^{\infty}a_{j,m}, \quad (6.77)$$

while $S_{n,m}$ and $\Sigma_{n,m}$ are defined as the *bounded solutions* of the difference equations

$$S_{n+1,m} - S_{n,m} = q_{n,m},$$

$$\Sigma_{n+1,m} - \Sigma_{n,m} = -p_{n+1,m}\frac{\pi_{n+2,m}}{\pi_{n+1,m+1}}. \tag{6.78}$$

The discrete evolution of the reflection coefficient turns out to be

$$R_{m+1} = \frac{1 - f_m^2(\lambda) - zf_m^1(\lambda)}{1 - f_m^2(\lambda) - f_m^1(\lambda)/z}R_m, \tag{6.79}$$

while the transmission coefficient T_m does not evolve in m. Comparing (6.79) with (6.62), we see that, identifying $R_m \equiv R(\lambda)$ and $R_{m+1} \equiv \widetilde{R}(\lambda)$, $f_m^1(\lambda) \equiv \delta(\lambda)$ and $f_m^2(\lambda) \equiv \gamma(\lambda)$, the two formulas are equal. So, the hierarchy of the Bäcklund transformations for the Toda Lattice is equal to the hierarchy for the discrete-time Toda Lattice. This result shows that we can use the Bäcklund Transformations as a discretizing tool [9, 10].

Choosing $f_m^2 = 0$ and $f_m^1 = \alpha$ in (6.75), we get the discrete Toda system:

$$a_{n,m+1} - a_{n,m} = \alpha(b_{n,m+1} - b_{n+1,m})\frac{\pi_{n,m+1}}{\pi_{n+1,m}}, \tag{6.80}$$

$$b_{n,m+1} - b_{n,m} = \alpha\left(\frac{\pi_{n-1,m+1}}{\pi_{n,m}} - \frac{\pi_{n,m+1}}{\pi_{n+1,m}}\right). \tag{6.81}$$

By a proper combination of (6.80), (6.81) we can get an expression of $b_{n,m}$ in terms of $a_{n,m}$

$$b_{n,m} = \alpha + \frac{1}{\alpha} - \alpha\frac{\pi_{n-1,m+1}}{\pi_{n,m}} - \frac{\pi_{n,m}}{\alpha\pi_{n,m+1}},$$

which, introduced in (6.81), gives the discrete–time Toda Lattice

$$\Delta_{\text{Toda}} = \pi_{n-1,m+2} - \frac{1}{\alpha^2}\pi_{n,m} - \pi_{n,m+1}^2\left(\frac{1}{\pi_{n+1,m}} - \frac{1}{\alpha^2\pi_{n,m+2}}\right) = 0.$$

Defining $\pi_{n,m} = e^{u_{n,m}}$, so that $a_{n,m} = e^{u_{n,m}-u_{n+1,m}}$, we get the alternative representation

$$e^{u_{n,m}-u_{n,m+1}} - e^{u_{n,m+1}-u_{n,m+2}} = \alpha^2(e^{u_{n-1,m+2}-u_{n,m+1}} - e^{u_{n,m+1}-u_{n+1,m}}), \tag{6.82}$$

i.e., the well-known discrete-time Toda Lattice equation or Hirota equation. This discrete nonlinear equation is a Bäcklund transformation of the Toda Lattice. In correspondence with the discrete-time Toda Lattice (6.82), we have the following discrete evolution of the spectrum:

$$R_{m+1} = \frac{1 - \alpha z}{1 - \alpha/z}R_m.$$

Its generalized symmetries are given by the *Toda Lattice hierarchy*, i.e., in the

- isospectral case:

$$\begin{pmatrix} a_{n,m} \\ b_{n,m} \end{pmatrix}_{,\epsilon_k} = (\mathcal{L})^k \begin{pmatrix} a_{n,m}(b_{n,m} - b_{n+1,m}) \\ a_{n-1,m} - a_{n,m} \end{pmatrix},$$

$$\frac{dR}{d\epsilon_k} = \mu \lambda^k R;$$

- non-isospectral case:

$$\begin{pmatrix} a_{n,m} \\ b_{n,m} \end{pmatrix}_{,\epsilon_k} = f_m^k(\mathcal{L}) \begin{pmatrix} a_{n,m}(b_{n,m} - b_{n+1,m}) \\ a_{n-1,m} - a_{n,m} \end{pmatrix}$$
$$+ \mathcal{L}^k \begin{pmatrix} a_{n,m}[(2n+3)b_{n+1,m} - (2n-1)b_{n,m}] \\ b_{n,m}^2 - 4 + 2[(n+1)a_{n,m} - (n-1)a_{n-1,m}] \end{pmatrix},$$

where

$$f_m^k(\lambda) = -2m\lambda^k \frac{2\alpha^2 - \alpha\lambda}{1 + \alpha^2 - \alpha\lambda},$$
$$\frac{dR_m(z, c_k)}{d\epsilon_k} = \mu f_m^k(\lambda) R_m(z, c_k), \qquad \lambda_{\epsilon_k} = \mu^2 \lambda^k.$$

The simplest isospectral generalized symmetry is obtained by choosing $k = 0$ and reads

$$a_{n,m,\epsilon} = a_{n,m}(b_{n,m} - b_{n+1,m}),$$
$$b_{n,m,\epsilon} = a_{n-1,m} - a_{n,m}.$$

Taking into account the definition of $a_{n,m}$ in terms of $u_{n,m}$, we get that this symmetry reduces to

$$u_{n,m,\epsilon} = b_{n,m} + c,$$

where c is an arbitrary constant. By a proper choice of the constant, we get the simplest generalized symmetry of the discrete-time Toda Lattice. This same result could be obtained directly by using the Lie point symmetry approach.

6.3 Formal symmetries and integrable lattice equations

In this section we present a technique based on the idea that the existence of generalized symmetries provides an integrability criteria for differential equations. In its discrete version it was developed mainly by R.I. Yamilov [25, 26], as part of a project developed by a group of researchers of Ufa (Russia) lead by A.B. Shabat [1, 17, 18]. Attempts to extend this approach, as it is, to PΔE have up to now failed. A modification of the approach adapted to treat PΔE is discussed in [12–14] (see also [19]).

The existence of infinite hierarchies of generalized symmetries and/or conservation laws is used by this method as an integrability criteria. This method enables one not only to test equations for integrability but also to classify integrable equations in classes defined by arbitrary functions of many variables. In the following we will present the main ideas needed to introduce the method in the case of differential–difference equations and a few theorems on integrability conditions. In this way we will be able to show the difference between S- and C-integrable systems [2] from the point of view of the existence of generalized symmetries and local conservation laws, and present a simple example of classification. More details of this theory can be found in [11, 26–28].

The class of lattice equations considered here is

$$\dot{u}_n = f(u_{n+1}, u_n, u_{n-1}), \qquad \frac{\partial f}{\partial u_{n+1}} \neq 0, \ \frac{\partial f}{\partial u_{n-1}} \neq 0, \qquad (6.83)$$

where $u_n = u_n(t)$. If we interpret $u_n(t)$ as an infinite set of functions $\{u_n(t) : n \in \mathbb{Z}\}$, then (6.83) is an infinite system of ordinary differential equations, defined by one arbitrary function of three variables: $f(z_1, z_2, z_3)$. The Volterra equation

$$\dot{u}_n = u_n(u_{n+1} - u_{n-1}) \qquad (6.84)$$

is an example of Nonlinear Differential-Difference Equation (NDΔE) in this class.

We will look here for generalized symmetries. As we have seen in the previous section in the case of the Toda Lattice, equations integrable by inverse scattering method (which sometimes are called S-integrable) have infinitely many generalized symmetries. Equations which can be transformed into a linear one, linearizable or C-integrable equations, also may possess this property. The generalized symmetries we consider here are

$$u_{n,\epsilon} = g(u_{n+m}, u_{n+m-1}, \ldots, u_{n+m'+1}, u_{n+m'}), \qquad \frac{\partial g}{\partial u_{n+m}} \frac{\partial g}{\partial u_{n+m'}} \neq 0, \qquad (6.85)$$

where $m \geq m'$ are two finite integers fixed for any given function g. Number m is called *the left order* (or *the order*) and m' is called *the right order* of generalized symmetry (6.85).

A symmetry is defined by one locally analytic function $g(z_1, z_2, z_3, \ldots z_{1+m-m'})$ depending on $m - m' + 1$ variables. Generalized symmetries are given by those (6.85) which are compatible with (6.83).

Let us introduce the following notations:

$$f_n = f(u_{n+1}, u_n, u_{n-1}), \qquad g_n = g(u_{n+m}, u_{n+m-1}, \ldots u_{n+m'}) \qquad (6.86)$$

for functions from the right hand sides of the NDΔE and the symmetry. If $u_n(t, \epsilon)$ is their common solution, they have to satisfy the following equation:

$$\frac{\partial^2 u_n}{\partial t \partial \epsilon} - \frac{\partial^2 u_n}{\partial \epsilon \partial t} = D_t g_n - D_\epsilon f_n = 0, \qquad (6.87)$$

where D_t and D_ϵ are the total derivative operators defined as

$$D_t g_n = \sum_{j=m'}^{m} \frac{\partial g_n}{\partial u_{n+j}} f_{n+j}, \qquad D_\epsilon f_n = \sum_{j=-1}^{1} \frac{\partial f_n}{\partial u_{n+j}} g_{n+j}, \qquad (6.88)$$

with

$$f_{n+j} = f(u_{n+1+j}, u_{n+j}, u_{n-1+j}),$$
$$g_{n+j} = g(u_{n+m+j}, u_{n+m-1+j}, \ldots, u_{n+m'+1+j}, u_{n+m'+j}).$$

As a result, we get the equation

$$D_t g_n = \frac{\partial f_n}{\partial u_{n+1}} g_{n+1} + \frac{\partial f_n}{\partial u_n} g_n + \frac{\partial f_n}{\partial u_{n-1}} g_{n-1}, \qquad (6.89)$$

which is to be satisfied for all the common solutions of (6.83), (6.85) and is an equation for the symmetries g_n when f_n is given or for the function f_n if the symmetries g_n are given.

In order to be able to find from (6.89) g_n for any given function $f(z_1, z_2, z_3)$, *we will require that (6.89) is identically satisfied for all values of the variables*

$$u_0, u_1, u_{-1}, u_2, u_{-2}, \ldots \qquad (6.90)$$

and for all $n \in \mathbb{Z}$.

As an example of the consequences of a difference relation which must be identically satisfied for all values of the dependent variables u_n, let us consider the equation

$$\phi_{n+1} - \phi_n = 0, \qquad (6.91)$$

where

$$\phi_n = \phi(u_{n+k}, u_{n+k-1}, \ldots, u_{n+k'}), \qquad k \geq k', k, k' \text{ finite}. \qquad (6.92)$$

Formula (6.92) means that ϕ_n may only depend on variables $u_{n+k}, u_{n+k-1}, \cdots u_{n+k'}$, for any n. In such a case we say that such ϕ_n is a *finite range function*. If ϕ_n is a non-constant function, one always can choose integers k, k' so that $\partial \phi_n / \partial u_{n+k} \neq 0$, $\partial \phi_n / \partial u_{n+k'} \neq 0$. Naturally, $\partial \phi_n / \partial u_{n+j}$ can be zero for some j in the interval (k, k'). If ϕ_n in (6.91) is a non-constant function, and k' is the minimum integer for which $\partial\phi_n/\partial u_{n+k'} \neq 0$, then by differentiating (6.91) with respect to $u_{n+k'}$, we find $\partial\phi_n/\partial u_{n+k'} = 0$ identically, i.e., a contradiction. So (6.91) implies that ϕ_n is a constant. Introducing the standard shift operator $T : n \rightarrow n + 1$, such that

$$T^j\phi_n = \phi_{n+j} = \phi(u_{n+j+k}, u_{n+j+k-1}, \ldots, u_{n+j+k'}), \qquad (6.93)$$

we can rewrite the previous results as the following property:

$$\ker(T - 1) = \mathbb{C}, \qquad (6.94)$$

as (6.91) can be rewritten as $(T - 1)\phi_n = 0$.

A local conservation law or, more shortly, a *conservation law* of the NDΔE is a relation of the form

$$D_t p_n = (T - 1)q_n, \qquad (6.95)$$

where p_n and q_n are functions of finite range, and D_t is the evolution differentiation corresponding to the NDΔE as given in (6.88). This relation must be satisfied identically for all values of the variables (6.90). The function p_n is called a *conserved density* of the NDΔE.

Any conservation law (6.95) of the NDΔE generates constant of motion $I = \sum p_n$ under appropriate boundary conditions. In fact,

$$\frac{dI}{dt} = \sum D_t p_n = \sum(q_{n+1} - q_n) = 0.$$

For any function (6.92) of finite range we can calculate *formal variational derivative* [4, 24, 26]

$$\frac{\delta\phi_n}{\delta u_n} = \sum_{i=k'}^{k} T^{-i}\frac{\partial\phi_n}{\partial u_{n+i}} = \sum_{j=-k}^{-k'}\frac{\partial\phi_{n+j}}{\partial u_n}. \qquad (6.96)$$

The operator $\delta/\delta u_n$ is the discrete analog of the standard variational derivative and possesses similar properties. Sometimes it is called Euler operator [7]. Using the formal variational derivative, we get that if $\phi_n = (T - 1)\psi_n$ then $\delta\phi_n/\delta u_n = 0$.

Thus the formal variational derivative helps us to check integrability conditions (see below) and to calculate *the order* of conservation laws ρ_n. If $\varrho_n = \delta p_n/\delta u_n = 0$, the conservation law is called trivial. If $\varrho_n = \varrho(u_n) \neq 0$, it has the order 0. In the case

$$\varrho_n = \varrho(u_{n+m}, u_{n+m-1}, \ldots u_{n-m}), \qquad \frac{\partial\varrho}{\partial u_{n+m}}\frac{\partial\varrho}{\partial u_{n-m}} \neq 0, \ m > 0,$$

the order equals m.

Let us obtain a *first integrability condition*. Let us consider g_n (6.86) with $m > 0$, $m' < 0$ and a given function f_n. Defining

$$f_n^{(i)} = \frac{\partial f_n}{\partial u_{n+i}}, \qquad g_n^{(i)} = \frac{\partial g_n}{\partial u_{n+i}}, \tag{6.97}$$

one can express the compatibility condition (6.89) in the form:

$$D_t g_n = \sum_{i=m'}^{m} g_n^{(i)} f_{n+i} = f_n^{(1)} g_{n+1} + f_n^{(0)} g_n + f_n^{(-1)} g_{n-1}. \tag{6.98}$$

If $m \geq 1$, applying the operator $\partial/\partial u_{n+m+1}$, we get

$$g_n^{(m)} f_{n+m}^{(1)} = g_{n+1}^{(m)} f_n^{(1)}. \tag{6.99}$$

Applying $\partial/\partial u_{n+m}$ and $\partial/\partial u_{n+m-1}$, two other relations can be derived:

$$D_t g_n^{(m)} + g_n^{(m)} f_{n+m}^{(0)} + g_n^{(m-1)} f_{n+m-1}^{(1)} = g_n^{(m)} f_n^{(0)} + g_{n+1}^{(m-1)} f_n^{(1)}, \quad m \geq 2; \tag{6.100}$$

$$D_t g_n^{(m-1)} + g_n^{(m)} f_{n+m}^{(-1)} + g_n^{(m-1)} f_{n+m-1}^{(0)} + g_n^{(m-2)} f_{n+m-2}^{(1)}$$
$$= g_{n-1}^{(m)} f_n^{(-1)} + g_n^{(m-1)} f_n^{(0)} + g_{n+1}^{(m-2)} f_n^{(1)}, \quad m \geq 3. \tag{6.101}$$

These relations allow us to obtain the partial derivatives $g_n^{(m)}$, $g_n^{(m-1)}$ and $g_n^{(m-2)}$.
   Defining

$$\Phi_n^{(N)} = f_n^{(1)} f_{n+1}^{(1)} \cdots f_{n+N}^{(1)}, \qquad N \geq 0, \tag{6.102}$$

and taking into account that $f_n^{(1)} \neq 0$, we can divide the previous relations by $\Phi_n^{(m)}, \Phi_n^{(m-1)}, \Phi_n^{(m-2)}$, respectively, and we get

$$(T - 1) \frac{g_n^{(m)}}{\Phi_n^{(m-1)}} = 0, \tag{6.103}$$

$$(T - 1) \frac{g_n^{(m-1)}}{\Phi_n^{(m-2)}} = \Theta_n^{(1)}(g_n^{(m)}), \tag{6.104}$$

$$(T - 1) \frac{g_n^{(m-2)}}{\Phi_n^{(m-3)}} = \Theta_n^{(2)}(g_n^{(m-1)}, g_n^{(m)}). \tag{6.105}$$

On the left-hand sides one has total differences, and on the right hand sides the functions $\Theta_n^{(i)}$ depending on already defined partial derivatives $g_n^{(j)}$. So, calculating $g_n^{(j)}$, we are led to the conditions that some known functions must be expressed as total differences.

The following theorem can be easily proved, using the relations (6.103), (6.104):

Theorem *If an equation of the form* (6.83) *possesses a generalized symmetry of an order* $m \geq 2$, *then there exists a function* $q_n^{(1)}$ *of finite range, such that*

$$\dot{p}_n^{(1)} = D_t p_n^{(1)} = (T - 1)q_n^{(1)}, \qquad p_n^{(1)} = \log \frac{\partial f_n}{\partial u_{n+1}}. \qquad (6.106)$$

6.3.1 Formal symmetries and further integrability conditions

Let us introduce the operator

$$h_n^* = \sum_{i=k'}^{k} \frac{\partial h_n}{\partial u_{n+i}} T^i, \qquad (6.107)$$

where h_n is any function of the form (6.92). Following the continuous case, we call the operator h_n^* the *Fréchet derivative* of the function h_n.

Let us define the "Lax equation"

$$\dot{L}_n = [f_n^*, L_n]. \qquad (6.108)$$

Here f_n^* is the operator

$$f_n^* = f_n^{(1)} T + f_n^{(0)} + f_n^{(-1)} T^{-1}, \qquad (6.109)$$

with coefficients defined in terms of the right hand side of (6.83). Solutions of (6.108) can be expressed as formal series in powers of the shift operator T,

$$L_n = \sum_{i=-\infty}^{N} l_n^{(i)} T^i, \qquad l_n^{(N)} \neq 0, \qquad (6.110)$$

whose coefficients $l_n^{(i)}$ are finite range functions of the form (6.92). In (6.108) $\dot{L}_n = \sum_{i=-\infty}^{N} \dot{l}_n^{(i)} T^i$, by [,] we denote the standard commutator and the multiplication is defined as: $(a_n T^i)(b_n T^j) = a_n b_{n+i} T^{i+j}$.

Introducing the operator

$$A(L_n) = \dot{L}_n - [f_n^*, L_n] \qquad (6.111)$$

for L_n given by (6.110), we easily can check the following two general formulas:

$$A(L_n^{-1}) = -L_n^{-1} A(L_n) L_n^{-1}, \qquad (6.112)$$

$$A(L_n \tilde{L}_n) = A(L_n) \tilde{L}_n + L_n A(\tilde{L}_n), \qquad (6.113)$$

which show that *for any two solutions* L_n *and* $\tilde{L}_n$ *of* (6.108), *the inverse* L_n^{-1} *and the product* $L_n \tilde{L}_n$ *also satisfy* (6.111), as well as any integer power L_n^i (here $L_n^0 = 1$ is to be interpreted as the operator of multiplication by unit).

The solution L_n of (6.108) is nothing but *the recursion operator* because it brings the right hand side g_n of a generalized symmetry into the right-hand side

$L_n g_n$ of a new generalized symmetry. As the compatibility condition (6.89) can be expressed, using the Fréchet derivative f_n^*, in the form

$$(D_t - f_n^*) g_n = 0, \tag{6.114}$$

applying D_t to $L_n g_n$ and using (6.108), (6.114), one easily checks that

$$D_t L_n g_n = \dot{L}_n g_n + L_n \dot{g}_n = (f_n^* L_n - L_n f_n^*) g_n + L_n f_n^* g_n = f_n^* L_n g_n.$$

It remains to prove that the new symmetry $u_{n,\epsilon'} = L_n g_n$ is local, i.e., the function $L_n g_n$ is of finite range. As any integer power L_n^i satisfies (6.108) and f_n is a trivial solution of (6.114) we obtain infinitely many, maybe non-local, generalized symmetries of (6.83):

$$u_{n,\epsilon_i} = L_n^i f_n, \tag{6.115}$$

where $i \in \mathbb{Z}$ and $\epsilon_0 = t$.

We call *residue* of the series (6.110) the coefficient of T^0, i.e.

$$\text{res } L_n = l_n^{(0)}. \tag{6.116}$$

It is clear that if $N < 0$, then $\text{res } L_n = 0$. If a series L_n of the form (6.110) satisfies the Lax equation and is such that $N > 0$, then

$$\log l_n^{(N)}, \tag{6.117}$$

and

$$\text{res } L_n^i, \qquad i \geq 1, \tag{6.118}$$

are conserved densities of (6.83).

In practice, it is difficult to construct recursion operator and difficult to prove that a generalized symmetry (6.115) is local. So we consider below approximate solutions of (6.108) which are easier to construct and which can be used to derive integrability conditions. These approximate solutions are approximate recursion operators, but we denote them as *formal symmetry* because of the close connection between formal and generalized symmetry.

Let us define N as *the order of an infinite series* of the form (6.110) and let us write $\text{ord } L_n = N$. For the operator f_n^*, $\text{ord } f_n^* = 1$, see (6.109). Let us also note that for any series L_n of order N, the series $A(L_n)$ (6.111) is given by

$$A(L_n) = a_n^{(N+1)} T^{N+1} + a_n^{(N)} T^N + a_n^{(N-1)} T^{N-1} + \cdots. \tag{6.119}$$

The series (6.110) is called a *formal symmetry of* (6.83) *of the length l* (we write $\text{len } L_n = l$) if the first l coefficients of the series $A(L_n)$ (6.119) vanish:

$$a_n^{(i)} = 0, \qquad N + 1 \geq i \geq N - l + 2. \tag{6.120}$$

We assume that $l \geq 1$ and $a_n^{(N-l+1)} \neq 0$. A recursion operator L_n of order N will be such that all the coefficients $a_n^{(i)}$ of the series $A(L_n)$ vanish.

If g_n is the right hand side of a generalized symmetry of order $m \geq 1$, then by applying the Frechét derivative to (6.114) we can show that $L_n = g_n^*$ is a formal symmetry, such that $\operatorname{ord} L_n = m$, $\operatorname{len} L_n \geq m$. This allows us to use (6.108) for deriving the integrability conditions instead of the compatibility condition (6.89). This simplifies the calculations. We can now just state the following theorem:

Theorem   *If (6.83) has a generalized symmetry of order m with $m \geq 4$ or a formal symmetry, such that $\operatorname{len} L_n \geq 4$, then f_n (6.86) has to satisfy the following conditions:*

$$\dot{p}_n^{(1)} = (T-1)q_n^{(1)}, \qquad p_n^{(1)} = \log \frac{\partial f_n}{\partial u_{n+1}},$$

$$\dot{p}_n^{(2)} = (T-1)q_n^{(2)}, \qquad p_n^{(2)} = q_n^{(1)} + \frac{\partial f_n}{\partial u_n},$$

$$\dot{p}_n^{(3)} = (T-1)q_n^{(3)}, \qquad p_n^{(3)} = q_n^{(2)} + \frac{1}{2}(p_n^{(2)})^2 + \frac{\partial f_n}{\partial u_{n+1}} \frac{\partial f_{n+1}}{\partial u_n},$$

where all involved function must be of finite range.

Let us define $\varrho_n = \delta p_n / \delta u_n$, the formal variational derivative of a conserved density p_n. If f_n^* is the Frechét derivative of f_n, one can define the operator $f_n^{*\dagger}$ as the corresponding adjoint operator:

$$(a_n T^i)^\dagger = T^{-i} \circ a_n = a_{n-i} T^{-i},$$

$$f_n^{*\dagger} = \sum_{i=-1}^{1} f_{n+i}^{(-i)} T^i = \sum_{i=-1}^{1} \frac{\partial f_{n+i}}{\partial u_n} T^i.$$

Then, for any conserved density p_n of (6.83), its variational derivative ϱ_n satisfies

$$(D_t + f_n^{*\dagger})\varrho_n = 0. \tag{6.121}$$

We can introduce a *second Lax-like equation*

$$\dot{S}_n + S_n f_n^* + f_n^{*\dagger} S_n = 0. \tag{6.122}$$

Here S_n is a formal series of the same type as L_n:

$$S_n = \sum_{i=-\infty}^{M} s_n^{(i)} T^i, \qquad s_n^{(M)} \neq 0, \tag{6.123}$$

and $s_n^{(i)}$ are functions of finite range. Let us introduce an operator B as before:

$$B(S_n) = \dot{S}_n + S_n f_n^* + f_n^{*\dagger} S_n. \tag{6.124}$$

It is easy to prove that the following identities take place:

$$B(S_n L_n) = B(S_n)L_n + S_n A(L_n), \tag{6.125}$$

$$A(S_n^{-1}\widetilde{S}_n) = S_n^{-1}B(\widetilde{S}_n) - S_n^{-1}B(S_n)S_n^{-1}\widetilde{S}_n, \tag{6.126}$$

for any formal series L_n, S_n, $\widetilde{S}_n$. The first identity (6.125) shows that, for any solutions S_n and L_n of (6.122) and (6.108), the series $S_n L_n$ will be a new solution of (6.122). On the other hand (6.126) tells us that if S_n and $\widetilde{S}_n$ are any two solutions of (6.122), then the series $S_n^{-1}\widetilde{S}_n$ will be a recursion operator.

As in the case of the Lax-like equation for the recursion operator, we are interested here in *approximate solutions* S_n. Such solutions will be called *formal conserved densities*, as they are strictly related to ϱ_n. If a series S_n (6.123) is such that first $l \geq 1$ coefficients of the series $B(S_n)$ vanish, i.e.

$$B(S_n) = b_n^{(M-l+1)}T^{M-l+1} + b_n^{(M-l)}T^{M-l} + \cdots, \qquad b_n^{(M-l+1)} \neq 0, \tag{6.127}$$

then S_n is called *formal conserved density of* (6.83) *of order M and length l,* and we can write: $\text{ord }S_n = M$, $\text{len }S_n = l$.

Two formal conserved densities S_n and $\widetilde{S}_n$ give a formal symmetry $L_n = S_n^{-1}\widetilde{S}_n$. Formal conserved density S_n together with formal symmetry L_n generate another formal conserved density $\widetilde{S}_n = S_n L_n$.

Applying the Fréchet derivative to (6.121), we can show that if (6.83) has a conservation law of order $m \geq 2$, then $S_n = \varrho_n^*$ is a formal conserved density, with $\text{ord }S_n = m$, $\text{len }S_n \geq m - 1$. So, calculating coefficients of a formal conserved density, we can derive some additional integrability conditions from (6.122) in the same way as we derived the first integrability condition from (6.108). The result of this calculation is presented in the following Theorem.

Theorem *If* (6.83) *has a conservation law of order $m \geq 3$ and a generalized symmetry of order $m \geq 2$, then there exist functions $\sigma_n^{(1)}$ and $\sigma_n^{(2)}$ of finite range which satisfy the following relations:*

$$r_n^{(1)} = (T-1)\sigma_n^{(1)}, \qquad r_n^{(1)} = \log\left(-\frac{\partial f_n/\partial u_{n+1}}{\partial f_n/\partial u_{n-1}}\right), \tag{6.128}$$

$$r_n^{(2)} = (T-1)\sigma_n^{(2)}, \qquad r_n^{(2)} = \dot{\sigma}_n^{(1)} + 2\frac{\partial f_n}{\partial u_n}. \tag{6.129}$$

6.3.2 Why integrable equations on the lattice must be symmetric

Let us present here a theorem which proves that only equations depending *symmetrically* on the lattice variables may possess higher-order local conservation laws. Then, on the example of the discrete Burgers equation, we will point out the difference between S- and C-integrable equations.

Let us consider an n- and t-independent equation of the form

$$\dot{u}_n = f_n = f(u_{n+N}, u_{n+N-1}, \ldots u_{n+M}), \qquad N \geq M, \quad \frac{\partial f_n}{\partial u_{n+N}} \frac{\partial f_n}{\partial u_{n+M}} \neq 0, \quad (6.130)$$

and let us prove the following theorem:

Theorem *If an equation of the form* (6.130) *possesses a conservation law of order m, such that*

$$m > \min(|N|, |M|), \tag{6.131}$$

then $M < 0$ and $N = -M$ (in particular $N \geq 0$).

Proof We have $m > 0$. The variational derivative ϱ_n of a conserved density p_n will satisfy (6.121), i.e.,

$$\sum_{i=-m}^{m} \frac{\partial \varrho_n}{\partial u_{n+i}} f_{n+i} + \sum_{k=-N}^{-M} \frac{\partial f_{n+k}}{\partial u_n} \varrho_{n+k} = 0, \tag{6.132}$$

and will be such that

$$\varrho_n = \varrho(u_{n+m}, u_{n+m-1}, \ldots u_{n-m}), \qquad \frac{\partial \varrho_n}{\partial u_{n+m}} \frac{\partial \varrho_n}{\partial u_{n-m}} \neq 0. \tag{6.133}$$

Taking into account the definition of ϱ_n and f_n we get for any i, k with $-m \leq i \leq m$ and $-N \leq k \leq -M$ the following table:

$$
\begin{aligned}
\frac{\partial}{\partial u_{n+j}} \frac{\partial \varrho_n}{\partial u_{n+i}} = 0, && j < -m, && j > m, \\[2mm]
\frac{\partial}{\partial u_{n+j}} f_{n+i} = 0, && j < M - m, && j > N + m, \\[2mm]
\frac{\partial}{\partial u_{n+j}} \frac{\partial f_{n+k}}{\partial u_n} = 0, && j < M - N, && j > N - M, \\[2mm]
\frac{\partial}{\partial u_{n+j}} \varrho_{n+k} = 0, && j < -m - N, && j > m - M.
\end{aligned}
\tag{6.134}
$$

Let us assume

$$N > 0, \qquad N > -M, \qquad m > -M. \tag{6.135}$$

Differentiating (6.132) with respect to u_{n+N+m} and using the table, we obtain:

$$\frac{\partial}{\partial u_{n+N+m}} \left(\frac{\partial \varrho_n}{\partial u_{n+m}} f_{n+m} \right) = \frac{\partial \varrho_n}{\partial u_{n+m}} \frac{\partial f_{n+m}}{\partial u_{n+N+m}} = \frac{\partial \varrho_n}{\partial u_{n+m}} T^m \frac{\partial f_n}{\partial u_{n+N}} = 0.$$

This last relation contradicts the assumptions given in (6.130), (6.133) on the functions f_n, ϱ_n and thus (6.135) is false.

We can consider, as a second hypothesis:

$$-M > 0, \qquad -M > N, \qquad m > N. \tag{6.136}$$

If we differentiate (6.132) with respect to u_{n+m-M}, we get

$$\frac{\partial}{\partial u_{n+m-M}}\left(\frac{\partial f_{n-M}}{\partial u_n}\varrho_{n-M}\right) = \frac{\partial f_{n-M}}{\partial u_n}\frac{\partial \varrho_{n-M}}{\partial u_{n+m-M}} = T^{-M}\left(\frac{\partial f_n}{\partial u_{n+M}}\frac{\partial \varrho_n}{\partial u_{n+m}}\right) = 0.$$

This again contradicts the assumptions done on f_n and ϱ_n. So, also (6.136) is not correct.

The remaining relations between N and M are possible:

$$\text{Case 1: } N > -M, \qquad \text{Case 2: } -M > N.$$

We can show that both are impossible. In Case 1, as $N \geq M$, we must have

$$N > -M \geq -N, \qquad N \geq M > -N; \tag{6.137}$$

hence $N > 0$ and thus $|N| = N \geq |M| \geq -M$. As $m > \min(|N|, |M|)$, then $m > -M$. So, we obtain the relations (6.135) which lead to a contradiction. Case 2 is dealt with in a very similar way and leads to the relations (6.136), which is also contraddictory. So, by necessity, $N = -M$. $\qquad\square$

This theorem is very important as it gives necessary conditions on the dependence of a given NDΔE from shifted values of u_n such that will admit conserved quantities of sufficiently high order.

Let us consider a linearizable equation, for example, *the discrete Burgers equation*

$$\dot{u}_n = u_n(u_{n+1} - u_n). \tag{6.138}$$

We can show that it has an infinite hierarchy of generalized symmetries, but no local conservation laws of positive order. Equation (6.138) is associated to the linear equation

$$\dot{v}_n = v_{n+1} \tag{6.139}$$

by the discrete Hopf–Cole transformation $u_n = v_{n+1}/v_n$. For any integer k, the equations

$$v_{n,\epsilon_k} = v_{n+k} \tag{6.140}$$

are the symmetries of the linear equation. The same discrete Hopf–Cole transformation allows us to obtain symmetries for the Burgers equation from the symmetries of the linear equation. In the case $k = 0$, we have a trivial symmetry: $u_{n,\epsilon_0} = 0$. If $k > 1$ or $k < 0$, then we are led to nontrivial generalized symmetries, for example

$$u_{n,\epsilon_2} = u_n u_{n+1}(u_{n+2} - u_n), \qquad u_{n,\epsilon_{-1}} = 1 - u_n/u_{n-1}.$$

One can check directly that these equations are generalized symmetries of the discrete Burgers equation.

From the previous theorem (6.138), being non-symmetric, cannot have conservation laws of order $m > 0$. It can have a conservation law of order $m = 0$;

for example the function $\log u_n$ is conserved density for the discrete Burgers equation and its symmetries:

$$(\log u_n)_t = (T - 1)u_n, \qquad (\log u_n)_{\epsilon_1} = (T - 1)(1/u_{n-1}).$$

In the case of linear equations (6.140) with $k \neq 0$, the theorem we proved above tells us just that there is no conservation law of order $m > |k|$. For the discrete Burgers equation the result is stronger: $m > 0$.

6.3.3 Example of classification problem

Let us consider the following class of lattice equations

$$\dot{u}_n = P(u_n)(u_{n+1} - u_{n-1}), \qquad P'(u_n) \neq 0, \tag{6.141}$$

depending on one function of one variable. This class is not empty, as it contains the Volterra equation (6.84). There is only one undefined function $P(z)$ here, and the aim is to find all equations of the form (6.141) satisfying the integrability condition (6.106).

As the right-hand side of the integrability condition is a total difference, we have just to check that $(\delta/\delta u_n)\dot{p}_n^{(1)} = 0$. As

$$p_n^{(1)} = \log P(u_n), \qquad \dot{p}_n^{(1)} = P'(u_n)(u_{n+1} - u_{n-1}),$$

we have

$$\frac{\delta}{\delta u_n}\dot{p}_n^{(1)} = P''(u_n)(u_{n+1} - u_{n-1}) + P'(u_{n-1}) - P'(u_{n+1}) = 0. \tag{6.142}$$

Equation (6.142) must be satisfied for all values of the three variables u_{n+1}, u_n, u_{n-1}. Applying the operator $\partial^2/(\partial u_n \partial u_{n+1})$, we see that $P'''(u_n) = 0$, i.e., P must be a quadratic polynomial:

$$P(u_n) = \alpha u_n^2 + \beta u_n + \gamma. \tag{6.143}$$

Condition (6.142) is satisfied together with the integrability condition we started from, as in this case we have

$$\dot{p}_n^{(1)} = (T - 1)(2\alpha u_n u_{n-1} + \beta u_n + \beta u_{n-1}).$$

So, the polynomial (6.143) describes all equations of the form (6.141) satisfying the integrability condition (6.106). This integrability condition provides any equation (6.141), (6.143) with a conserved density of order 0.

It is easy to see that, using the linear transformation $\tilde{u}_n = c_1 u_n + c_2$ with c_1 and c_2 constants, $c_1 \neq 0$, one can transform any equation of the form (6.141), (6.143) into the Volterra equation or into the modified Volterra equation

$$\dot{u}_n = (c^2 - u_n^2)(u_{n+1} - u_{n-1}).$$

So, up to linear transformations, the resulting list of integrable equations (6.141) consists of the Volterra and modified Volterra equations.

There exists a discrete Miura transformation $\tilde{u}_n = (c + u_n)(c - u_{n+1})$ which transforms any solution u_n of the modified Volterra equation into a solution $\tilde{u}_n$ of the Volterra equation. For this reason, the list of integrable equations (6.141) consists, up to Miura type transformations, just of the Volterra equation.

Both the Volterra and modified Volterra equations are known to be integrable and to have infinite hierarchies of generalized symmetries and local conservation laws.

The problem we have considered is an example of complete (exhaustive) classification of integrable differential-difference equations. In [26] one can find the complete classification of the equations of the class (6.83).

Acknowledgments

R.I.Y. has been partially supported by the Russian Foundation for Basic Research (Grant number 10-01-00088-a and 11-01-97005-r-povolzhie-a). L.D. has been partly supported by the Italian Ministry of Education and Research, PRIN "Nonlinear waves: integrable finite dimensional reductions and discretizations" from 2007 to 2009 and PRIN "Continuous and discrete nonlinear integrable evolutions: from water waves to symplectic maps" from 2010.

References

[1] Adler, V. É., Shabat, A. B., and Yamilov, R. I. 2000. The symmetry approach to the integrability problem. *Theoret. and Math. Phys.*, **125**(3), 1603–1661.

[2] Calogero, F. 1991. Why are certain nonlinear PDEs both widely applicable and integrable? Pages 1–62 of: *What is integrability?* Springer Ser. Nonlinear Dynam. Berlin: Springer.

[3] Chiu, S. C., Ladik, J. F. 1977. Generating exactly soluble nonlinear discrete evolution equations by a generalized Wronskian technique. *J. Mathematical Phys.* **18**(4), 690–700.

[4] Dorodnitsyn, V. A. 1993. A finite-difference analogue of Noether's theorem. *Phys. Dokl.*, **38**(2), 66–68.

[5] Flaschka, H. 1974. On the Toda lattice. II. Inverse-scattering solution. *Progr. Theoret. Phys.* **51**, 703–716.

[6] Fokas, A. S. 1980. A symmetry approach to exactly solvable evolution equations. *J. Math. Phys.*, **21**(6), 1318–1325.

[7] Hydon, P. E., and Mansfield, E. L. 2004. A variational complex for difference equations. *Found. Comput. Math.*, **4**(2), 187–217.

[8] Inonu, E., and Wigner, E. P. 1953. On the contraction of groups and their representations. *Proc. Nat. Acad. Sci. U.S.A.*, **39**, 510–524.

[9] Levi, D. 1981. Nonlinear differential-difference equations as Bäcklund transformations. *J. Phys. A*, **14**(5), 1083–1098.

[10] Levi, D., and Benguria, R. 1980. Bäcklund transformations and nonlinear differential difference equations. *Proc. Nat. Acad. Sci. U.S.A.*, **77**(9, part 1), 5025–5027.

[11] Levi, D., and Yamilov, R. I. 1997. Conditions for the existence of higher symmetries of evolutionary equations on the lattice. *J. Math. Phys.*, **38**(12), 6648–6674.

[12] Levi, D., and Yamilov, R. I. 2009. The generalized symmetry method for discrete equations. *J. Phys. A*, **42**(45), 454012.

[13] Levi, D. Yamilov R.I. 2010. Integrability test for discrete equations via generalized symmetries. *Symmetries in Nature* (AIP Conference Proceedings 1323) ed. Benet L., Hess P. O., Torres J. M., Wolf K. B. (Melville, New York: AIP), 203–214.

[14] Levi D., Yamilov R. I. 2011. Generalized symmetry integrability test for discrete equations on the square lattice. *J. Phys. A*, **44**(14) 145207 (22 pp)

[15] Levi, D., Petrera, M., Scimiterna, C., and Yamilov, R. I. 2008. On Miura transformations and Volterra-type equations associated with the Adler-Bobenko-Suris equations. *SIGMA Symmetry Integrability Geom. Methods Appl.*, **4**, Paper 077.

[16] Levi, D., Winternitz, P. 2006. Continuous symmetries of difference equations. *J. Phys. A* **39**(2), R1–R63.

[17] Mikhailov, A. V., Shabat, A. B., and Yamilov, R. I. 1987. A symmetric approach to the classification of nonlinear equations. Complete lists of integrable systems. *Russian Math. Surveys*, **42**(4), 1–63.

[18] Mikhailov, A. V., Shabat, A. B., and Sokolov, V. V. 1991. The symmetry approach to classification of integrable equations. Pages 115–184 of: *What is integrability?* Springer Ser. Nonlinear Dynam. Berlin: Springer.

[19] Mikhailov, A. V., Wang, J. P., and Xenitidis, P. 2010. *Recursion operators, conservation laws and integrability conditions for difference equations.* arXiv:1004.5346.

[20] Noether, E. 1918. Invariante Variationsprobleme. *Nachr. v. d. Ges. d. Wiss. zu Göttingen*, 235–257. See Noether E. 1971. Invariant variation problems. *Transport Theory Statist. Phys.*, **1**(3), 186–207.

[21] Olver, P. J. 1993. *Applications of Lie groups to Differential Equations.* 2nd edn. Grad. Texts in Math., vol. 107. New York: Springer.

[22] Rasin, O. G., and Hydon, P. E. 2007. Symmetries of integrable difference equations on the quad-graph. *Stud. Appl. Math.*, **119**(3), 253–269.

[23] Toda, M. 1989. *Theory of nonlinear lattices.* Second edition. Springer Series in Solid-State Sciences, 20. Springer-Verlag, Berlin, x+225 pp.

[24] Yamilov, R. I. 1980. Conservation laws for the discrete Korteweg–de Vries equation. *Comput. Methods Appl. Mech. Engrg.*, **44**, 164–173. in Russian.

[25] Yamilov, R. I. 1983. Classification of discrete evolution equations. *Uspekhi Mat. Nauk*, **38**(6), 155–156.

[26] Yamilov, R. I. 2006. Symmetries as integrability criteria for differential difference equations. *J. Phys. A*, **39**(45), R541–R623.

[27] Yamilov, R. I. 2007. Integrability conditions for analogues of the relativistic Toda chain. *Theoret. and Math. Phys.*, **151**(1), 492–504.

[28] Yamilov, R. I., and Levi, D. 2004. Integrability conditions for n and t dependent dynamical lattice equations. *J. Nonlinear Math. Phys.*, **11**(1), 75–101.

7

Four Lectures on Discrete Systems

Sergey P. Novikov

7.1 Introduction

In this chapter we present a review of some work done in collaboration with
Veselov, Krichever, Dynnikov, Grinevich and Boyle on discrete systems. In
particular the Sections 7.2, 7.3, 7.4 are based on the articles nos. 136, 137,
140, 148, 159, 163 in the author's list of publications contained in the home-
page www.mi.ras.ru/~snovikov. The most recent results are described in
Section 7.5.

Mike Boyle from the University of Maryland helped me to use the methods
of symbolic dynamics to obtain the results presented in Section 7.5.

7.2 Discrete symmetries and completely integrable systems

Well known completely integrable systems like the Korteweg–de Vries equation (KdV) are associated with linear operators (see [16]). The strong and
weak factorization properties of second-order operators in one and two space
dimensions

(1D)

$$L = -\partial_x^2 + u(x) = QQ^+ + c, \qquad L\psi = \lambda\psi, \qquad Q = \partial_x + a(x), \quad (7.1)$$

(2D,a) hyperbolic

$$L = -Q_1 Q_2 + V = -(\partial_x + A)(\partial_y + B)) + V, \qquad L\psi = 0 \quad (7.2)$$

(2D,b) elliptic

$$L = QQ^+ + V = -(\partial + A)(\bar\partial + B) + V, \qquad L\psi = 0 \quad (7.3)$$

play a fundamental role in the integrability of the associated systems. Such a factorization in one dimension requires the solution of the Riccati equation

$$a_x + a^2 = u - c. \tag{7.4}$$

The "Darboux transformation," introduced by Euler in the 1740s to solve what later has been called the Schrödinger equation, follows from the strong factorization in one dimension,

$$L \to \tilde{L} = Q^+ Q + c, \psi \to \tilde{\psi} = Q^+ \psi. \tag{7.5}$$

It preserves the solutions for all λ (except may be for one such that $Q^+\psi_0 = 0$).

The isospectral deformations $dL/dt = [A, L]$ of the L-operator lead to the KdV and other famous systems.

In the case of partial differential equations (two-dimensional systems) the Laplace transformation follows from the weak factorization. In the hyperbolic case we have

$$\psi \to Q_2 \psi, \qquad L \to VQ_2 V^{-1} Q_1 + V, \tag{7.6}$$

and similarly in the elliptic case, replacing

$$Q_1 \to Q, \qquad Q_2 \to Q^+ = -\bar{\partial} + B. \tag{7.7}$$

The chain L_n of Laplace transformations $L_n \to L_{n+1}$ with potentials $V_n = \exp\{f_{n+1} - f_n\}$ is equivalent to the two-dimensional Toda lattice (Toda–Gordon equation)

$$f_{n,xy} = f_{n,z\bar{z}} = \exp(f_{n+1} - f_n) - \exp(f_n - f_{n-1}). \tag{7.8}$$

Remark For partial differential equations in $2 + 1$ dimensions one can introduce an analog of the isospectral deformations as

$$\frac{dL}{dt} = [A, L] + BL. \tag{7.9}$$

This deformation equation gives rise to the 2D analogs of the KdV hierarchy of equations like the so-called Novikov–Veselov hierarchy [17]. This hierarchy has better properties than those of the Kadomtsev–Petviashvili hierarchy (KP) [6]. Let us remember here that such deformations of the two-dimensional Schrödinger operators (L-A-B-triple) were discovered first by Manakov in 1976 [13] and developed in the same year by Dubrovin–Krichever–Novikov [3]. In particular, in [3] the inverse spectral problem (ISP) for 2-periodic two-dimensional Schrödinger operators has been introduced in terms of Riemann surfaces ("the complex Fermi curves"). A really interesting hierarchy was found in 1984 by Novikov and Veselov [19, 20] when the zero magnetic field condition was formulated in terms of the inverse spectral data (as a Riemann surface with selected pole divisor of the Bloch function). Later the ISP for rapidly decreasing two-dimensional potentials was also partly solved in an effective way in terms of one energy level data by Grinevich, R. Novikov, Manakov and the present author [8, 9, 11].

7.3 Discretization of linear operators

When looking for discrete integrable systems we may ask ourselves, which is the best discretization?

We may look for discretizations of the linear operators which preserve the discrete symmetries described above. In the one-dimensional case we introduce the shift operator

$$T: n \to n+1, \tag{7.10}$$

and consider the class of linear difference operators

$$L = c_n T + T^{-1} c_n + v_n. \tag{7.11}$$

This class can be factorized in the same way as has been done earlier for differential operators:

$$L = QQ^+ + c \tag{7.12}$$

where

$$Q = u_n T + b_n, \qquad Q^* = T^{-1} a_n + b_n. \tag{7.13}$$

The isospectral deformations

$$\frac{dL}{dt} = [A, L], \tag{7.14}$$

give rise to the "Toda lattice" and "discrete KdV" equation when $v_n = 0$.

In the two-dimensional case there are two different discretizations. Given the basic shifts $T_1(m, n) = (m+1, n)$ and $T_2(m, n) = (m, n+1)$ we have:

- In the hyperbolic case, see Figure 7.1, we take a square lattice and the equations $L\psi = 0$ on it. In this case we choose

$$L = a_{m,n} + b_{m,n} T_1 + c_{m,n} T_2 + d_{m,n} T_1 T_2. \tag{7.15}$$

This class of equations admits the following gauge transformations:

$$L \to fLg, \qquad \psi \to g^{-1}\psi \tag{7.16}$$

characterized by the functions f and g, which must be nonzero everywhere.

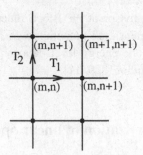

Figure 7.1

Thus L depends on two specific gauge invariant functions. We can always present L in the form

$$L = f((1 + uT_1)(1 + vT_2) + w) = f(Q_1Q_2 + w) \qquad (7.17)$$

implying the validity of the same gauge-invariant Laplace transformation as presented above in the continuous case.

• In the elliptic case, see Figure 7.2, real self-adjoint operators were first studied in [14] and developed in [16].

They are described on an equilateral triangle lattice by 7-point difference operators of the form

$$L = a + bT_1 + cT_2 + dT_1^{-1}T_2 + T_1^{-1}b + T_2^{-1}c + T_2^{-1}T_1d. \qquad (7.18)$$

It is easy to prove that we can always weakly factorize them as

$$L = QQ^+ + V, \qquad Q = u + vT_1 + wT_2. \qquad (7.19)$$

So the Laplace transformations are well-defined. Here the functions a, b, c, d, u, v, w are real valued functions of the vertices of the triangles.

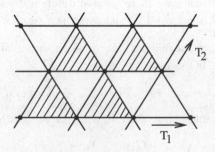

Figure 7.2

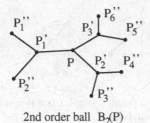

2nd order ball $B_2(P)$

Figure 7.3 2nd-order ball $B_2(P)$

Definition We call the first-order difference operators defined on the equilateral triangle lattice (see Figure7.2)

$$Q = u + vT_1 + wT_2 \tag{7.20}$$

"black triangle operator" and the difference operators

$$Q^+ = u + T_1^{-1}v + T_2^{-1}w = u + v'T_1^{-1} + w'T_2^{-1} \tag{7.21}$$

"white triangle operator."
We call

$$Q\psi = 0, \qquad Q^+\psi = 0, \tag{7.22}$$

respectively a "black triangle equation" and a "white triangle equation."

Another interesting class, considered in [12], is outlined in the following.
Let us consider a trivalent tree (see Figure 7.3) and on it any real self-adjoint 10 point operator

$$L\psi(P) = \sum_{i=1}^{6} b_{PP_i''}\psi(P_i'') + \sum_{j=1}^{3} b_{PP_j'}\psi(P_j') + V(P)\psi(P) \tag{7.23}$$

$$L = QQ^+ + v \tag{7.24}$$

The operators (7.23) can be factorized in terms of some second-order operators Q. In this way we get some completely integrable systems described by L-A-B-triples defined on this graph. Nothing like that exists for the second-order L operators on this graph. This result has been extended to higher trees by Chekhov and Puzyrnikova [2].

7.4 Discrete GL_n connections and triangle equation

The ideas presented in this section were developed in the references [4, 15, 16].
Let us recall some basic definitions from topology (see [18]):

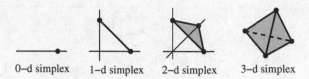

| 0–d simplex | 1–d simplex | 2–d simplex | 3–d simplex |

Figure 7.4 0-simplex in a point, 1-simplex is a line segment, 2-simplex in a triangle, 3-simplex is a tetrahedron.

The standard n-simplex $\sigma^{(n)}$ is the subset of the $n+1$-dimensional Euclidian space defined by the equations:

$$\begin{cases} x^1 + x^2 + \cdots + x^{n+1} = 1 \\ x^i \geq 0, \quad i = 1, \ldots, n+1. \end{cases}$$

A k-dimensional face of $\sigma^{(n)}$ is subset of $\sigma^{(n)}$ defined by equations $x^{i_1} = \cdots = x^{i_{n-k}} = 0$ for a fixed set $(i_1, \ldots, i_{n-k})$. The 0-dimnensional faces are called vertices, the 1-dimensional faces are called edges.

An n-simplex in a topological space K is a subset of K homeomorphic to the standard n-simplex.

A toplogical space K is called *a simplicial complex* if it is represented as an union of simplexes $K = \bigcup \sigma_j$ such, that for every pair σ_i, σ_j their intersection $\sigma_i \cap \sigma_j$ is either empty or a face for both of them [21].

Let K be a simplicial complex (n-manifold) with a selected family of n-simplices X and set of coefficients $b_{T:P}$ such that $b_{T:P} \neq 0$ for every n-simplex $T \in X$ and its vertex $P \in T$. Every vertex should belong to at least 2 simplices of the selected family X.

A Triangle Operator

$$Q^X \psi(T) = \sum_{P \in T} b_{T:P} \psi(P) \tag{7.25}$$

is defined in the space of scalar functions of the vertices.

Three families of triangle operators will be considered; black denoted by $X = b$, white by $X = w$ and all when no superscript X is written. The n–simplices of K corresponding to b or w will be colored in black and white colors, respectively. The operators Q^b (or Q^w) are those operators composed of the set of all black (or all white) simplices. When no superscript X is present, i.e., if we write just the operator Q then this means the set of all n-simplices $T \in K$.

Definitions   1. We call the triangle equation $Q\psi = 0$ the "discrete GL_n connection," while the ratios $\mu_{PP'}^T = b_{T:P}/b_{T:P'}$ are called the "connection

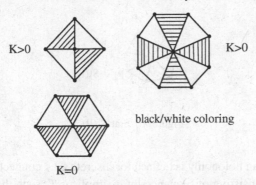

Figure 7.5

coefficients." The connection coefficients are defined up to an abelian gauge transformation defined by a scalar function f_P such that

$$\mu_{PP'}^T \rightarrow \mu_{PP'}^T f_P / f_{P'} \qquad (7.26)$$

However, curvature and holonomy are nonabelian.

Our approach is different from the standard one used (after K. Wilson [22]) since 30 years ago in the theory of Yang–Mills fields [7]. The GL_n-connection for us is an over-determined linear operator. It is a good object in the theory of linear operators but not for the selection of compact Lie groups, as needed in the theory of elementary particles.

2. We call the solutions to the triangle equation $Q^b\psi = 0$ for $b_{T:P} = 1$ "discrete holomorphic functions" and those of the triangle equation $Q^w\psi = 0$ "discrete antiholomorphic functions" (see Figure 7.5).

What do we mean by nonabelian curvature and holonomy? Let us work only with scalar-valued functions of the vertices. Figure 7.6 explains how one can define a nontrivial nonabelian curvature C_P on a vertex P in the case of discrete connections with $n = 2$. For every simplex P we start from the vertex P_1 in its star. Knowing the functions $\psi(P)$ and $\psi(P_1)$ we can calculate all the remaining $\psi(P_i)$ "along the circle" in the star. However, a contradiction might appear after returning to the original point P_1 in the form of a nonunit triangle matrix C_P. We call C_P the "curvature operator."

Important example Let us consider the case when $n = 2$ and $b_{T:P} = 1$. We call this connection "canonical." "The zero curvature" case appears when $C_P = 1$ and it simply means that an even number of edges (triangles) enters P.

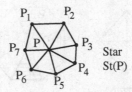

Figure 7.6 Star St(P)

The nonabelian holonomy is defined for discrete GL_n connections along the "thick paths" consisting of sequences of n-simplices T_j such that $T_{j+1} \cap T_j$ is exactly one face Δ_j of dimension $n - 1$. For the canonical connection we have $b_{T:P} = 1$ and $C_P = 1$ and the holonomy along the cyclic paths belongs to the permutation group S_{n+1}.

For every "closed framed path" $\gamma = [P_0, P_1, \dots, P_k = P_0]$ consisting of "framed edges" $\gamma_j = P_{j-1}, P_j$ (i.e., with selected n-simplices $\gamma_j \subset T_j$), such that

$$\gamma \equiv \langle \gamma_1, T_1; \gamma_2, T_2; \dots; \gamma_k, T_k \rangle \tag{7.27}$$

we define "abelian holonomy" $A(\gamma)$ as the product of the corresponding connection coefficients

$$A(\gamma) = \mu_{P_0, P_1}^{T_1} \times \cdots \times \mu_{P_{k-1}, P_k}^{T_k}. \tag{7.28}$$

In particular, we can introduce nontrivial "local abelian holonomy coefficients" as those ratios of connection coefficients

$$\rho_{PP'}^{TT'} = \mu_{PP'}^{T} / \mu_{P'P}^{T'} \tag{7.29}$$

that are gauge invariant.

The relation between abelian and nonabelian holonomies has been investigated in [15]. The "Inverse Problem" necessary to reconstruct the connections starting from the holonomy was solved in the same paper.

In [15] one can also find the theory of curvature while the theory of the characteristic classes for the GL_n-connections has not been constructed yet.

7.5 New discretization of complex analysis

The classical discretization of complex analysis was carried out in the case of square lattices by Ferrand, in 1940 [5]. Many people worked on this approach, see for example the results presented in [1]. Our ideas are instead based on the properties of equilateral triangle lattice (see [4]).

Discrete holomorphic functions do not form a ring in the square lattice discretization or in the triangular one since the Leibnitz rule is never satisfied. Thus we have to abandon the multiplication operation.

For every $n = 2$ simplician complex with black/white triangulation and $b_{T:P} = 1$ we can define discrete holomorphic (d-holomorphic) functions as those real functions ψ which satisfy the equations $Q^b\psi = 0$ and d-antiholomorphic functions are defined as those real functions ψ which satisfy the equations $Q^w\psi = 0$. We define as "covariant constants" those functions ψ that satisfy the equations $Q\psi = 0$, i.e., are such that

$$Q^b\psi = 0, \qquad Q^w\psi = 0. \tag{7.30}$$

For the standard Laplace–Beltrami operator $L_0 = \partial\partial^*$ we have

$$-2L_0 + 3m_P = Q^+Q = 2Q^{b+}Q^b = 2Q^{w+}Q^w \tag{7.31}$$

where m_P is equal to the number of edges (triangles) entering P. So for $m_P =$ const the zero modes of Q^+Q coincide with maximal modes of the Laplace–Beltrami operator L_0.

Theorem 7.1 *For compact manifolds every d-holomorphic function is a covariant constant.*

Proof $Q^b\psi = 0$ implies $Q^{b+}Q^b\psi = 0$ implies $Q^+Q\psi = 0$ implies $(Q\psi, Q\psi) = 0$ and therefore $Q\psi = 0$ because inner product is well-defined and positive. $\square$

We call the content of the previous theorem the Liouville principle for discrete equations.

Let us assume that the space of covariant constants is exactly 2-dimensional. We may have to go to a finite covering (at most 6 sheets), to be able to satisfy this condition for a non-simply-connected surface.

How can we carry out the continuous limit on triangles? Take a covariant constant f_0 whose values in every triangle are $1, \zeta, \zeta^2$ where $\zeta^3 = 1$. Let us perform the gauge transformation

$$L \to f_0^{-1}Lf_0, \qquad \psi \to f_0^{-1}\psi.$$

After that our theory should be considered over the complex field $\mathbb{C}$ as one of the covariant constants is now a real constant. In this gauge form the continuous limit can be easily defined: one half of our theory converges to ordinary complex analysis but the second half of this discrete theory is divergent for small scales.

We are working here in the most symmetric purely real gauge form over the field $\mathbb{R}$ imitating all of complex analysis.

Maximum principle

Let ψ be a holomorphic function in a finite domain D consisting of black triangles whose vertices all belong to D. A *boundary triangle* is such that at least one of its vertices belongs to some black triangle outside of D.

We can introduce an *evaluation map* $E_\psi : T \to R^2$ which assigns to any black triangle with vertices P, P', P'' a vector in the space of covariant constants R^2 defined by $\psi(P), \psi(P'), \psi(P'')$. We can then introduce the following theorem:

Theorem 7.2   *The image $E_\psi(D)$ coincides with the convex hull of the image $E_\psi(\partial D)$, where by the symbol ∂D we mean the boundary of D.*

D-holomorphic polynomials and Taylor series

How can we define polynomials without having defined the multiplication operation in our space? In the case of the Euclidean plane we can provide a positive answer.

Let us consider an equilateral triangle lattice on the plane with shifts T_1, T_2 and natural b/w coloring as we discussed above (see Figure 7.7). Then our operators $Q^b = 1 + T_1 + T_2$ and $Q^w = 1 + T_1^{-1} + T_2^{-1}$ map the space of functions of vertices into itself.

Definition   We call a d-holomorphic function ψ a d-holomorphic polynomial P_k of degree k, if $(Q^w)^{k+1} \psi = 0$.

Let us consider a big equilateral triangle T_k whose edges are black from inside and which contains exactly $2k + 2$ vertices (see Figure 7.7).

Theorem 7.3 (The Taylor approximation)   *For every d-holomorphic function ψ and big triangle T_k there exists exactly one d-holomorphic polynomial P_k of degree k such that $\psi - P_k = 0$ in the triangle T_k. The space H_k of d-holomorphic polynomials has dimension $2k + 2$ over the reals $\mathbb{R}$.*

The choice of the basis of d-holomorphic polynomials depends on T_k. There are 3 functions $P_k^\alpha(T_k), \alpha = 1, 2, 3$, which are equal to zero in T_k except for one boundary edge with value α. Along this edge P_k^α is equal to $1, -1, 1, -1, \ldots, -1$ (see Figure 7.7). A linear combination of these "boundary d-polynomials" which equals to zero inside, gives us the natural analog of the real space $\mathbb{R}^2$ generated by the real and imaginary parts of z^k. The following lemma is true:

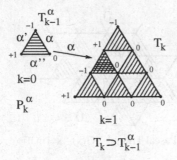

$$T_k \supset T^\alpha_{k-1}$$

Figure 7.7

Lemma 7.4 *The space of boundary d-polynomials is two-dimensional mod-ulo d-holomorphic polynomials of degree $k - 1$:*

$$P^1_k + P^2_k + P^3_k = P_{k-1} \in H_{k-1}, \qquad \alpha = 1, 2, 3 \qquad (7.32)$$

In (7.32) $P_{k-1} = P^\alpha_{k-1}$ belongs to the triangle T^α_{k-1} (see Figure 7.7).

Cauchy formula

It is well known that the Cauchy kernel $1/z$ plays a double role in complex algebra and analysis:

1. It generates all rational functions using differentiation ∂_z and translations:

$$\partial^k(1/z) = \text{const} \times 1/z^{k+1}, \qquad f(z) = \sum a_j/(z - z_j)^{k_j} \qquad (7.33)$$

2. It generates a boundary formula expressing a function in terms of its bound-ary values as $\bar{\partial}(1/z) = 2\pi i \delta(z)$.

What is the correct analog of the Cauchy Kernel in the discrete setting?

Let us start from the boundary formulas. Let ψ be a d-holomorphic function in the bounded domain D on an equilateral triangle lattice. We can easily construct some fundamental solutions $G(x - y)$ such that $Q^b G(x - y) = \delta_y(x)$. Here $x = (m, n)$ and $\delta_y(x) = 1$, if $y = x$ and zero otherwise. For $y = 0$ such a function is given in Figure 7.8. It is equal to zero for all $x = (m, n)$ where $m > 0$ or $n > 0$. Its values at the boundary are $(-1)^m$ in the points $(-m, 0)$ and $(-1)^n$ in the points $(0, -n)$ and $G = (-1)^{m+n} \frac{(m+n)!}{m!n!}$ for $m < 0, n < 0$ (the Pascal triangle).

Let us take a function which is $\tilde{\psi} = \psi$ in D and zero outside. Then the function $Q^b \tilde{\psi}$ is concentrated along the boundary ∂D, a "strip." We can then formulate the following theorem:

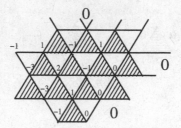

"Pascal Triangle" G(x)   x=(m,n)

Figure 7.8

Theorem 7.5   *The following Cauchy formula will be valid for $x \in D$:*

$$\sum_{y}(Q^{b}\tilde{\psi}(y))G(x-y) = \psi(x) \qquad (7.34)$$

To carry out this calculation any Green's function can be used. However our function looks more like a hyperbolic than an elliptic one. It cannot be considered as the correct analog of the kernel $1/z$.

Recently Grinevich and R. Novikov presented an essentially unique genuinely elliptic function $G(x - y)$ decreasing for $|x - y| \to \infty$ [10]. Such a Green's function provides an essentially unique Cauchy kernel. It is defined simply in terms of the Fourier transform, which in our case is convergent

$$G_{m,n} = \int_{0}^{2\pi} \int_{0}^{2\pi} dk_{1}\,dk_{2} \times \exp\{imk_{1} + ink_{2}\}/(1 + e^{ik_{1}} + e^{ik_{2}}). \qquad (7.35)$$

This result implies that every d-holomorphic function in R^{2} with polynomial growth at infinity is a d-polynomial.

Our conclusion is that the discrete analogs of rational functions appear naturally in the Euclidean plane.

Hyperbolic (Lobachevsky) plane

Recently we started to develop a discrete analog of complex analysis on equilateral lattices on a hyperbolic plane. Neither the analog of Taylor polynomials nor the Grinevich–R. Novikov-type Green's function are known in this case. We have negative curvature if the number of edges entering into every vertex is an even number with $m_P > 6$. For homogeneous triangulations with $m_P = 8, 10, 12, \ldots$ we have a big group preserving triangulations. Let us concentrate on the minimal one which appears when $m_P = 8$.

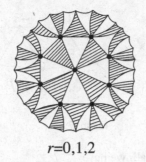

$$r=0,1,2$$

Figure 7.9

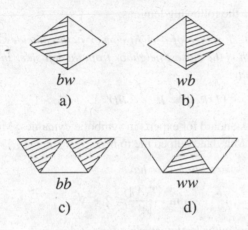

bw   *wb*

a)   b)

bb   *ww*

c)   d)

Figure 7.10

How can we describe the boundary of an *r*-ball for every integer *r*? A picture is presented below for $r = 0, 1, 2$. Let us define a class of right-convex oriented simplicial paths – see Figure 7.10a,b,c,d. These pictures follow directly from the definitions.

We codify the right-convex oriented paths by words formed in terms of two symbols *b*, *w* characterizing the kind of triangle we are considering. We assign the word *bw* to Figure 7.10a, *wb* to Figure 7.10b, *bb* to Figure 7.10c and *ww* to Figure 7.10d. Let us introduce a *structural transformation T* in the space of infinite periodic right-convex oriented paths by the formulas

$$T: \quad bw \to bwbw, \quad wb \to wbwb, \quad bb \to bwb, \quad ww \to wbw. \quad (7.36)$$

We apply *T* to every pair of neighboring letters in a word and then delete the old letters. For example in the case of the word $R_1 = \ldots bwbwbwbw \ldots$, which

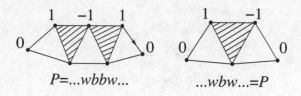

$$P=...wbbw...\qquad\qquad ...wbw...=P$$

Figure 7.11

is at the boundary of a 1-ball, we have

$$R_2 = T(R_1) = \ldots bwbwwbwbbwbwwbwbbwbwwbwbbwbwwbwb\ldots \quad (7.37)$$

We can then write the following lemmas:

Lemma 7.6 *The T-image of a right-convex path coincides with the right-convex path which is the closest neighbor from the left side. In particular, for an r-ball D_r we have*

$$[T(R_1)]^r = R_{r+1} = \partial D^{r+1}, \qquad r \geq 1 \qquad (7.38)$$

Such maps are standard for experts in symbolic dynamics. Mike Boyle from the University of Maryland helped me to investigate them.

Lemma 7.7 *For every word A we have*:

$$\lim_{|A|\to\infty} \frac{|T(A)|}{|A|} = 2 + \sqrt{3}. \qquad (7.39)$$

This asymptotic behavior is almost exact for $r \geq 4$, $A = R_r$.

In particular we have $|R_1| = 8$, $|R_2| = 32$, $|R_3| = 120$, $|R_4| = 448$, $|R_5| = 1672, \ldots$

Let us consider the problem of constructing a basis of d-holomorphic functions $z_P^r(x)$ such that $z_P^r = 0$ for all points x in R_k, $k < r$ and for all points in R_r except for those satisfying $P \subset R_r$. Here we consider the case when $P = wbbw$ or $P = wbw$ (see Figure 7.11 for the values of these functions in P).

Let us introduce the following conjecture:

Conjecture *There exists a basis of d-holomorphic functions z_P^r which are globally bounded in the hyperbolic plane. Their linear combinations are equivalent in the continuous case to the polynomials $\sum_{k=0}^{n} a_k z^k$ on the unit disc.*

We can consider another basis, defined by fixing the zero point and a right-convex line $l = \ldots bbbbbb \ldots$ passing through 0. We can easily construct a specific d-holomorphic function $h^{0,l}$ which is equal to zero on the right side of l and equal to ± 1 along the line l (see Figure 7.12).

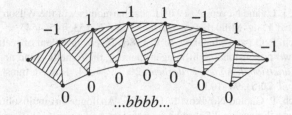

Figure 7.12

Its continuation to the left side of l is nonunique. We can make it unique by using an "optimal" continuation to the left-hand side (i.e., such that it has the minimal possible growth). What kind of growth is it? One can easily construct a basis of d-holomorphic functions using the specific function $h^{0,l}$ and its group shifts.

We present the following theorem:

Theorem 7.8 *The space of d-holomorphic functions in the r-ball D_r has dimension $1 + |R_r|/2$.*

This result is quite similar to the one we obtain in the continuous case.

We can construct a similar basis with d-antiholomorphic functions $\bar{z}_p^r(x)$ by replacing b by w in the previous definition and in Figure 7.11. On the boundary R_r these two spaces generate the space of all functions. Their intersection is exactly the space of covariant constants.

References

[1] Bobenko, A. I., Mercat, C., and Suris, Yu. B. 2005. Linear and nonlinear theories of discrete analytic functions. Integrable structure and isomonodromic Green's function. *J. Reine Angew. Math.*, **583**, 117–161.

[2] Chekhov, L. O., and Puzyrnikova, N. V. 2000. Integrable systems on graphs. *Uspekhi Mat. Nauk*, **55**(5(335)), 181–182. English transl., *Russian Math. Surveys* **55**(5), 992–994.

[3] Dubrovin, B. A., and Krichever, I. M. 1976. The Schrödinger equation in a periodic field and Riemann surfaces. *Dokl. Akad. Nauk SSSR*, **229**(1), 15–18. Russian.

[4] Dynnikov, I. A., and Novikov, S. P. 2003. Geometry of the triangle equation on two-manifolds. *Mosc. Math. J.*, **3**(2), 419–438.

[5] Ferrand, J. 1944. Fonctions préharmoniques et fonctions préholomorphes. *Bull. Sci. Math. (2)*, **68**, 152–180.

[6] Fokas, A. S., and Santini, P. M. 1988. Bi-Hamiltonian formulation of the Kadomtsev–Petviashvili and Benjamin–Ono equations. *J. Math. Phys.*, **29**(3), 604–617.

[7] Gervais, J. L., and Neveu, A. 1979. Local harmonicity of the Wilson loop integral in classical Yang–Mills theory. *Nuclear Phys. B*, **153**, 445–454.

[8] Grinevich, P. G., and Manakov, S. V. 1986. Inverse problem of scattering theory for the two-dimensional Schrödinger operator, the $\bar{\partial}$-method and nonlinear equations. *Funktsional. Anal. i Prilozhen.*, **20**(2), 14–24. English transl., *Functional Anal. Appl.* **20**(2), 94–103.

[9] Grinevich, P. G., and Novikov, R. G. 1986. Analogues of multisoliton potentials for the two-dimensional Schrödinger operator, and a nonlocal Riemann problem. *Dokl. Akad. Nauk SSSR*, **286**(1), 19–22. English transl., *Soviet Math. Dokl.* **33**(1), 9–12.

[10] Grinevich, P. G., and Novikov, R. G. 2007. The Cauchy kernel for the Novikov–Dynnikov DN-discrete complex analysis on a triangular lattice. *Uspekhi Mat. Nauk*, **62**(4), 155–156. English transl., *Russian Math. Surveys* **62**(4), 799–801.

[11] Grinevich, P. G., and Novikov, S. P. 1988. A two-dimensional "inverse scattering problem" for negative energies, and generalized-analytic functions. I. Energies lower than the ground state. *Funktsional. Anal. i Prilozhen.*, **22**(1), 23–33. English transl., *Funct. Anal. Appl.* **22**(1), 19–27.

[12] Krichever, I. M., and Novikov, S. P. 1999. Trivalent graphs and solitons. *Uspekhi Mat. Nauk*, **54**(6), 149–150. English transl., *Russian Math. Surveys* **54**(6), 1248–1249.

[13] Manakov, S. V. 1976. The method of the inverse scattering problem, and two-dimensional evolution equations. *Uspehi Mat. Nauk*, **31**(5), 245–246. English transl., *Russian Math. Surveys* **31**(5), 245–246.

[14] Novikov, S. P. 1997. Algebraic properties of two-dimensional difference operators. *Uspekhi Mat. Nauk*, **52**(1), 225–226. English transl., *Russian Math. Surveys* **52**(1), 226–227.

[15] Novikov, S. P. 2004. Discrete connections and linear difference equations. *Tr. Mat. Inst. Steklova*, **247**(Geom. Topol. i Teor. Mnozh.), 186–201. English transl., *Proc. Steklov Inst. Math.* **2004**(4), 168–183.

[16] Novikov, S. P., and Dynnikov, I. A. 1997. Discrete spectral symmetries of small-dimensional differential operators and difference operators on regular lattices and two-dimensional manifolds. *Uspekhi Mat. Nauk*, **52**(5), 175–234. English transl., *Russian Math. Surveys* **52**(5), 1057–1116.

[17] Novikov, S. P., and Veselov, A. P. 1986. Two-dimensional Schrödinger operator: inverse scattering transform and evolutional equations. *Phys. D*, **18**(1-3), 267–273.

[18] Seifert, H., and Threlfall, W. 1980. *Seifert and Threlfall: a textbook of topology*. Pure Appl. Math., vol. 89. New York: Academic Press.

[19] Veselov, A. P., and Novikov, S. P. 1984a. Finite-gap two-dimensional potential Schrödinger operators. Explicit formulas and evolution equations. *Dokl. Akad. Nauk SSSR*, **279**(1), 20–24. English transl., *Soviet Math. Dokl.* **30**(3), 588–591.

[20] Veselov, A. P., and Novikov, S. P. 1984b. Finite-gap two-dimensional Schrödinger operators. Potential operators. *Dokl. Akad. Nauk SSSR*, **279**(4), 784–788.

[21] Wikipedia. 2009. *Simplicial complex*. http://en.wikipedia.org/wiki/Simplicial_complex.

[22] Wilson, K. G. 1974. Confinement of quarks. *Phys. Rev. D*, **10**(8), 2445–2459.

8

Lectures on Moving Frames

Peter J. Olver[a]

Abstract

This chapter presents the equivariant method of moving frames for finite-dimensional Lie group actions, surveying a variety of applications, including geometry, differential equations, computer vision, numerical analysis, the calculus of variations, and invariant flows.

8.1 Introduction

According to Akivis [1], the method of moving frames originates in work of the Estonian mathematician Martin Bartels (1769–1836), a teacher of both Gauss and Lobachevsky. The field is most closely associated with Élie Cartan [22], who forged earlier contributions by Darboux, Frenet, Serret, and Cotton into a powerful tool for analyzing the geometric properties of submanifolds and their invariants under the action of transformation groups. In the 1970s, several researchers, cf. [25, 37, 38, 49], began the process of developing a firm theoretical foundation for the method. The final crucial step [32], is to define a moving frame simply as an equivariant map from the manifold back to the transformation group. All classical moving frames can be reinterpreted in this manner. Moreover, the equivariant approach is completely algorithmic, and applies to very general group actions.

Cartan's normalization construction of a moving frame can be interpreted as the choice of a cross-section to the group orbits. This enables one to algorithmically construct an equivariant moving frame along with a complete systems of invariants through the induced invariantization process. The existence of an equivariant moving frame requires freeness of the underlying group action, i.e.,

[a] Supported in part by NSF Grant DMS 08-07317

the isotropy subgroup of any single point is trivial. Classically, non-free actions are made free by prolonging to jet space, leading to differential invariants and the solution to equivalence and symmetry problems via the differential invariant signature. Alternatively, applying the moving frame method to Cartesian product actions leads to the classification of joint invariants and joint differential invariants [78]. Finally, an amalgamation of jet and Cartesian product actions dubbed *multi-space* was proposed in [77] to serve as the basis for the geometric analysis of numerical approximations, and systematic construction of invariant numerical algorithms [54].

With the basic moving frame machinery in hand, a plethora of new, unexpected, and significant applications soon appeared. In [7, 55, 56, 75], the theory was applied to produce new algorithms for solving the basic symmetry and equivalence problems of polynomials that form the foundation of classical invariant theory. In [2, 6, 10, 21, 72, 91], the characterization of submanifolds via their differential invariant signatures was applied to the problem of object recognition and symmetry detection, [17, 18, 31, 87]. Applications to the classification of joint invariants and joint differential invariants appear in [11, 32, 78]. In computer vision, joint differential invariants have been proposed as noise-resistant alternatives to the standard differential invariant signatures [29, 69]. The approximation of higher order differential invariants by joint differential invariants and, generally, ordinary joint invariants leads to fully invariant finite difference numerical schemes [10, 20, 21, 54, 77]. The all-important recurrence formulae lead to a complete characterization of the differential invariant algebra of group actions, and lead to new results on minimal generating invariants, even in very classical geometries [44, 45, 48, 79, 82]. The general problem from the calculus of variations of directly constructing the invariant Euler-Lagrange equations from their invariant Lagrangians was solved in [57]. Applications to the evolution of differential invariants under invariant submanifold flows, leading to integrable soliton equations and signature evolution in computer vision, can be found in [50, 80].

Applications of equivariant moving frames that are being developed by other research groups include the computation of symmetry groups and classification of partial differential equations [60, 70]; geometry of curves and surfaces in homogeneous spaces, with applications to integrable systems [61, 62, 64]; symmetry and equivalence of polygons and point configurations [12, 13], recognition of DNA supercoils [90], recovering structure of three-dimensional objects from motion [6], classification of projective curves in visual recognition [42]; construction of integral invariant signatures for object recognition in 2D and 3D images [33]; determination of invariants and covariants of Killing tensors, with applications to general relativity, separation of variables, and

Hamiltonian systems [28, 66, 67]; further developments in classical invariant theory [7, 55, 56]; computation of Casimir invariants of Lie algebras and the classification of subalgebras, with applications in quantum mechanics [14, 15]. A rigorous, algebraically-based reformulation of the method, suitable for symbolic computations, has been proposed by Hubert and Kogan [46, 47].

Finally, in recent work with Pohjanpelto [83–85], the theory and algorithms have recently been extended to the vastly more complicated case of infinite-dimensional Lie pseudo-groups. Applications to infinite-dimensional symmetry groups of partial differential equations can be found in [23, 24, 71, 94], and to the classification of Laplace invariants and factorization of linear partial differential operators in [92].

8.2 Equivariant moving frames

We begin by describing the general equivariant moving frame construction. Let G be an r-dimensional Lie group acting smoothly on an m-dimensional manifold M.

Definition 8.1 A *moving frame* is a smooth, G-equivariant map $\rho : M \to G$.

There are two principal types of equivariance:

$$\rho(g \cdot z) = \begin{cases} g \cdot \rho(z) & \text{left moving frame} \\ \rho(z) \cdot g^{-1} & \text{right moving frame} \end{cases} \tag{8.1}$$

If $\rho(z)$ is any right-equivariant moving frame then $\tilde{\rho}(z) = \rho(z)^{-1}$ is left-equivariant and conversely. All classical moving frames are left-equivariant, but the right versions are often easier to compute. In classical geometrical situations, one can identify left-equivariant moving frames with the usual frame-based versions, cf. [40].

It is not difficult to establish the basic requirements for the existence of an equivariant moving frame.

Theorem 8.2 *A moving frame exists in a neighborhood of a point $z \in M$ if and only if G acts freely and regularly near z.*

Recall that G acts *freely* if the isotropy subgroup $G_z = \{g \in G \mid g \cdot z = z\}$ of each point $z \in M$ is trivial: $G_z = \{e\}$. This implies *local freeness*, meaning that the isotropy subgroups G_z are all discrete, or, equivalently, that the orbits all have the same dimension, r, as G itself. *Regularity* requires that, in addition, the orbits form a regular foliation.

The explicit construction of a moving frame relies on the choice of a (local) *cross-section* to the group orbits, meaning an $(m - r)$-dimensional submanifold $\mathcal{K} \subset M$ that intersects each orbit transversally and at most once.

Theorem 8.3 *Let G act freely and regularly on M, and let* $\mathcal{K} \subset M$ *be a cross-section. Given* $z \in M$, *let* $g = \rho(z)$ *be the unique group element that maps z to the cross-section:* $g \cdot z = \rho(z) \cdot z \in \mathcal{K}$. *Then* $\rho \colon M \to G$ *is a right moving frame.*

Given local coordinates $z = (z_1, \ldots, z_m)$ on M, suppose the cross-section $\mathcal{K}$ is defined by the r equations

$$Z_1(z) = c_1, \ldots, Z_r(z) = c_r, \tag{8.2}$$

where $Z_1, \ldots, Z_r$ are scalar-valued functions, while $c_1, \ldots, c_r$ are suitably chosen constants. In many applications, the Z_λ are merely coordinate functions. The associated right moving frame $g = \rho(z)$ is obtained by solving the *normalization equations*

$$Z_1(g \cdot z) = c_1, \ldots, Z_r(g \cdot z) = c_r, \tag{8.3}$$

for the group parameters $g = (g_1, \ldots, g_r)$ in terms of the coordinates $z = (z_1, \ldots, z_m)$. Transversality combined with the Implicit Function Theorem implies the existence of a local solution to these algebraic equations.

The specification of a moving frame by choice of a cross-section induces an invariantization process that maps functions to invariants.

Definition 8.4 The *invariantization* of a function $F \colon M \to \mathbb{R}$ is the unique invariant function $I = \iota(F)$ that agrees with F on the cross-section: $I|\mathcal{K} = F|\mathcal{K}$.

In practice, the invariantization of a function $F(z)$ is obtained by first transforming it according to the group, $F(g \cdot z)$ and then replacing the group parameters by their moving frame formulae $g = \rho(z)$, so that $\iota[F(z)] = F(\rho(z) \cdot z)$. In particular, invariantization of the coordinate functions yields the *fundamental invariants*: $I_\nu(z) = \iota(z_\nu)$. Once these have been computed, the invariantization of a general function $F(z)$ is simply given by

$$\iota[F(z_1, \ldots, z_n)] = F(I_1(z), \ldots, I_r(z)). \tag{8.4}$$

In particular, the functions defining the cross-section (8.2) have constant invariantization, $\iota(Z_\nu) = c_\nu$, and are known as the *phantom invariants*. Thus, there are precisely $m - r$ functionally independent fundamental invariants. Moreover, if $I(z)$ is any invariant, then clearly $\iota(I) = I$, which implies the elegant and powerful *Replacement Rule*

$$I(z_1, \ldots, z_n) = I(I_1(z), \ldots, I_n(z)), \tag{8.5}$$

that can be used to immediately rewrite $I(z)$ in terms of the fundamental invariants.

Of course, most interesting group actions are *not* free, and therefore do not admit moving frames in the sense of Definition 8.1. There are two well-known methods that convert a nonfree (but effective) action into a free action. The first is to look at the Cartesian product action of G on several copies of M, which leads to joint invariants. The second is to prolong the group action to jet space, which is the natural setting for the traditional moving frame theory, leading to differential invariants. Combining the two methods of jet prolongation and Cartesian product results in joint differential invariants. In applications of symmetry constructions to numerical approximations of derivatives and differential invariants, one requires a unification of these different actions into a common framework, called multispace [54, 77]. These are discussed in turn in the following sections.

8.3 Moving frames on jet space and differential invariants

Traditional moving frames are obtained by prolonging the group action to the nth order submanifold jet bundle $J^n = J^n(M, p)$, which is defined as the set of equivalence classes of p-dimensional submanifolds $S \subset M$ under the equivalence relation of nth order contact at a single point; see [73, chapter 3] for details. Since G preserves the contact equivalence relation, it induces an action on the jet space J^n, known as its nth order *prolongation* and denoted by $G^{(n)}$. The formulas for the prolonged group action are found by implicit differentiation.

We can assume, without significant loss of generality, that G acts effectively on open subsets of M, meaning that the only group element that fixes *every* point in any open $U \subset M$ is the identity element. This implies [76] that the prolonged action is locally free on a dense open subset $\mathcal{V}^n \subset J^n$ for $n \gg 0$ sufficiently large. In all known examples, the prolonged action is, in fact, free on such a $\mathcal{V}^n$ although there is, frustratingly, no general proof of this property. The points $z^{(n)} \in \mathcal{V}^n$ are known as *regular jets*.

The normalization construction based on a choice of local cross-section $\mathcal{K}^n \subset \mathcal{V}^n$ to the prolonged group orbits can be used to produce a (locally defined) nth order *equivariant moving frame* $\rho^{(n)} : J^n \to G$ in a neighborhood of any regular jet. Once the moving frame is established, the induced invariantization process will map general differential functions $F(x, u^{(n)})$ to differential invariants $I = \iota(F)$. The *fundamental differential invariants* are obtained by invariantization of the coordinate functions:

$$H^i = \iota(x^i), \qquad I_J^\alpha = \iota(u_J^\alpha), \qquad \alpha = 1, \ldots, q, \quad \# J \geq 0. \tag{8.6}$$

These naturally split into two classes: The $r = \dim G$ combinations defining the cross-section will be constant, and are known as the *phantom differential invariants*. The remainder, called the *basic differential invariants*, form a complete system of functionally independent differential invariants. Indeed, if $I(x, u^{(n)}) = I(\ldots x^i \ldots u_J^\alpha \ldots)$ is any differential invariant, then the Replacement Rule (8.5) allows one to immediately rewrite $I = I(\ldots H^i \ldots I_J^\alpha \ldots)$ in terms of the fundamental differential invariants. The moving frame also produces p independent invariant differential operators by invariantizing the usual total derivative operators, $\mathcal{D}_1 = \iota(D_1), \ldots, \mathcal{D}_p = \iota(D_p)$, which can be iteratively applied to lower order differential invariants to generate the higher order differential invariants; see below for full details.

Example 8.5 The paradigmatic example is the action of the orientation-preserving Euclidean group SE(2) on plane curves $C \subset M = \mathbb{R}^2$. The group transformation $g \in$ SE(2) maps the point $z = (x, u)$ to the point $w = (y, v) = g \cdot z$, given by

$$y = x \cos \phi - u \sin \phi + a, \qquad v = x \sin \phi + u \cos \phi + b, \tag{8.7}$$

where $g = (\phi, a, b) \in$ SE(2) are the group parameters. The prolonged group transformations are obtained by successively applying the implicit differentiation operator

$$D_y = (\cos \phi - u_x \sin \phi)^{-1} D_x \tag{8.8}$$

to v, producing

$$v_y = \frac{\sin \phi + u_x \cos \phi}{\cos \phi - u_x \sin \phi}, \qquad v_{yy} = \frac{u_{xx}}{(\cos \phi - u_x \sin \phi)^3}, \qquad \ldots \tag{8.9}$$

Observe that the prolonged action is locally free on the first order jet space J^1. (To simplify the exposition, we gloss over the remaining discrete ambiguity caused by a 180° rotation; see [78] for a more precise development.) The classical moving frame is based on the cross-section

$$\mathcal{K}^1 = \{x = u = u_x = 0\}. \tag{8.10}$$

Solving the corresponding normalization equations $y = v = v_y = 0$ for the group parameters produces the right moving frame

$$\phi = -\tan^{-1} u_x, \qquad a = -\frac{x + u u_x}{\sqrt{1 + u_x^2}}, \qquad b = \frac{x u_x - u}{\sqrt{1 + u_x^2}}. \tag{8.11}$$

The classical left-equivariant Frenet frame [40] is obtained by inverting the Euclidean group element given by (8.11). Invariantization of the coordinate functions, which is done by substituting the moving frame formulae (8.11) into the prolonged group transformations (8.9), produces the fundamental normalized differential invariants:

$$\iota(x) = H = 0, \qquad \iota(u) = I_0 = 0, \qquad \iota(u_x) = I_1 = 0,$$
$$\iota(u_{xx}) = I_2 = \kappa, \qquad \iota(u_{xxx}) = I_3 = \kappa_s, \qquad \iota(u_{xxxx}) = I_4 = \kappa_{ss} + 3\kappa^3, \tag{8.12}$$

and so on. The first three are the *phantom invariants*. The lowest order basic differential invariant is the Euclidean curvature $I_2 = \kappa = (1 + u_x^2)^{-3/2} u_{xx}$. The corresponding invariant differential operator is the arc length derivative,

$$\mathcal{D} = D_s = \frac{1}{\sqrt{1 + u_x^2}} D_x \tag{8.13}$$

which is obtained by invariantizing (8.8). Using the general recursion formulae, that relate the normalized and differentiated differential invariants, to be presented in detail below, we can readily prove that the curvature and its successive derivatives with respect to arc length, $\kappa, \kappa_s, \kappa_{ss}, \ldots$, form a complete system of differential invariants.

8.4 Equivalence and signatures

A motivating application of the moving frame method is to solve problems of equivalence and symmetry of submanifolds under group actions. Given a group action of G on M, two submanifolds $S, \overline{S} \subset M$ are said to be *equivalent* if $\overline{S} = g \cdot S$ for some $g \in G$. A *symmetry* of a submanifold is a self-equivalence, that is a group transformation $g \in G$ that maps S to itself: $S = g \cdot S$. The solution to the equivalence and symmetry problems for submanifolds is based on the functional interrelationships among the fundamental differential invariants restricted to the submanifold.

Suppose we have constructed an nth order moving frame $\rho^{(n)} \colon J^n \to G$ defined on an open subset of jet space. A submanifold S is called *regular* if its n-jet $j_n S$ lies in the domain of definition of the moving frame. For any $k \geq n$, we use $J^{(k)} = I^{(k)} \mid j_k S$, where $I^{(k)} = (\ldots H^i \ldots I_J^\alpha \ldots)$, $\#J \leq k$, to denote the kth order *restricted differential invariants*.

Definition 8.6 The kth order *signature* $S^{(k)} = S^{(k)}(S)$ is the set parametrized by the restricted differential invariants $J^{(k)} \colon j_k S \to \mathbb{R}^{n-k}$, where $n_k = p + q\binom{p+k}{k} = \dim J^k$.

The submanifold S is called *fully regular* if $J^{(k)}$ has constant rank $0 \le t_k \le p = \dim S$ for all $k \ge n$. In this case, $S^{(k)}$ forms a submanifold of dimension t_k – perhaps with self-intersections. In the fully regular case,

$$t_n < t_{n+1} < t_{n+2} < \cdots < t_s = t_{s+1} = \cdots = t \le p, \qquad (8.14)$$

where t is the *differential invariant rank* and s the *differential invariant order* of S.

Theorem 8.7 *Two fully regular p-dimensional submanifolds $S, \overline{S} \subset M$ are (locally) equivalent if and only if they have the same differential invariant order s and their signature manifolds of order $s + 1$ are identical:* $S^{(s+1)}(\overline{S}) = S^{(s+1)}(S)$.

Since symmetries are merely self-equivalences, the signature also determines the symmetry group of the submanifold.

Theorem 8.8 *If $S \subset M$ is a fully regular p-dimensional submanifold of differential invariant rank t, then its symmetry group G_S is an $(r-t)$-dimensional subgroup of G that acts locally freely on S.*

A submanifold with maximal differential invariant rank $t = p$, and hence only a discrete symmetry group, is called *nonsingular*. The number of symmetries of a nonsingular submanifold is determined by its *index*, which is defined as the number of points in S map to a single generic point of its signature:

$$\operatorname{ind} S = \min\{\#(J^{(s+1)})^{-1}\{\zeta\} \mid \zeta \in S^{(s+1)}\}. \qquad (8.15)$$

Theorem 8.9 *If S is a nonsingular submanifold, then its symmetry group is a discrete subgroup of cardinality $\operatorname{ind} S$.*

At the other extreme, a rank 0 or *maximally symmetric* submanifold [81] has all constant differential invariants, and so its signature degenerates to a single point.

Theorem 8.10 *A regular p-dimensional submanifold S has differential invariant rank 0 if and only if its symmetry group is a p-dimensional subgroup $H \subset G$ and hence S is an open submanifold of an H-orbit: $S \subset H \cdot z_0$.*

Remark "Totally singular" submanifolds may have even larger, non-free symmetry groups, but these are not covered by the preceding results. See [76] for details, including Lie algebraic characterizations.

Remark See [72] for some counterexamples when one tries to relax the regularity assumptions in the above results.

Example 8.11 The *Euclidean signature* for a curve $C \subset M = \mathbb{R}^2$ is the planar curve $\mathcal{S}(C) = \{(\kappa, \kappa_s)\}$ parametrized by the curvature invariant κ and its first derivative with respect to arc length. Two fully regular planar curves are equivalent under an oriented rigid motion if and only if they have the same signature curve. The maximally symmetric curves have constant Euclidean curvature, and so their signature curve degenerates to a single point. These are the circles and straight lines, and, in accordance with Theorem 8.10, each is the orbit of its one-parameter symmetry subgroup of SE(2). The number of Euclidean symmetries of a nonsingular curve is equal to its index – the number of times the Euclidean signature is retraced as we go around the curve.

In Figure 8.1 we display some signature curves computed from the left ventricle of a gray-scale digital MRI scan of a canine heart. The boundary of the ventricle has been automatically segmented through use of the conformally Riemannian snake flow proposed in [51, 96]. The ventricle boundary curve is then smoothed with the Euclidean-invariant curve shortening flow (see the final section for details) and the Euclidean signatures of the resulting curves computed. As the progressively smoothed curves approach circularity, their signatures exhibit less variation in curvature and wind more and more tightly around a single point, which is the signature of a circle of area equal to the area inside the evolving curve. Despite the rather extensive smoothing involved, except for an overall shrinking as the contour approaches circularity, the basic qualitative features of the different signature curves appear to be remarkably robust.

8.5 Joint invariants and joint differential invariants

As always, the starting point is the action of a Lie group G on a manifold M. Consider the *joint action*

$$g \cdot (z_0, \ldots, z_n) = (g \cdot z_0, \ldots, g \cdot z_n), \qquad g \in G, \quad z_0, \ldots, z_n \in M. \qquad (8.16)$$

on the $(n + 1)$-fold Cartesian product $M^{\times(n+1)} = M \times \cdots \times M$. An invariant $I(z_0, \ldots, z_n)$ of (8.16) is an $(n + 1)$-*point joint invariant* of the original transformation group. In most cases of interest (although not in general), if G acts effectively on M, then, for $n \gg 0$ sufficiently large, the product action is free and regular on an open subset of $M^{\times(n+1)}$, cf. [78]. Consequently, the equivariant moving frame method can be applied to such joint actions, and thereby establish complete classifications of joint invariants and, via prolongation to Cartesian products of jet spaces, joint differential invariants.

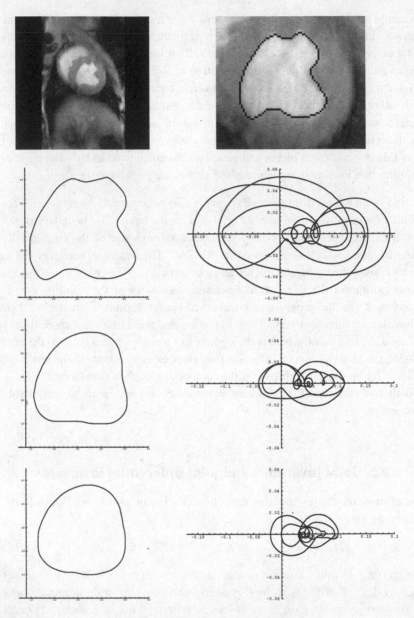

Figure 8.1 Signature of a canine ventricle

Example 8.12   Consider the Euclidean group SE(2) acting on curves $C \subset M = \mathbb{R}^2$. For the Cartesian product action on $M^{\times n}$, we take the simplest cross-section $\mathcal{K} = \{x_0 = u_0 = x_1 = 0, u_1 > 0\}$ leads to the normalization equations

$$y_0 = x_0 \cos\phi - u_0 \sin\phi + a = 0, \qquad v_0 = x_0 \sin\phi + u_0 \cos\phi + b = 0,$$
$$y_1 = x_1 \cos\phi - u_1 \sin\phi + a = 0. \tag{8.17}$$

Solving, we obtain a right moving frame

$$\phi = \tan^{-1}\left(\frac{x_1 - x_0}{u_1 - u_0}\right),$$
$$a = -x_0 \cos\phi + u_0 \sin\phi, \qquad b = -x_0 \sin\phi - u_0 \cos\phi, \tag{8.18}$$

along with the fundamental interpoint distance invariant

$$I = \iota(u_1) = \|z_1 - z_0\|. \tag{8.19}$$

Substituting (8.18) into the prolongation formulae (8.9) leads to the the normalized first and second order joint differential invariants

$$J_k = \iota\left(\frac{du_k}{dx}\right) = -\frac{(z_1 - z_0) \cdot \dot{z}_k}{(z_1 - z_0) \wedge \dot{z}_k},$$
$$K_k = \iota\left(\frac{d^2 u_k}{dx^2}\right) = -\frac{\|z_1 - z_0\|^3 (\dot{z}_k \wedge \ddot{z}_k)}{[(z_1 - z_0) \wedge \dot{z}_0]^3}, \tag{8.20}$$

where the dots indicate derivatives of $z_k(t_k)$ with respect to the curve parameter t_k.

Theorem 8.13 *If $n \geq 2$, then every n-point joint Euclidean differential invariant is a function of the interpoint distances $\|z_i - z_j\|$ and their iterated derivatives with respect to the invariant differential operators*

$$\mathcal{D}_k = \iota(D_{t_k}) = -\frac{\|z_1 - z_0\|}{(z_1 - z_0) \wedge \dot{z}_k} D_{t_k}.$$

Consequently, to create a Euclidean signature based entirely on joint invariants, we can take four points z_0, z_1, z_2, z_3 on our curve $C \subset \mathbb{R}^2$. As illustrated in Figure 8.2, there are six different interpoint distance invariants

$$a = \|z_1 - z_0\|, \qquad b = \|z_2 - z_0\|, \qquad c = \|z_3 - z_0\|,$$
$$d = \|z_2 - z_1\|, \qquad e = \|z_3 - z_1\|, \qquad f = \|z_3 - z_2\|, \tag{8.21}$$

which parametrize the joint signature $\widehat{S} = \widehat{S}(C)$ that uniquely characterizes the curve C up to Euclidean motion. Since this signature avoids any differentiation, it is insensitive to noisy image data. There are two local syzygies

$$\Phi_1(a, b, c, d, e, f) = 0, \qquad \Phi_2(a, b, c, d, e, f) = 0, \tag{8.22}$$

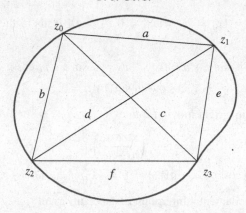

Figure 8.2 Four-point Euclidean curve invariants

among the the six interpoint distances. One of these is the universal *Cayley–Menger syzygy*

$$\det \begin{vmatrix} 2a^2 & a^2 + b^2 - d^2 & a^2 + c^2 - e^2 \\ a^2 + b^2 - d^2 & 2b^2 & b^2 + c^2 - f^2 \\ a^2 + c^2 - e^2 & b^2 + c^2 - f^2 & 2c^2 \end{vmatrix} = 0, \qquad (8.23)$$

which is valid for all possible planar configurations of the four points, cf. [8]. The second syzygy in (8.22) is curve-dependent and serves to effectively characterize the joint invariant signature.

A variety of additional examples, including curves and surfaces in two- and three-dimensional space under the Euclidean, equi-affine, affine and projective groups, are investigated in detail in [78].

8.6 Invariant numerical approximations

In modern numerical analysis, the development of numerical schemes that incorporate additional structure enjoyed by the problem being approximated, e.g., symmetries, conservation laws, symplectic structure, etc., is now known as *geometric numerical integration* [19, 30, 41, 65]. In practical applications of invariant theory to computer vision, group-invariant numerical schemes to approximate differential invariants have been applied to the problem of symmetry-based object recognition [10, 20, 21]. In this section, I discuss the use of moving frame methods to construct symmetry-preserving numerical approximations.

The first step is to construct a suitable manifold that incorporates both the differential equation under consideration and its numerical approximations.

Currently, only the case of ordinary differential equations, involving $p = 1$ independent variables, is completely understood, and so we restrict ourselves to this context.

Finite difference approximations to the derivatives of a function $u = f(x)$ rely on its values $u_0 = f(x_0), \ldots, u_n = f(x_n)$ at several distinct points $z_i = (x_i, u_i) = (x_i, f(x_i))$ on the graph. Thus, discrete approximations to jet coordinates on J^n are functions $F(z_0, \ldots, z_n)$ defined on the $(n + 1)$-fold Cartesian product space $M^{\times(n+1)}$. As the points $z_0, \ldots, z_n$ coalesce, the approximation $F(z_0, \ldots, z_n)$ will not be well-defined unless we specify the "direction" of convergence. Thus, strictly speaking, F is not defined on all of $M^{\times(n+1)}$, but, rather, on the "off-diagonal" part

$$M^{\diamond(n+1)} = \{(z_0, \ldots, z_n) \mid z_i \neq z_j \text{ for all } i \neq j\} \subset M^{\times(n+1)}.$$

As two or more points come together, the limiting value of $F(z_0, \ldots, z_n)$ will be governed by the derivatives (or jet) of the appropriate order governing the direction of convergence. This motivates our construction of the *nth order multi-space $M^{(n)}$*.

Definition 8.14 An $(n + 1)$-*pointed curve* $\mathbf{C} = (z_0, \ldots, z_n; C)$ consists of a smooth curve C and $n + 1$ not necessarily distinct points $z_0, \ldots, z_n \in C$ thereon. Two $(n + 1)$-pointed curves $\mathbf{C} = (z_0, \ldots, z_n; C)$, $\widetilde{\mathbf{C}} = (\tilde{z}_0, \ldots, \tilde{z}_n; \tilde{C})$, have *nth order multi-contact* if and only if $z_i = \tilde{z}_i$, and $j_{\#i-1}C|_{z_i} = j_{\#i-1}\widetilde{C}|_{z_i}$, where $\#i = \#\{j \mid z_j = z_i\}$, for each $i = 0, \ldots, n$.

Definition 8.15 The *nth order multi-space*, denoted $M^{(n)}$ is the set of equivalence classes of $(n + 1)$-pointed curves in M under the equivalence relation of *nth order multi-contact*. The equivalence class of an $(n + 1)$-pointed curves $\mathbf{C}$ is called its *nth order multi-jet*, and denoted $\mathbf{j}_n\mathbf{C} \in M^{(n)}$.

We can identify the subset of multi-jets of multi-pointed curves having distinct points with the off-diagonal Cartesian product space $M^{\diamond(n+1)} \subset J^n$. On the other hand, the multi-space equivalence relation reduces to the ordinary jet space equivalence relation on the set of coincident multi-pointed curves, and in this way $J^n \subset M^{(n)}$. Intermediate cases, when some but not all points coincide, correspond to "off-diagonal" Cartesian products of jet spaces

$$J^{k_1} \diamond \cdots \diamond J^{k_l} \equiv \{(z_0^{(k_1)}, \ldots, z_i^{(k_l)}) \in J^{k_1} \times \cdots \times J^{k_l} \mid \pi(z_\nu^{(k_\nu)}) \text{ are distinct}\}, \quad (8.24)$$

where $\sum k_\nu = n$ and $\pi \colon J^k \to M$ is the usual jet space projection.

Theorem 8.16 *If M is a smooth m-dimensional manifold, then its nth order multi-space $M^{(n)}$ is a smooth manifold of dimension $(n + 1)m$, which contains the off-diagonal part $M^{\diamond(n+1)}$ of the Cartesian product space as an open, dense submanifold, and the nth order jet space J^n as a smooth submanifold.*

Just as local coordinates on J^n are provided by the coefficients of Taylor polynomials, local coordinates on $M^{(n)}$ are provided by the coefficients of interpolating polynomials, which are the classical divided differences of numerical interpolation theory [77].

Definition 8.17 Given an $(n+1)$-pointed graph $\mathbf{C} = (z_0, \ldots, z_n; C)$, its divided differences are defined by $[z_j]_C = f(x_j)$, and

$$[z_0 z_1 \ldots z_{k-1} z_k]_C = \lim_{z \to z_k} \frac{[z_0 z_1 z_2 \ldots z_{k-2} z]_C - [z_0 z_1 z_2 \ldots z_{k-2} z_{k-1}]_C}{x - x_{k-1}}. \qquad (8.25)$$

When taking the limit, the point $z = (x, f(x))$ must lie on the curve C, and take limiting values $x \to x_k$ and $f(x) \to f(x_k)$.

It is not hard to show that two $(n + 1)$-pointed graphs $\mathbf{C}, \widetilde{\mathbf{C}}$ have nth order multi-contact if and only if they have the same divided differences: $[z_0 \ldots z_k]_C = [z_0 z_1 \ldots z_k]_{\widetilde{C}}$ for all $k = 0, \ldots, n$. Therefore, the required local coordinates on multi-space $M^{(n)}$ consist of the independent variables along with all the divided differences

$$\begin{aligned} x_0, \ldots, x_n, \qquad & u^{(0)} = u_0 = [z_0]_C, \\ u^{(1)} = [z_0 z_1]_C, \qquad \cdots \qquad & u^{(n)} = n![z_0 z_1 \ldots z_n]_C. \end{aligned} \qquad (8.26)$$

The $n!$ factor is included so that $u^{(n)}$ agrees with the usual derivative coordinate when restricted to J^n.

In general, implementation of a finite difference numerical solution scheme for a system of ordinary differential equations

$$\Delta_1(x, u, u^{(1)}, \ldots, u^{(n)}) = \cdots = \Delta_k(x, u, u^{(1)}, \ldots, u^{(n)}) = 0, \qquad (8.27)$$

requires suitable discrete approximations to each of its defining differential functions Δ_ν. This requires extending the differential functions from the jet space to the associated multi-space, in accordance with the following definition.

Definition 8.18 An $(n+1)$-point *numerical approximation of order k* to a differential function $\Delta \colon J^n \to \mathbb{R}$ is an function $F \colon M^{(n)} \to \mathbb{R}$ that, when restricted to the jet space, agrees with Δ to order k.

Now let us consider an r-dimensional Lie group G which acts smoothly on M. Since G evidently maps multi-pointed curves to multi-pointed curves while preserving the multi-contact equivalence relation, it induces an action on the multi-space $M^{(n)}$ that will be called the nth *multi-prolongation* of G and denoted by $G^{(n)}$. On the jet subset $J^n \subset M^{(n)}$ the multi-prolonged action reduces to the usual jet space prolongation. On the other hand, on the off-diagonal part

$M^{\circ(n+1)} \subset M^{(n)}$ the action coincides with the $(n + 1)$-fold Cartesian product action of G on $M^{\times(n+1)}$.

We define a *multi-invariant* to be a function $K: M^{(n)} \to \mathbb{R}$ on multi-space which is invariant under the multi-prolonged action of $G^{(n)}$. The restriction of a multi-invariant K to jet space will be a differential invariant, $I = K|J^n$, while restriction to $M^{\circ(n+1)}$ will define a joint invariant $J = K|M^{\circ(n+1)}$. Restriction to intermediate multi-jet subspaces (8.24) will produce joint differential invariants. Smoothness of K will imply that the joint invariant J is an *invariant nth order numerical approximation to the differential invariant I*. Moreover, every invariant finite difference numerical approximation arises in this manner. Thus, the theory of multi-invariants *is* the theory of invariant numerical approximations!

Assuming regularity and freeness of the multi-prolonged action on an open subset of $M^{(n)}$, we can apply the equivariant moving frame construction. The resulting *multi-frame* $\rho^{(n)}: M^{(n)} \to G$ will lead us immediately to the required multi-invariants and hence a general, systematic construction for invariant numerical approximations to differential invariants through its induced invariantization procedure. The basic multi-invariants are

$$(H_i, K_i) = I_i \equiv \iota(z_i) = (\iota(r_i), \iota(u_i)), \qquad i = 0, \ldots, n, \tag{8.28}$$

and their divided differences

$$I^{(k)} = \iota([z_0 z_1 \ldots z_k]) = [I_0 \ldots I_k] = \frac{[I_0 \ldots I_{k-2} I_k] - [I_0 \ldots I_{k-2} I_{k-1}]}{H_k - H_{k-1}}. \tag{8.29}$$

Example 8.19 For the planar Euclidean action (8.7), the multi-prolonged action is locally free on $M^{(n)}$ for $n \geq 1$. We can thereby determine a first order multi-frame and use it to completely classify Euclidean multi-invariants. The first order transformation formulae are

$$y_0 = x_0 \cos \phi - u_0 \sin \phi + a, \qquad v_0 = x_0 \sin \phi + u_0 \cos \phi + b,$$

$$y_1 = x_1 \cos \phi - u_1 \sin \phi + a, \qquad v^{(1)} = \frac{\sin \phi + u^{(1)} \cos \phi}{\cos \phi - u^{(1)} \sin \phi}, \tag{8.30}$$

where $u^{(1)} = [z_0 z_1] = (u_1 - u_0)/(x_1 - x_0)$. Normalization based on the cross-section $y_0 = v_0 = v_1 = 0$ results in the right moving frame

$$a = -x_0 \cos \phi + u_0 \sin \phi = -\frac{x_0 + u^{(1)} u_0}{\sqrt{1 + (u^{(1)})^2}},$$

$$\tan \phi = -u^{(1)}. \tag{8.31}$$

$$b = -x_0 \sin \phi - u_0 \cos \phi = \frac{x_0 u^{(1)} - u_0}{\sqrt{1 + (u^{(1)})^2}},$$

Substituting the moving frame formulae (8.31) into the lifted divided differences produces a complete system of (oriented) Euclidean multi-invariants. These are easily computed by beginning with the fundamental joint invariants

$$H_k = \iota(x_k) = \frac{(x_k - x_0) + u^{(1)}(u_k - u_0)}{\sqrt{1 + (u^{(1)})^2}} = (x_k - x_0)\frac{1 + [z_0z_1][z_0z_k]}{\sqrt{1 + [z_0z_1]^2}},$$

$$K_k = \iota(u_k) = \frac{(u_k - u_0) - u^{(1)}(x_k - x_0)}{\sqrt{1 + (u^{(1)})^2}} = (x_k - x_0)\frac{[z_0z_k] - [z_0z_1]}{\sqrt{1 + [z_0z_1]^2}}.$$

The higher order multi-invariants are obtained by forming divided difference quotients

$$[I_0 I_k] = \frac{K_k - K_0}{H_k - H_0} = \frac{K_k}{H_k} = \frac{(x_k - x_1)[z_0z_1z_k]}{1 + [z_0z_k][0z_1]}$$

where, in particular, $I^{(1)} = [I_0 I_1] = 0$. The second order multi-invariant

$$I^{(2)} = 2[I_0 I_1 I_2] = 2\frac{[I_0 I_2] - [I_0 I_1]}{H_2 - H_1} = \frac{2[z_0z_1z_2]\sqrt{1 + [z_0z_1]^2}}{(1 + [z_0z_1][z_1z_2])(1 + [z_0z_1][z_0z_2])}$$

$$= \frac{u^{(2)}\sqrt{1 + (u^{(1)})^2}}{[1 + (u^{(1)})^2 + u^{(1)}u^{(2)}(x_2 - x_0)/2][1 + (u^{(1)})^2 + u^{(1)}u^{(2)}(x_2 - x_1)/2]}$$

provides a Euclidean-invariant numerical approximation to the Euclidean curvature:

$$\lim_{z_1, z_2 \to z_0} I^{(2)}2 = \kappa = \frac{u^{(2)}}{(1 + (u^{(1)})^2)^{3/2}}.$$

Similarly, the third order multi-invariant

$$I^{(3)} = 6[I_0 I_1 I_2 I_3] = 6\frac{[I_0 I_1 I_3] - [I_0 I_1 I_2]}{H_3 - H_2}$$

will form a Euclidean-invariant approximation for the normalized differential invariant $\kappa_s = \iota(u_{xxx})$, the derivative of curvature with respect to arc length [21, 32]. In [27], my undergraduate student Derek Dalle makes detailed comparisons between the various divided difference approximations to differential invariants, and shows a number of advantages of such moving frame-based approximations.

Given a symmetry group of an ordinary differential equation, we can construct a moving frame on the associated multispace and apply the induced invariantization procedure to standard numerical schemes, e.g., Runge–Kutta methods, to systematically derive invariantized schemes that respect the symmetries. As emphasized by Pilwon Kim [52–54], the key to the success of the invariantized numerical scheme lies in the intelligent choice of cross-section

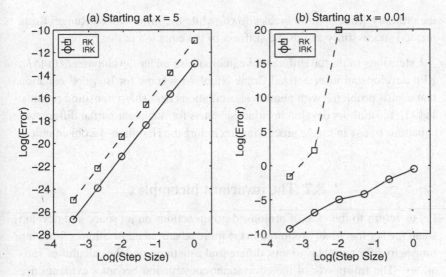

Figure 8.3 Invariantized Runge–Kutta schemes for Ames' equation

for the moving frame. Let us look at one simple illustrative example taken from [52].

Example 8.20 (Ames' equation)

$$u_{xx} = -\frac{u_x}{x} - e^u \qquad (8.32)$$

is a well-studied stiff ordinary differential equation that arises in a wide range of fields, including kinetics and heat transfer, vortex motion of incompressible fluids, and the mass distribution of gaseous interstellar material under influence of its own gravitational fields [3]. The infinitesimal generators

$$v_1 = -x\frac{\partial}{\partial x} + 2\frac{\partial}{\partial u}, \qquad v_2 = -\frac{1}{2}x\log x\frac{\partial}{\partial x} + (1 + \log x)\frac{\partial}{\partial u},$$

induce one-parameter symmetry groups mapping (x, u) to

$$(e^{\varepsilon_1}x, u + 2\varepsilon_1), \qquad (x^{e^{-\varepsilon_2/2}}, u + 2(1 - e^{-\varepsilon_2/2})\log x + \varepsilon_2),$$

respectively. Individually, neither group leads to a significant improvement in the integration scheme, but a suitable combination induces a moving frame that maps every point (x, u) to the cross-section $\{u = 0\}$. Figure 8.3 compares the Runge–Kutta and the invariantized Runge–Kutta schemes starting at $x = 5$. Even in this region, the invariantized scheme outperforms the standard scheme. A more dramatic effect appears when they are applied around $x = 0$, where the ordinary Runge–Kutta breaks down, while the invariantized Runge–Kutta method successfully avoids the stiffness of the equation in that regime.

Extensions to partial differential equations are under development. In [53], Kim develops an invariantized Crank-Nicolson scheme for Burgers' equation that avoids problems with numerical oscillations near sharp transition regions. In [95], the authors develop invariant schemes for nonlinear partial differential equations of use in image processing, including the Hamilton–Jacobi equation.

8.7 The invariant bicomplex

Let us return to the case of prolonged group actions on jet space and develop some further machinery required in the more advanced applications of moving frames to differential invariants, differential equations, and the calculus of variations. The full power of the equivariant construction becomes evident once we incorporate the contact structure and induced variational bicomplex on the infinite order jet bundle $J^\infty = J^\infty(M, p)$, which we now review, [4, 74].

Separating the local coordinates $(x, u) = (x^1, \ldots, x^p, u^1, \ldots, u^q)$ on M into independent and dependent variables naturally splits[1] the differential one-forms on J^∞ into *horizontal forms*, spanned by $dx^1, \ldots, dx^p$, and *vertical forms*, spanned by the *basic contact one-forms*

$$\theta_J^\alpha = du_J^\alpha - \sum_{i=1}^p u_{J,i}^\alpha \, dx^i, \qquad \alpha = 1, \ldots, q, \quad \# J \geq 0. \tag{8.33}$$

Let π_H and π_V denote the projections mapping one-forms on J^∞ to their horizontal and vertical (contact) components, respectively. We accordingly decompose the differential $d = \pi_H \circ d + \pi_V \circ d = d_H + d_V$, which results in the *variational bicomplex* on J^∞. If $F(x, u^{(n)})$ is any differential function, its horizontal differential is

$$d_H F = \sum_{i=1}^p (D_i F) \, dx^i, \tag{8.34}$$

in which $D_i = D_{x^i}$ denote the usual total derivatives with respect to the independent variables. Thus, $d_H F$ can be identified with the "total gradient" of F. Similarly, its vertical differential is

$$d_V F = \sum_{\alpha, J} \frac{\partial F}{\partial u_J^\alpha} \theta_J^\alpha = \sum_{\alpha, J} \frac{\partial F}{\partial u_J^\alpha} D_J \theta^\alpha = D_F(\theta), \tag{8.35}$$

---

[1] The splitting, which depends on the choice of local coordinates, only works at infinite order, which is the reason we work on J^∞.

in which the total derivatives act as Lie derivatives on the contact forms $\theta = (\theta^1, \ldots, \theta^q)^T$, and D_F denotes the *formal linearization operator* or *Fréchet derivative* of the differential function F. Thus, the vertical differential $d_V F$ can be identified[2] with the (first) *variation*, hence the name "variational bicomplex."

Let $\pi_n \colon J^\infty \to J^n$ be the natural jet space projections. Choosing a cross-section $\mathcal{K}^n \subset \mathcal{V}^n \subset J^n$, we extend the induced nth order moving frame $\rho^{(n)}$ to the infinite jet bundle by setting $\rho(x, u^{(\infty)}) = \rho^{(n)}(x, u^{(n)})$ whenever $(x, u^{(n)}) = \pi_n(x, u^{(\infty)})$ lies in the domain of definition of $\rho^{(n)}$. We will employ our moving frame to *invariantize* the variational bicomplex. As before, the invariantization of a differential form is the unique invariant differential form that agrees with its progenitor on the cross-section. In particular, the invariantization process does not affect invariant differential forms. In practice, one determines the invariantization by first transforming the differential form by the prolonged group action and then substituting the moving frame formulae for the group parameters.

As in (8.6), the fundamental differential invariants are obtained by invariantizing the jet coordinates: $H^i = \iota(x^i)$, $I_J^\alpha = \iota(u_J^\alpha)$. Let

$$\varpi^i = \iota(dx^i) = \omega^i + \eta^i, \quad \text{where} \quad \omega^i = \pi_H(\varpi^i), \quad \eta^i = \pi_V(\varpi^i), \quad (8.36)$$

denote the *invariantized horizontal one-forms*. Their horizontal components $\omega^1, \ldots, \omega^p$ prescribe, in the language of [74], a contact-invariant coframe for the prolonged group action, while the contact forms $\eta^1, \ldots, \eta^p$ are required to make $\varpi^1, \ldots, \varpi^n$ fully G-invariant. Finally, the *invariantized basis contact forms* are denoted by

$$\vartheta_J = \iota(\theta_J^\alpha), \qquad \alpha = 1, \ldots, q, \quad \#J \geq 0. \tag{8.37}$$

Invariantization of more general differential forms relies on the fact that it preserves the exterior algebra structure, and so

$$\iota(\Omega + \Psi) = \iota(\Omega) + \iota(\Psi), \qquad \iota(\Omega \wedge \Psi) = \iota(\Omega) \wedge \iota(\Psi), \tag{8.38}$$

for any differential forms (or functions) Ω, Ψ on J^∞.

As in the ordinary bicomplex construction, the decomposition of invariant one-forms on J^∞ into invariant horizontal and invariant contact components induces a decomposition of the differential. However, now $d = d_{\mathcal{H}} + d_V + d_W$ splits into three constituents, where $d_{\mathcal{H}}$ adds an invariant horizontal form, d_V adds a invariant contact form, while d_W replaces an invariant horizontal one-form with a combination of wedge products of two invariant contact forms.

---

[2] This becomes clearer when you rewrite $\theta_j^\alpha = \delta u_j^\alpha$.

They satisfy the "quasi-tricomplex" identities

$$d_{\mathcal{H}}^2 = 0, \qquad d_{\mathcal{H}}\,d_{\mathcal{V}} + d_{\mathcal{V}}\,d_{\mathcal{H}} = 0,$$
$$d_{\mathcal{W}}^2 = 0, \qquad d_{\mathcal{V}}\,d_{\mathcal{W}} + d_{\mathcal{W}}\,d_{\mathcal{V}} = 0, \qquad d_{\mathcal{V}}^2 + d_{\mathcal{H}}\,d_{\mathcal{W}} + d_{\mathcal{W}}\,d_{\mathcal{H}} = 0. \quad (8.39)$$

Fortunately, the third, anomalous component $d_{\mathcal{W}}$ plays no role (to date) in the applications; in particular, $d_{\mathcal{W}}F = 0$ for any differential function F. Even better, if the group acts projectably, $d_{\mathcal{W}} \equiv 0$. The corresponding dual invariant differential operators $\mathcal{D}_1, \ldots, \mathcal{D}_p$ are then defined so that

$$d_{\mathcal{H}}F = \sum_i^p (\mathcal{D}_i F)\varpi^i, \quad d_{\mathcal{H}}\Omega = \sum_i^p \varpi^i \wedge \mathcal{D}_i\,\Omega, \qquad (8.40)$$

for any differential function F and, more generally, differential form Ω, on which the $\mathcal{D}_i$ act via Lie differentiation. Keep in mind that, in general, the invariant differential operators do not commute; see (8.49) below.

The most important fact underlying the moving frame construction is that, while it does preserve algebraic structure, the invariantization map ι does *not* respect the differential. The *recurrence formulae* [32, 57] which we now review, provide the missing "correction terms", i.e., $d\iota(\Omega) - \iota(d\Omega)$. Remarkably, they can be explicitly and algorithmically constructed using merely linear differential algebra – without knowing the explicit formulas for either the differential invariants or invariant differential forms, the invariant differential operators, or even the moving frame!

Let $v_1, \ldots, v_r$ be a basis for the infinitesimal generators of our transformation group. For conciseness, we will retain the same notation for the corresponding prolonged vector fields on J^∞ which, in local coordinates, take the form

$$v_\kappa = \sum_i^p \xi_\kappa^i(x, u)\frac{\partial}{\partial x^i} + \sum_{\alpha=1}^q \sum_{j=\#J\geq 0} \varphi_{J,\kappa}^\alpha(x, u^{(j)})\frac{\partial}{\partial u_J^\alpha}, \qquad \kappa = 1, \ldots, r. \quad (8.41)$$

The coefficients $\varphi_{J,\kappa}^\alpha = v_\kappa(u_J^\alpha)$ can be successively constructed by Lie's recursive prolongation formula [73, 74]:

$$\varphi_{Ji,\kappa}^\alpha = D_i\varphi_{J,\kappa}^\alpha - \sum_{j=1}^p u_{Jj}^\alpha D_i\xi_\kappa^j. \qquad (8.42)$$

With this in hand, we can formulate the *universal recurrence formula*.

Theorem 8.21 *If Ω is any differential function or form on J^∞, then*

$$d\iota(\Omega) = \iota(d\Omega) + \sum_{\kappa=1}^r \nu^\kappa \wedge \iota[v_\kappa(\Omega)], \qquad (8.43)$$

where $v^1, \ldots, v^r$ are the invariantized Maurer–Cartan forms *dual to the infinitesimal generators* $v_1, \ldots, v_r$, *while* $v_\kappa(\Omega)$ *denotes the corresponding Lie derivative of* Ω.

In general, the invariantized Maurer–Cartan forms are obtained by pulling back the dual Maurer–Cartan forms $\mu^1, \ldots, \mu^r$ on G via the moving frame map: $v^\kappa = \rho^* \mu^\kappa$. The full details [57] are, fortunately, not required thanks to the following marvelous result that allows us to compute them directly without reference to their underlying definition:

Proposition 8.22 *Let* $\mathcal{K} = \{Z_1(x, u^{(n)}) = c_1, \ldots, Z_r(x, u^{(n)}) = c_r\}$ *be the cross section defining our moving frame, so that* $c_\lambda = \iota(Z_\lambda)$ *are the phantom differential invariants. Then the corresponding* phantom recurrence formulae

$$0 = d\iota(Z_\lambda) = \iota(dZ_\lambda) + \sum_{kappa=1}^{r} v^\kappa \wedge \iota[v_\kappa(Z_\lambda)], \qquad \lambda = 1, \ldots, r, \qquad (8.44)$$

can be uniquely solved for the invariantized Maurer–Cartan forms:

$$nu^\kappa = \sum_{i=1}^{p} R_i^\kappa \varpi^i + \sum_{\alpha, J} S_\alpha^{\kappa, J} \vartheta_J, \qquad (8.45)$$

where $R_i^\kappa, S_\alpha^{\kappa, J}$ *are certain differential invariants.*

The R_i^κ are called the *Maurer–Cartan invariants* [44, 79]. In the case of curves, $p = 1$, they are the entries of the Frenet–Serret matrix $\mathcal{D}\rho^{(n)}(x, u^{(n)}) \cdot \rho^{(n)}(x, u^{(n)})^{-1}$, cf. [40].

Substituting (8.45) into the universal formula (8.43) produces a complete system of explicit recurrence relations for all the differentiated invariants and invariant differential forms. In particular, taking Ω to be any one of the individual jet coordinate functions x^i, u_j^α, results in the recurrence formulae for the fundamental differential invariants (8.6):

$$\mathcal{D}_i H^j = \delta_i^j + \sum_{\kappa=1}^{r} R_i^\kappa \iota(\xi_\kappa^i), \mathcal{D}_i I_J^\alpha = I_{Ji}^\alpha + \sum_{\kappa=1}^{r} R_i^\kappa \iota(\varphi_{J,\kappa}^\alpha), \qquad (8.46)$$

where δ_i^j is the usual Kronecker delta, and $\xi_\kappa^i, \varphi_{J,\kappa}^\alpha$ are the coefficients of the prolonged infinitesimal generators (8.41). Owing to the functional independence of the non-phantom differential invariants, these formulae, in fact, serve to completely prescribe the structure of the non-commutative differential invariant algebra engendered by G [32, 45, 79].

Similarly, the recurrence formulae (8.43) for the invariant horizontal forms are

$$d\varpi^i = d[\iota(dx^i)] = \iota(d^2x^i) + \sum_{\kappa=1}^{r} v^\kappa \wedge \iota[v_\kappa(dx^i)]$$

$$= \sum_{\kappa=1}^{r}\sum_{k=1}^{p} \iota(D_k\xi_\kappa^i)v^\kappa \wedge \varpi^k + \sum_{\kappa=1}^{r}\sum_{\alpha=1}^{q} \iota\left(\frac{\partial\xi_\kappa^i}{\partial u^\alpha}\right)v^\kappa \wedge \vartheta^\alpha. \quad (8.47)$$

The terms in (8.47) involving wedge products of two horizontal forms are

$$d_{\mathcal{H}}\varpi^i = -\sum_{j<k} Y^i_{jk}\varpi^j \wedge \varpi^k,$$

where

$$Y^i_{jk} = -Y^i_{kj} = \sum_{\kappa=1}^{r}\sum_{j=1}^{p} R^\kappa_j\iota(D_j\xi_\kappa^i) - R^\kappa_k\iota(D_k\xi_\kappa^i) \quad (8.48)$$

are called the *commutator invariants*, since they prescribe the commutators of the invariant differential operators:

$$[\mathcal{D}_j, \mathcal{D}_k] = \sum_{i=1}^{p} Y^i_{jk}\mathcal{D}_i. \quad (8.49)$$

The terms in (8.47) involving wedge products of a horizontal and a contact form yield

$$d_{\mathcal{V}}\varpi^i = \sum_{\kappa=1}^{r}\left[\sum_{\alpha=1}^{q} \iota\left(\frac{\partial\xi_\kappa^i}{\partial u^\alpha}\right)R^\kappa_j\varpi^i \wedge \vartheta^\alpha - \sum_{k=1}^{p} \iota(D_k\xi_\kappa^i)S^{\kappa,J}_\alpha\varpi^k \wedge \vartheta_J\right]. \quad (8.50)$$

Finally, the remaining terms, involving wedge products of two contact forms, provide the formulas for the anomalous third component of the differential:

$$d_{\mathcal{W}}\varpi^i = \sum_{\kappa=1}^{r}\sum_{\alpha=1}^{q} \iota\left(\frac{\partial\xi_\kappa^i}{\partial u^\alpha}\right)S^{\kappa,J}_\alpha\vartheta_J \wedge \vartheta^\alpha. \quad (8.51)$$

In a similar fashion, we derive the recurrence formulae (8.43) for the differentiated invariant contact forms: In particular, the horizontal components

$$\mathcal{D}_i\vartheta_J = \vartheta_{Ji} + \sum_{\kappa=1}^{r} R^\kappa_i\iota(v_\kappa(\theta^\alpha_J)). \quad (8.52)$$

can be inductively solved to express the higher order invariantized contact forms as certain invariant derivatives of those of order 0:

$$\vartheta_J = \mathcal{E}^\alpha_J(\vartheta) = \sum_{\beta=1}^{q} \mathcal{E}^\alpha_{J,\beta}(\vartheta^\beta), \quad (8.53)$$

in which $\vartheta = (\vartheta^1, \ldots, \vartheta^q)^T$ denotes the column vector containing the order zero invariantized contact forms, while $\mathcal{E}_J^\alpha = (\mathcal{E}_{J,1}^\alpha, \ldots, \mathcal{E}_{J,q}^\alpha)$ is a row vector of invariant differential operators, i.e., each $\mathcal{E}_{J,\alpha} = \sum A_{J,\alpha}^K \mathcal{D}^K$ for certain differential invariants $A_{J,\alpha}^K$.

Combining these formulae allows us to express the invariant vertical derivative or *invariant variation* of any differential invariant K in the form

$$d_\mathcal{V} K = \mathcal{A}_K(\vartheta), \tag{8.54}$$

in which $\mathcal{A}_K$ is a row vector of invariant differential operators. Formula (8.54) can be viewed as the invariant version of the vertical differentiation formula (8.35), and so will refer to $\mathcal{A}_K$ as the *invariant linearization operator* associated with the differential invariant K. Similarly, we derive the recurrence formulae for the vertical differentials of the invariant horizontal forms:

$$d_\mathcal{V} \varpi^i = \sum_{j=1}^{p} \sum_{\alpha=1}^{q} \mathcal{B}_{j\alpha}^i(\vartheta^\alpha) \wedge \varpi^j = \sum_{j=1}^{p} \mathcal{B}_j^i(\vartheta) \wedge \varpi^j \tag{8.55}$$

in which $\mathcal{B}_j^i = (\mathcal{B}_{j1}^i, \ldots, \mathcal{B}_{jq}^i)$ is a family of p^2 row-vector-valued invariant differential operators, known, collectively, as the *invariant Hamiltonian operator complex*, stemming from its role in the calculus of variations, cf [57, 88].

Example 8.23   Let us return to the Euclidean group acting on plane curves initiated in Example 8.5. The basic invariant horizontal one-form $\varpi = \iota(dx)$ is obtained by first transforming dx by a general group element:

$$dx \mapsto dy = (\cos \phi - u_x \sin \phi) \, dx + (\sin \phi)\theta, \tag{8.56}$$

where

$$\theta = du - u_x \, dx, \qquad \theta_x = du_x - u_{xx} \, dx, \qquad \ldots, \tag{8.57}$$

are the ordinary basis contact forms. Substituting the moving frame formulae (8.11) for the group parameters into (8.56) yields the basic invariant horizontal one-form

$$\varpi = \iota(dx) = \frac{dx + u_x \, du}{\sqrt{1 + u_x^2}} = \sqrt{1 + u_x^2} \, dx + \frac{u_x}{\sqrt{1 + u_x^2}} \theta. \tag{8.58}$$

Its (noninvariant) horizontal component is the contact-invariant arc length form

$$\omega = \pi_H(\varpi) = ds = \sqrt{1 + u_x^2} \, dx,$$

and so the corresponding invariant differential operator is the usual arc length derivative $\mathcal{D} = D_s$. In the same manner we obtain the basis invariant contact forms

$$\vartheta = \iota(\theta) = \frac{\theta}{\sqrt{1 + u_x^2}}, \qquad \vartheta_1 = \iota(\theta_x) = \frac{(1 + u_x^2)\theta_x - u_x u_{xx}\theta}{(1 + u_x^2)^2}, \qquad \cdots \quad (8.59)$$

To construct the recurrence formulae for the differentiated functions and forms, we begin with the prolonged infinitesimal generators of SE(2):

$$v_1 = \partial_x, \qquad v_2 = \partial_u,$$
$$v_3 = -u\partial_x + x\partial_u + (1 + u_x^2)\partial_{u_x} + 3u_x u_{xx}\partial_{u_{xx}} + (4u_x u_{xxx} + 3u_{xx}^2)\partial_{u_{xxx}} + \cdots .$$

The pulled back dual Maurer–Cartan forms v^1, v^2, v^3 are found by applying the universal recurrence formulae (8.43) to the phantom invariants:

$$0 = dH = \iota(dx) + \iota(v_1(x))v^1 + \iota(v_2(x))v^2 + \iota(v_3(x))v^3 = \varpi + v^1,$$
$$0 = dI_0 = \iota(du) + \iota(v_1(u))v^1 + \iota(v_2(u))v^2 + \iota(v_3(u))v^3 = \vartheta + v^2,$$
$$0 = dI_1 = \iota(du_x) + \iota(v_1(u_x))v^1 + \iota(v_2(u_x))v^2 + \iota(v_3(u_x))v^3 = \kappa\varpi + \vartheta_1 + v^3,$$

since $du = u_x\,dx + \theta$, $du_x = u_{xx}\,dx + \theta_x$. Therefore,

$$v^1 = -\varpi, \qquad v^2 = -\vartheta, \qquad v^3 = -\kappa\varpi - \vartheta_1. \qquad (8.60)$$

We are now ready to substitute the nonphantom invariants into (8.43):

$$d\kappa = d\iota(u_{xx}) = \iota(du_{xx}) + \iota(v_1(u_{xx}))v^1 + \iota(v_2(u_{xx}))v^2 + \iota(v_3(u_{xx}))v^3$$
$$= \iota(u_{xxx}\,dx + \theta_{xx}) - \iota(3u_x u_{xx})(\kappa\varpi + \vartheta_1) = I_3\varpi + \vartheta_2,$$
$$dI_3 = d\iota(u_{xxx}) = \iota(du_{xxx}) + \iota(v_1(u_{xxx}))v^1 + \iota(v_2(u_{xxx}))v^2 + \iota(v_3(u_{xxx}))v^3$$
$$= \iota(u_{xxxx}\,dx + \theta_{xxx}) - \iota(4u_x u_{xxx} + 3u_{xx}^2)(\kappa\varpi + \vartheta_1)$$
$$= (I_4 - 3\kappa^3)\varpi + \vartheta_3 - 3\kappa^2\vartheta_1,$$

and so on. Breaking these formulas into their horizontal and vertical components yields

$$I_3 = \mathcal{D}\kappa = \kappa_s, \qquad\qquad d_V\kappa = \vartheta_2,$$
$$I_4 = \mathcal{D}I_3 + 3\kappa^3 = \kappa_{ss} + 3\kappa^3, \qquad d_V I_3 = d_V\kappa_s = \vartheta_3 - 3\kappa^2\vartheta_1. \qquad (8.61)$$

To proceed further, we compute the differentials of the invariant contact forms, again using (8.43), (8.60):

$$d\vartheta = \iota(d\theta) + v^1 \wedge \iota(v_1(\theta)) + v^2 \wedge \iota(v_2(\theta)) + v^3 \wedge \iota(v_3(\theta))$$
$$= \iota(dx \wedge \theta_x) - (\kappa\varpi + \vartheta_1) \wedge \iota(u_x\theta) = \varpi \wedge \vartheta_1,$$
$$d\vartheta_1 = \iota(d\theta_x) + v^1 \wedge \iota(v_1(\theta_x)) + v^2 \wedge \iota(v_2(\theta_x)) + v^3 \wedge \iota(v_3(\theta_x))$$
$$= \iota(dx \wedge \theta_{xx}) - (\kappa\varpi + \vartheta_1) \wedge \iota(2u_x\theta_x + u_{xx}\theta)$$
$$= \varpi \wedge (\vartheta_2 - \kappa^2\vartheta) - \kappa\vartheta_1 \wedge \vartheta,$$
$$d\vartheta_2 = \iota(d\theta_{xx}) + v^1 \wedge \iota(v_1(\theta_{xx})) + v^2 \wedge \iota(v_2(\theta_{xx})) + v^3 \wedge \iota(v_3(\theta_{xx}))$$
$$- \iota(dx \wedge \theta_{xxx}) - (\kappa\varpi + \vartheta_1) \wedge \iota(3u_x\theta_{xx} + 3u_{xx}\theta_x + u_{xxx}\theta)$$
$$= \varpi \wedge (\vartheta_3 - 3\kappa^2\vartheta_1 - \kappa\kappa_s\vartheta) - \kappa_s\vartheta_1 \wedge \vartheta,$$

and so on. Concentrating on the terms involving the invariant horizontal form and comparing with (8.40), we deduce

$$\vartheta_1 = \mathcal{D}\vartheta, \qquad \vartheta_2 = \mathcal{D}\vartheta_1 + \kappa^2\vartheta = (\mathcal{D}^2 + \kappa^2)\vartheta,$$
$$\vartheta_3 = \mathcal{D}\vartheta_2 + 3\kappa^2\vartheta_1 + \kappa\kappa_s\vartheta = (\mathcal{D}^3 + 4\kappa^2\mathcal{D} + 3\kappa\kappa_s)\vartheta.$$

Substituting back into (8.61), we find

$$d_V\kappa = (\mathcal{D}^2 + \kappa^2)\vartheta, \qquad d_V\kappa_s = (\mathcal{D}^3 + \kappa^2\mathcal{D} + 3\kappa\kappa_s)\vartheta.$$

Thus, the invariant linearization operators for the curvature and its arc length derivative are

$$\mathcal{A}_\kappa = \mathcal{D}^2 + \kappa^2, \qquad \mathcal{A}_{\kappa_s} = \mathcal{D}^3 + \kappa^2\mathcal{D} + 3\kappa\kappa_s. \qquad (8.62)$$

Finally, applying (8.43) and (8.60) to the invariant arc length form $\varpi = \iota(dx)$ yields

$$d\varpi = \iota(d^2 x) + v^1 \wedge \iota(v_1(dx)) + v^2 \wedge \iota(v_2(dx)) + v^3 \wedge \iota(v_3(dx))$$
$$= (\kappa\varpi + \vartheta_1) \wedge \iota(u_x\,dx + \theta) = \kappa\varpi \wedge \vartheta + \vartheta_1 \wedge \vartheta.$$

Therefore,

$$d_V\varpi = -\kappa\vartheta \wedge \varpi, \qquad \text{and so} \quad \mathcal{B} = -\kappa \qquad (8.63)$$

is the invariant Hamiltonian operator.

8.8 Generating differential invariants

Let us now apply the recurrence formulae to study the structure of the differential invariant algebra associated with the prolonged group action. A set of differential invariants $\mathcal{I} = \{I_1, \ldots, I_k\}$ is said to be *generating* if, locally, every differential invariant can be expressed as a function of the generators and their iterated invariant derivatives $\mathcal{D}_J I_\nu$. Let

$$\mathcal{I}^{(n)} = \{H^1, \ldots, H^p\} \cup \{I_J^\alpha \mid \alpha = 1, \ldots, q, \#J \leq n\} \tag{8.64}$$

denote the entire set of fundamental differential invariants (8.6) of order $\leq n$. In particular, assuming we choose a cross-section that projects to a cross-section on M, then $\mathcal{I}^{(0)} = \{H^1, \ldots, H^p, I^1, \ldots I^q\}$ are the ordinary invariants for the action of G on M. If, as in the examples treated here, G acts transitively on M, the normalized order 0 invariants are all constant, and hence are superfluous in any generating systems.

The first result is a direct consequence of the recurrence formulae (8.46) for the fundamental differential invariants and the fact that the Maurer–Cartan invariants, being solutions to the phantom recurrence relations, have order bounded by that of the moving frame.

Theorem 8.24 *If the moving frame has order n, then the set of normalized differential invariants $\mathcal{I}^{(n+1)}$ of order $\leq n + 1$ forms a generating set.*

Almost all applications rely on a cross-section $\mathcal{K}^n \subset J^n$ of *minimal order*, which means that its projections $\mathcal{K}^k = \pi_k^n(\mathcal{K}^n) \subset J^k$ form cross-sections for all $0 \leq k < n$. In this case, one can significantly reduce the set of required generators [45, 79]:

Theorem 8.25 *If $\mathcal{K}^n = \{Z_1(x, u^{(n)}) = c_1, \ldots, Z_r(x, u^{(n)}) = c_r\}$ is a minimal order cross-section, then $\mathcal{I}(0) \cup \mathcal{Z}^{(1)}$, where $\mathcal{Z}^{(1)} = \{\iota(\mathcal{D}_i(Z_j)) \mid 1 \leq i \leq p, 1 \leq j \leq r\}$, form a generating set of differential invariants.*

The result is false in general if the cross-section is not minimal [79]. An alternative interesting generating system was found in [44]; again, the proof is entirely based on the recurrence formulae.

Theorem 8.26 *Let $\mathcal{R} = \{R_a^i \mid 1 \leq i \leq p, 1 \leq a \leq r\}$ be the Maurer–Cartan invariants. Then $\mathcal{I}^{(0)} \cup \mathcal{R}$ form a generating system.*

In both cases, the $\mathcal{I}^{(0)}$ constituent can be omitted if G acts transitively on M. The preceding generating sets are rarely minimal. For curves, where $p = 1$, under mild restrictions on the group action (specifically transitivity and no pseudo-stabilization under prolongation), there are exactly $m - 1$ independent generating differential invariants, and any other differential invariant is a function of the generating invariants and their successive derivatives with respect to the G-invariant arc length element. Thus, for instance, the differential invariants of a space curve $C \subset \mathbb{R}^3$ under the standard action of the Euclidean group $SE(3) = SO(3) \ltimes \mathbb{R}^3$ are generated by $m - 1 = 2$ differential invariants, namely its curvature and torsion.

For higher-dimensional submanifolds, the minimal number of generating differential invariants cannot be fixed a priori, but depends the particularities of the group action and, in fact, can be arbitrarily large, even for surfaces in three-dimensional space [79]. Even in very well-studied, classical situations, there are interesting subtleties that have not been noted before [48, 82].

Example 8.27 Consider the standard action of the special Euclidean group $SE(3)$ on surfaces $S \subset \mathbb{R}^3$. The classical moving frame construction [40, chapter 10], or its equivariant reformulation [57, example 9.9], relies on the cross-section

$$x = y = u = u_x = u_y = u_{xy} = 0, \qquad u_{xx} \neq u_{yy}. \tag{8.65}$$

The two basic differential invariants are the principal curvatures

$$\kappa_1 = \iota(u_{xx}), \qquad \kappa_2 = \iota(u_{yy}), \tag{8.66}$$

or, equivalently, the *mean curvature* and *Gauss curvature*

$$H = \tfrac{1}{2}(\kappa_1 + \kappa_2), \qquad K = \kappa_1 \kappa_2. \tag{8.67}$$

The surface admits a classical moving frame provided we are at a nonumbilic point, where $\kappa_1 \neq \kappa_2$. (At a nondegenerate umbilic, one could, in principle, employ a higher order moving frame.) The corresponding invariant horizontal coframe $\varpi^1 = \iota(dx)$, $\varpi^2 = \iota(dy)$, can be identified with the diagonalizing Frenet frame on the surface, [40]. We let $\mathcal{D}_1, \mathcal{D}_2$ denote the dual invariant differential operators.

Let $I_{jk} = \iota(u_{jk})$ denote the higher-order normalized differential invariants, so $I_{20} = \kappa_1$, $I_{11} = 0$, $I_{02} = \kappa_2$. The third-order recurrence relations are readily found:

$$
\begin{aligned}
I_{30} &= \mathcal{D}_1\kappa_1 = \kappa_{1,1}, & I_{21} &= \mathcal{D}_2\kappa_1 = \kappa_{1,2}, \\
I_{12} &= \mathcal{D}_1\kappa_2 = \kappa_{2,1}, & I_{03} &= \mathcal{D}_2\kappa_2 = \kappa_{2,2}.
\end{aligned}
\tag{8.68}
$$

The two fourth-order recurrence relations for

$$I_{22} = \mathcal{D}_2 I_{21} + \frac{I_{30}I_{12} - 2I_{12}^2}{\kappa_1 - \kappa_2} + \kappa_1\kappa_2^2 = \mathcal{D}_1 I_{12} - \frac{I_{21}I_{03} - 2I_{21}^2}{\kappa_1 - \kappa_2} + \kappa_1^2\kappa_2$$

imply the celebrated *Codazzi syzygy*

$$\kappa_{1,22} - \kappa_{2,11} + \frac{\kappa_{1,1}\kappa_{2,1} + \kappa_{1,2}\kappa_{2,2} - 2\kappa_{2,1}^2 - 2\kappa_{1,2}^2}{\kappa_1 - \kappa_2} - \kappa_1\kappa_2(\kappa_1 - \kappa_2) = 0. \tag{8.69}$$

The well-known fact that the principal curvatures κ_1, κ_2, or, equivalently, the Gauss and mean curvatures H, K, form a generating system follows from Theorem 8.24 combined with (8.68). Remarkably, as we now show, neither is a minimal generating set!

Applying the moving frame machinery, the recurrence relations for the invariant horizontal forms are found to be

$$
\begin{aligned}
d_{\mathcal{H}} \varpi^1 &= Y_2 \varpi^1 \wedge \varpi^2, \\
d_{\mathcal{H}} \varpi^2 &= Y_1 \varpi^1 \wedge \varpi^2,
\end{aligned}
\qquad \text{where} \quad Y_1 = \frac{\kappa_{2,1}}{\kappa_1 - \kappa_2}, \ Y_2 = \frac{\kappa_{1,2}}{\kappa_2 - \kappa_1}, \qquad (8.70)
$$

are the commutator invariants. The invariant differential operators therefore satisfy the commutation relation

$$
[\mathcal{D}_1, \mathcal{D}_2] = \mathcal{D}_1 \mathcal{D}_2 - \mathcal{D}_2 \mathcal{D}_1 = Y_2 \mathcal{D}_1 - Y_1 \mathcal{D}_2. \qquad (8.71)
$$

An easy computation shows that the Codazzi syzygy (8.69) can be written compactly as

$$
K = \kappa_1 \kappa_2 = -(\mathcal{D}_1 + Y_1) Y_1 - (\mathcal{D}_2 + Y_2) Y_2, \qquad (8.72)
$$

which is the key identity employed by Guggenheimer [40], for a short proof of Gauss' Theorema Egregium.

Let us now show how, for suitably nondegenerate surfaces, we can write the Gauss curvature K as a universal rational combination of the invariant derivatives of the mean curvature H. In view of the Codazzi formula (8.72), it suffices to write the commutator invariants Y_1, Y_2 in terms of the mean curvature. To this end, we note that the commutator identity (8.71) can be applied to any differential invariant. In particular,

$$
\mathcal{D}_1 \mathcal{D}_2 H - \mathcal{D}_2 \mathcal{D}_1 H = Y_2 \mathcal{D}_1 H - Y_1 \mathcal{D}_2 H, \qquad (8.73)
$$

and, furthermore, for $j = 1$ or 2,

$$
\mathcal{D}_1 \mathcal{D}_2 \mathcal{D}_j H - \mathcal{D}_2 \mathcal{D}_1 \mathcal{D}_j H = Y_2 \mathcal{D}_1 \mathcal{D}_j H - Y_1 \mathcal{D}_2 \mathcal{D}_j H. \qquad (8.74)
$$

Provided the nondegeneracy condition

$$
(\mathcal{D}_1 H)(\mathcal{D}_2 \mathcal{D}_j H) \neq (\mathcal{D}_2 H)(\mathcal{D}_1 \mathcal{D}_j H), \qquad \text{for } j = 1 \text{ or } 2, \qquad (8.75)
$$

holds, we can solve (8.73)–(8.74) to write the commutator invariants Y_1, Y_2 as rational functions of invariant derivatives of H. Plugging these expressions into the right-hand side of the Codazzi identity (8.72) produces an explicit formula for the Gauss curvature as a rational function of the invariant derivatives, of order ≤ 4, of the mean curvature, valid for all surfaces satisfying the nondegeneracy condition (8.75).

In [82] it was also proved that, for suitably generic surfaces in $\mathbb{R}^3$, the algebra of equi-affine differential invariants is generated by the third order Pick invariant alone through invariant differentiation. In [48] it was proved that the algebras of conformal and projective differential invariants are also both generated by a single differential invariant.

8.9 Invariant variational problems

As first recognized by Sophus Lie [59], every invariant variational problem can be written in terms of the differential invariants of the symmetry group. The associated Euler-Lagrange equations automatically inherit the symmetry group of the variational problem, and so can also be written in terms of the differential invariants [73]. The formula for directly constructing the differential invariant form of the Euler–Lagrange equations from that of the variational problem was only known in a handful of particular cases [4, 39] until, applying the invariant variational bicomplex machinery, the general version was established in [57]. Recent applications to the equilibrium configurations of flexible Mobius bands can be found in [93].

Let us begin by recalling how variational problems $\mathcal{L}[u] = \int L(x, u^{(n)}) \, dx$ appear in the variational bicomplex, [4]. The integrand or *Lagrangian form*

$$\lambda = L(x, u^{(n)}) \, dx = L(x, u^{(n)}) \, dx^1 \wedge \cdots \wedge dx^p, \qquad (8.76)$$

is a differential form on J^∞ of type $(p, 0)$, meaning that it involves p horizontal forms and no contact forms. Classically, to compute the associated Euler–Lagrange equations, one begins with the first variation, followed by an integration by parts. According to (8.35), we identify the first variation with the vertical differential $d_V \lambda = d_V L \wedge dx$ of the Lagrangian form, which is a form of type $(p, 1)$. Integration by parts can be viewed as quotienting out by the image of the horizontal differential, so $\omega \equiv \widetilde{\omega}$ whenever $\omega - \widetilde{\omega} = d_H \psi$ for some differential form ψ. The induced equivalence classes are represented by *source forms*

$$\omega = \sum_{\alpha=1}^q \Delta_\alpha(x, u^{(n)}) \theta^\alpha \wedge dx, \qquad (8.77)$$

whose vanishing defines a system of differential equations: $\Delta_\alpha(x, u^{(n)}) = 0$. In the case of a variational problem, $\Delta_\alpha = E_\alpha(L) = 0$ are the classical Euler–Lagrange equations.

The Lagrangian of a G-invariant variational problem can be written in the invariant form

$$\lambda = \tilde{L}(I^{(n)})\omega^1 \wedge \cdots \wedge \omega^p,$$

where $\omega^1, \ldots, \omega^p$ denote the contact invariant coframe induced by the moving frame, (8.36), while $\tilde{L}(I^{(n)})$ is a function of the generating differential invariants $I = (I^1, \ldots, I^l)$ and their invariant derivatives $\mathcal{D}_J I^\kappa$ up to some finite order $\#J \leq k$. Since they differ by contact forms (which vanish when evaluated on submanifold jets), we do not affect anything by replacing the ω^i by their fully invariant counterparts ϖ^i, and so will use the fully *invariant Lagrangian form*

$$\tilde{\lambda} = \tilde{L}(I^{(n)})\varpi^1 \wedge \cdots \wedge \varpi^p \tag{8.78}$$

in our subsequent computations. To find the invariant form of the Euler–Lagrange equations, we first compute the invariant variation $d_V\tilde{\lambda}$, followed by an invariant integration by parts. Two new complications arise: first, whereas the ordinary vertical derivative does not affect the basis horizontal forms dx^i, formula (8.50) shows that this is not true for the invariant vertical derivatives of the invariant horizontal forms ϖ^i. Secondly, invariant integration by parts, which amounts to working modulo the image of the invariant horizontal differential $d_{\mathcal{H}}$, also introduces new terms owing to (8.48). As a result, the invariant Euler–Lagrange equation expressions are considerably more complicated.

For simplicity, let's just work out the case of curves, so we have only $p = 1$ independent variable, and $q \geq 1$ dependent variables. (The higher-dimensional case has some extra twists; see [57] for details.) Consider an invariant Lagrangian form $\tilde{\lambda} = \tilde{L}(I^{(n)})\varpi$ depending on the generating differential invariants $I = (I^1, \ldots, I^l)$, their invariant derivatives $I^\alpha_{,i} = \mathcal{D}^i I^\alpha$, and the fully G-invariant arc length form $\varpi = \iota(dx)$. Its first variation is computed as follows:

$$d_V\tilde{\lambda} = d_V(\tilde{L}\varpi) = d_V\tilde{L} \wedge \varpi + \tilde{L}\, d_V\varpi = \sum_{i,\alpha} \frac{\partial\tilde{L}}{\partial I^\alpha_{,i}}\, d_V I^\alpha_{,i} \wedge \varpi + \tilde{L}\, d_V\varpi. \tag{8.79}$$

We then invariantly integrate by parts by applying the basic identity

$$F\, d_V(\mathcal{D}H) \wedge \varpi \equiv -\mathcal{D}F\, d_V H \wedge \varpi - F(\mathcal{D}H)\, d_V\varpi, \tag{8.80}$$

where we work modulo the image of $d_{\mathcal{H}}$. We eventually arrive at the formula

$$d_V\tilde{\lambda} \equiv \mathcal{E}(\tilde{L})\, d_V I \wedge \varpi - \mathcal{H}(\tilde{L})\, d_V\varpi, \tag{8.81}$$

where $\mathcal{E}(\tilde{L})$, the *invariantized Eulerian* of $\tilde{L}$, has components

$$\mathcal{E}_\alpha(\tilde{L}) = \sum_{i=0}^{\infty}(-\mathcal{D})^i \frac{\partial\tilde{L}}{\partial I^\alpha_{,i}}, \qquad \alpha = 1, \ldots, l, \tag{8.82}$$

while

$$\mathcal{H}(\tilde{L}) = \sum_{\alpha=1}^{m} \sum_{i>j} I_{,i-j}^{\alpha}(-\mathcal{D})^j \frac{\partial \tilde{L}}{\partial I_{,i}^{\alpha}} - \tilde{L} \tag{8.83}$$

is known as the *invariantized Hamiltonian*, being the invariant counterpart of the usual Hamiltonian associated with a higher-order Lagrangian $L(x, u^{(n)})$, cf. [4, 88].

In the second phase of the computation, we use the recurrence formulae (8.54), (8.55) to compute the vertical differentials

$$d_V I = \mathcal{A}(\vartheta), \qquad d_V \varpi = \mathcal{B}(\vartheta) \wedge \varpi, \tag{8.84}$$

of the differential invariants $I = (I^1, \ldots, I^l)$ and the invariant horizontal (arc length) form in terms of invariant derivatives of the zeroth-order invariant contact forms $\vartheta = (\vartheta^1, \ldots, \vartheta^q)$. Substituting (8.84) into (8.81) and performing one last integration by parts, we arrive at the key formula

$$d_V \tilde{\lambda} \equiv \mathcal{E}(\tilde{L})\mathcal{A}(\vartheta) \wedge \varpi - \mathcal{H}(\tilde{L})\mathcal{B}(\vartheta) \wedge \varpi \equiv [\mathcal{A}^* \mathcal{E}(\tilde{L}) - \mathcal{B}^* \mathcal{H}(\tilde{L})]\vartheta \wedge \varpi,$$

where * denotes the *formal invariant adjoint* of an invariant differential operator, so if

$$\mathcal{P} = \sum_n P_k \mathcal{D}^k, \quad \text{then} \quad \mathcal{P}^* = \sum_k (-\mathcal{D})^k \cdot P_k.$$

We conclude that the Euler–Lagrange equations for our invariant variational problem are equivalent to the invariant system of differential equations

$$\mathcal{A}^* \mathcal{E}(\tilde{L}) - \mathcal{B}^* \mathcal{H}(\tilde{L}) = 0. \tag{8.85}$$

Example 8.28 Any Euclidean-invariant variational problem corresponds to an invariant Lagrangian $\tilde{\lambda} = \tilde{L}(\kappa, \kappa_s, \kappa_{ss}, \ldots)\varpi$ depending on the arc length derivatives of the curvature, and the invariant arc length form (8.58). According to (8.62), (8.63), $\mathcal{A} = \mathcal{D}^2 + \kappa^2 = \mathcal{A}^*$, while $\mathcal{B} = -\kappa = \mathcal{B}^*$. The invariant Euler–Lagrange formula (8.85) reduces to the known formula

$$(\mathcal{D}^2 + \kappa^2)\mathcal{E}(\tilde{L}) + \kappa \mathcal{H}(\tilde{L}) = 0 \tag{8.86}$$

for the Euclidean-invariant Euler–Lagrange equation [4, 39].

Additional, more intricate examples can be found in [80].

8.10 Invariant curve flows

Finally, let us discuss some recent applications of the invariant variational bicomplex construction to invariant curve flows. (Extensions to higher-dimensional invariant submanifold flows can be found in [80].) Setting $p = 1$, let us single out the $m = 1 + q$ invariant one-forms

$$\varpi, \vartheta^1, \ldots, \vartheta^q \tag{8.87}$$

consisting of the invariant arc length form $\varpi = \iota(dx)$ and the order 0 invariant contact forms $\vartheta^\alpha = \iota(\theta^\alpha)$. Let $C \subset M$ be a curve. Evaluating the coefficients of (8.87) on the curve jet $(x, u^{(n)}) = j_n C|_z$ produces a G-equivariant coframe, i.e., a basis for the cotangent space $T^*M|_z$ at $z = (x, u) \in C$. Let $\mathbf{t}, \mathbf{n}_1, \ldots, \mathbf{n}_q$ denote the corresponding dual G-equivariant *frame* on C, with $\mathbf{t}$ tangent, while $\mathbf{n}_1, \ldots, \mathbf{n}_q$ form a basis for the complementary G-invariant *normal bundle* $N \to C$ induced by the moving frame.

In general, let

$$\mathbf{V} = \mathbf{V}_T + \mathbf{V}_N = I\mathbf{t} + \sum_{\alpha=1}^{q} J^\alpha \mathbf{n}_\alpha \tag{8.88}$$

be a G-equivariant section of $TM \to C$, where $\mathbf{V}_T, \mathbf{V}_N$ denote, respectively, its tangential and normal components, while $I, J^1, \ldots, J^q$ are differential invariants. We will, somewhat imprecisely, refer to $\mathbf{V}$ as a *vector field*, even though it depends on the underlying curve jet. Any $\mathbf{V}$ generates a G-invariant curve flow:

$$\frac{\partial C}{\partial t} = \mathbf{V}|_{C(t)}. \tag{8.89}$$

The tangential component $\mathbf{V}_T$ only affects the curve's internal parametrization, and hence can be ignored as far as the external curve geometry goes. For example, if we set $\mathbf{V}_T \doteq 0$, the resulting vector field $\mathbf{V}_N$ is said to generate a *normal flow*, since each point on the curve moves in the G-invariant normal direction.

Example 8.29 The most well-studied are the Euclidean-invariant plane curve flows. The dual frame vectors to the invariant one-forms (8.58), (8.59) are the usual Euclidean frame vectors[3] – the unit tangent and unit normal:

$$\mathbf{t} = \frac{1}{\sqrt{1 + u_x^2}} \left(\frac{\partial}{\partial x} + u_x \frac{\partial}{\partial u} \right), \qquad \mathbf{n} = \frac{1}{\sqrt{1 + u_x^2}} \left(-u_x \frac{\partial}{\partial x} + \frac{\partial}{\partial u} \right). \tag{8.90}$$

A Euclidean-invariant normal flow is generated by a vector field of the form $\mathbf{V} = \mathbf{V}_N = J\mathbf{n}$, in which $J(\kappa, \kappa_s, \ldots)$ is any differential invariant. Particular cases include:

- $\mathbf{V} = \mathbf{n}$: the geometric optics or grassfire flow, [9, 89];

---

[3] For simplicity, we are assuming the curve is represented as the graph of a function $u = u(x)$; generalizing the formulas to arbitrarily parametrized curves is straightforward [80].

- $\mathbf{V} = \kappa\mathbf{n}$: the celebrated curve shortening flow [34, 36], also used to great effect in image processing [86, 89];
- $\mathbf{V} = \kappa^{1/3}\mathbf{n}$: the induced flow is equivalent, modulo reparametrization, to the equi-affine invariant curve shortening flow, also used in image processing [5, 86, 89];
- $\mathbf{V} = \kappa_s\mathbf{n}$: this flow induces the modified Korteweg–de Vries equation for the curvature evolution, and is the simplest example of a soliton equation arising in a geometric curve flow [26, 35, 63];
- $\mathbf{V} = \kappa_{ss}\mathbf{n}$: this flow models thermal grooving of metals [16].

A key question is how the differential invariants evolve under an invariant curve flow.

Theorem 8.30 *Let* $\mathbf{V}_N = \sum J^\alpha \mathbf{n}_\alpha$ *generate an invariant normal curve flow. If* K *is any differential invariant, then*

$$\frac{\partial K}{\partial t} = \mathbf{V}(K) = \mathcal{A}_K(J), \tag{8.91}$$

where S_K *is the corresponding invariant linearization operator.*

Example 8.31 For any of the Euclidean invariant normal plane curve flows $\mathbf{C}_t = J\mathbf{n}$ listed in Example 8.29, we have, according to (8.62),

$$\frac{\partial \kappa}{\partial t} = (\mathcal{D}^2 + \kappa^2)J, \qquad \frac{\partial \kappa_s}{\partial t} = (\mathcal{D}^3 + \kappa^2\mathcal{D} + 3\kappa\kappa_s)J. \tag{8.92}$$

For instance, for the grassfire flow $J = 1$, and so

$$\frac{\partial \kappa}{\partial t} = \kappa^2, \qquad \frac{\partial \kappa_s}{\partial t} = 3\kappa\kappa_s. \tag{8.93}$$

The first equation immediately implies finite time blow-up at a caustic for a convex initial curve segment, where $\kappa > 0$. For the curve shortening flow, $J = \kappa$, and

$$\frac{\partial \kappa}{\partial t} = \kappa_{ss} + \kappa^3, \qquad \frac{\partial \kappa_s}{\partial t} = \kappa_{sss} + 4\kappa^2\kappa_s, \tag{8.94}$$

thereby recovering formulas used in Gage and Hamilton's analysis [34]; see also [68]. Finally, for the modified Korteweg-deVries flow, $J = \kappa_s$,

$$\frac{\partial \kappa}{\partial t} = \kappa_{sss} + \kappa^2\kappa_s, \qquad \frac{\partial \kappa_s}{\partial t} = \kappa_{ssss} + \kappa^2\kappa_{ss} + 3\kappa\kappa_s^2. \tag{8.95}$$

Warning Normal flows *do not preserve arc length*, and so the arc length parameter s will vary in time. Or, to phrase it another way, time differentiation ∂_t and arc length differentiation $\mathcal{D} = D_s$ *do not commute* – as can easily be seen in the preceding examples. Thus, one must be very careful *not* to interpret the resulting evolutions (8.93)–(8.95) as partial differential equations in the usual sense. Rather, one should regard the differential invariants $\kappa, \kappa_s, \kappa_{ss}, \ldots$ as satisfying an infinite-dimensional dynamical system of coupled ordinary differential equations.

A second important class are the invariant curve flows that preserve arc length, which requires $[\mathbf{V}, \mathcal{D}] = 0$, or, equivalently that the Lie derivative $\mathbf{V}(\varpi) \equiv 0$ is a contact form. Applying the Cartan formula and (8.55) to the latter characterization, we conclude that arc length preservation under (8.88) requires

$$\mathcal{D}I = \mathcal{B}(J) = \sum_{\alpha=1}^{q} \mathcal{B}_\alpha(J^\alpha), \qquad (8.96)$$

where $\mathcal{D}$ is the arc length derivative, while $\mathcal{B} = (\mathcal{B}_1, \ldots, \mathcal{B}_q)$ is the *invariant Hamiltonian operator* (8.55).

Theorem 8.32   *Under an arc-length preserving flow,*

$$\kappa_t = \mathcal{R}_\kappa(J) \qquad where \ \mathcal{R}_\kappa = \mathcal{A}_\kappa - \kappa_s \mathcal{D}^{-1} \mathcal{B}. \qquad (8.97)$$

More generally, the time evolution of $\kappa_n = \mathcal{D}^n \kappa$ is given by arc length differentiation: $\partial \kappa_n / \partial t = \mathcal{D}^n \mathcal{R}_\kappa(J)$.

Here, the arc length and time derivatives commute, and hence the arc-length preserving flow (8.97) is an ordinary evolution equation – albeit possibly with nonlocal terms. Moreover, when (8.97) is a local evolution equation, it often turns out to be integrable, with $\mathcal{R}_\kappa$ the associated recursion operator [73]. However, as yet, there is no general explanation for this phenomenon.

Example 8.33   For the Euclidean action on plane curves, the condition (8.96) that a curve flow generated by the vector field $\mathbf{V} = I\mathbf{t} + J\mathbf{n}$ preserve arc length is that

$$\mathcal{D}I = -\kappa J. \qquad (8.98)$$

Most of the curve flows listed in Example 8.29 have *nonlocal* arc length preserving counterparts owing to the non-invertibility of the arc length derivative operator on κJ. An exception is the modified Korteweg-deVries flow, where $J = \kappa_s$, and so $I = -\frac{1}{2}\kappa^2$. For such flows, the evolution of the curvature is given by (8.97), where

$$\mathcal{R}_\kappa = \mathcal{A}_\kappa - \kappa_s \mathcal{D}^{-1} \mathcal{B} = \mathcal{D}^2 + \kappa^2 + \kappa_s \mathcal{D}^{-1} \cdot \kappa = D_s^2 + \kappa^2 + \kappa_s D_s^{-1} \cdot \kappa \qquad (8.99)$$

is the modified Korteweg–de Vries recursion operator [73]. In particular, when $J = \kappa_s$, (8.97) is the modified Korteweg–de Vries equation

$$\kappa_t = \mathcal{R}_\kappa(\kappa_s) = \kappa_{sss} + \frac{3}{2}\kappa^2\kappa_s.$$

Example 8.34 In the case of space curves $C \subset \mathbb{R}^3$, under the usual action of the Euclidean group $G = SE(3)$, the coordinate cross-section

$$\mathcal{K}^2 = \{x = u = v = u_x = v_x = v_{xx} = 0\}$$

produces the classical moving frame [40, 57]. There are two generating differential invariants: the curvature $\kappa = \iota(u_{xx})$ and the torsion $\iota = \iota(v_{xxx}/u_{xx})$. According to [57], the relevant moving frame formulae are

$$d_{\mathcal{V}}\kappa = \mathcal{A}_\kappa(\vartheta), \qquad d_{\mathcal{V}}\tau = \mathcal{A}_\iota(\vartheta), \qquad d_{\mathcal{V}}\varpi = \mathcal{B}(\vartheta) \wedge \varpi,$$

where $\vartheta = (\vartheta_1, \vartheta_2)^T$ are the order 0 invariant contact forms, while

$$\mathcal{A}_\kappa = (\mathcal{D}^2 + (\kappa^2 - \tau^2), -2\tau\mathcal{D} - \tau_s),$$

$$\mathcal{A}_\tau = \left(\frac{2\tau}{\kappa}\mathcal{D}^2 + \frac{3\kappa\tau_s - 2\kappa_s\tau}{\kappa^2}\mathcal{D} + \frac{\kappa\tau_{ss} - \kappa_s\tau_s + 2\kappa^3\tau}{\kappa^2},\right.$$

$$\left. \frac{1}{\kappa}\mathcal{D}^3 - \frac{\kappa_s}{\kappa^2}\mathcal{D}^2 + \frac{\kappa^2 - \tau^2}{\kappa}\mathcal{D} + \frac{\kappa_s\tau^2 - 2\kappa\tau\tau_s}{\kappa^2}\right),$$

$$\mathcal{B} = (\kappa, 0).$$

Thus, under an arc length preserving flow with normal component $\mathbf{V}_N = J\mathbf{n}_1 + K\mathbf{n}_2$, the curvature and torsion evolve according to

$$\begin{pmatrix}\kappa_t \\ \tau_t\end{pmatrix} = \mathcal{R}\begin{pmatrix}J \\ K\end{pmatrix}, \qquad \text{where } \mathcal{R} = \begin{pmatrix}\mathcal{R}_\kappa \\ \mathcal{R}_\tau\end{pmatrix} = \begin{pmatrix}\mathcal{A}_\kappa \\ \mathcal{A}_\tau\end{pmatrix} - \begin{pmatrix}\kappa_s\mathcal{D}^{-1}\kappa & 0 \\ \tau_s\mathcal{D}^{-1}\kappa & 0\end{pmatrix}$$

is the recursion operator for the integrable vortex filament flow, which corresponds to the choice $J = \kappa_s$, $K = \tau_s$. The latter flow can be mapped to the nonlinear Schrödinger equation via the Hasimoto transformation [43, 58].

Further developments, including applications to image processing and object recognition, can be found in Kenney's thesis [50].

References

[1] Akivis, M. A., and Rosenfeld, B. A. 1993. *Élie Cartan (1869–1951)*. Transl. Math. Monogr., vol. 123. Providence, RI: Amer. Math. Soc.

[2] Ames, A. D., Jalkio, J. A., and Shakiban, C. 2002. Three-dimensional object recognition using invariant Euclidean signature curves. Pages 13–23 of: *Analysis, Combinatorics and Computing*. Hauppauge, NY: Nova Sci. Publ.

[3] Ames, W. F. 1968. *Nonlinear Ordinary Differential Equations in Transport Processes*. Math. Sci. Engrg., vol. 42. New York: Academic Press.

[4] Anderson, I. M. 1989. *The Variational Bicomplex*. Tech. rept. Utah State University.

[5] Angenent, S., Sapiro, G., and Tannenbaum, A. 1998. On the affine heat equation for non-convex curves. *J. Amer. Math. Soc.*, **11**(3), 601–634.

[6] Bazin, P.-L., and Boutin, M. 2004. Structure from motion: a new look from the point of view of invariant theory. *SIAM J. Appl. Math.*, **64**(4), 1156–1174.

[7] Berchenko, I., and Olver, P. J. 2000. Symmetries of polynomials. *J. Symbolic Comput.*, **29**(4-5), 485–514.

[8] Blumenthal, L. M. 1953. *Theory and Applications of Distance Geometry*. Oxford: Oxford Univ. Press.

[9] Born, M., and Wolf, E. 1970. *Principles of Optics*. New York: Pergamon Press.

[10] Boutin, M. 2000. Numerically invariant signature curves. *Int. J. Comput. Vision*, **40**(3), 235–248.

[11] Boutin, M. 2002. On orbit dimensions under a simultaneous Lie group action on *n* copies of a manifold. *J. Lie Theory*, **12**(1), 191–203.

[12] Boutin, M. 2003. Polygon recognition and symmetry detection. *Found. Comput. Math.*, **3**(3), 227–271.

[13] Boutin, M., and Kemper, G. 2004. On reconstructing *n*-point configurations from the distribution of distances or areas. *Adv. in Appl. Math.*, **32**(4), 709–735.

[14] Boyko, V., Patera, J., and Popovych, R. 2006. Computation of invariants of Lie algebras by means of moving frames. *J. Phys. A*, **39**(20), 5749–5762.

[15] Boyko, V., Patera, J., and Popovych, R. 2008. Invariants of solvable Lie algebras with triangular nilradicals and diagonal nilindependent elements. *Linear Algebra Appl.*, **428**(4), 834–854.

[16] Broadbridge, P., and Tritscher, P. 1994. An integrable fourth-order nonlinear evolution equation applied to thermal grooving of metal surfaces. *IMA J. Appl. Math.*, **53**(3), 249–265.

[17] Bruckstein, A. M., and Shaked, D. 1998. Skew symmetry detection via invariant signatures. *Pattern Recognition*, **31**(2), 181–192.

[18] Bruckstein, A. M., Holt, R. J., Netravali, A. N., and Richardson, T. J. 1993. Invariant signatures for planar shape recognition under partial occlusion. *Comput. Vision Graphics Image Process.*, **58**(1), 49–65.

[19] Budd, C. J., and Iserles, A. 1999. Geometric integration: numerical solution of differential equations on manifolds. *R. Soc. Lond. Philos. Trans. Ser. A Math. Phys. Eng. Sci.*, **357**(1754), 945–956.

[20] Calabi, E., Olver, P. J., and Tannenbaum, A. 1996a. Affine geometry, curve flows, and invariant numerical approximations. *Adv. Math.*, **124**(1), 154–196.

[21] Calabi, E., Olver, P. J., Shakiban, C., Tannenbaum, A., and Haker, S. 1996b. Differential and numerically invariant signature curves applied to object recognition. *Int. J. Comput. Vision*, **26**, 107–135.

[22] Cartan, É. 1935. *La méthode du repère mobile, la théorie des groupes continus, et les espaces généralisés*. Exposés de géométrie, vol. 5. Paris: Hermann.

[23] Cheh, J., Olver, P. J., and Pohjanpelto, J. 2005. Maurer–Cartan equations for Lie symmetry pseudogroups of differential equations. *J. Math. Phys.*, **46**(2), 023504.

[24] Cheh, J., Olver, P. J., and Pohjanpelto, J. 2008. Algorithms for differential invariants of symmetry groups of differential equations. *Found. Comput. Math.*, **8**(4), 501–532.

[25] Chern, S. S. 1985. Moving frames. *Astérisque*, numéro hors série, 67–77.

[26] Chou, K.-S., and Qu, C.-Z. 2003. Integrable equations arising from motions of plane curves. II. *J. Nonlinear Sci.*, **13**(5), 487–517.

[27] Dalle, D. 2006. Comparison of numerical techniques for Euclidean curvature. *Rose-Hulman Undergraduate Math. J.*, **7**(1).

[28] Deeley, R J., Horwood, J. T., McLenaghan, R. G., and Smirnov, R. G. 2004. Theory of algebraic invariants of vector spaces of Killing tensors: methods for computing the fundamental invariants. Pages 1079–1086 of: *Symmetry in Nonlinear Mathematical Physics. Part 1, 2, 3*. Pr. Inst. Mat. Nats. Akad. Nauk Ukr. Mat. Zastos., 50, Part 1, vol. 2. Kiev: Natsīonal. Akad. Nauk Ukraïni Īnst. Mat.

[29] Dhooghe, P. F. 1996. Multilocal invariants. Pages 121–137 of: *Geometry and Topology of Submanifolds. VIII*. Brussels, 1995/Nordfjordeid, 1995: World Sci. Publ., River Edge, NJ.

[30] Dorodnitsyn, V. A. 1994. Finite difference models entirely inheriting continuous symmetry of original differential equations. *Internat. J. Modern Phys. C*, **5**(4), 723–734.

[31] Faugeras, O. D. 1994. Cartan's moving frame method and its application to the geometry and evolution of curves in the Euclidean, affine and projective planes. Pages 11–46 of: *Applications of Invariance in Computer Vision*. Lecture Notes In Comput. Sci., vol. 825. Berlin: Springer.

[32] Fels, M., and Olver, P. J. 1999. Moving coframes. II. Regularization and theoretical foundations. *Acta Appl. Math.*, **55**(2), 127–208.

[33] Feng, S., Kogan, I., and Krim, H. Classification of curves in 2D and 3D via affine integral signatures. *Acta. Appl. Math.*, **109**(3), 903–937.

[34] Gage, M., and Hamilton, R. S. 1986. The heat equation shrinking convex plane curves. *J. Differential Geom.*, **23**(1), 69–96.

[35] Goldstein, R. E., and Petrich, D. M. 1991. The Korteweg–de Vries hierarchy as dynamics of closed curves in the plane. *Phys. Rev. Lett.*, **67**(23), 3203–3206.

[36] Grayson, M. A. 1987. The heat equation shrinks embedded plane curves to round points. *J. Differential Geom.*, **26**(2), 285–314.

[37] Green, M. L. 1978. The moving frame, differential invariants and rigidity theorems for curves in homogeneous spaces. *Duke Math. J.*, **45**(4), 735–779.

[38] Griffiths, P. 1974. On Cartan's method of Lie groups and moving frames as applied to uniqueness and existence questions in differential geometry. *Duke Math. J.*, **41**, 775–814.

[39] Griffiths, P. A. 1983. *Exterior differential systems and the calculus of variations*. Progr. Math., vol. 25. Boston, MA: Birkhäuser.

[40] Guggenheimer, Heinrich W. 1963. *Differential geometry*. New York: McGraw-Hill.

[41] Hairer, E., Lubich, C., and Wanner, G. 2002. *Geometric numerical integration*. Springer Ser. Comput. Math., vol. 31. Berlin: Springer.

[42] Hann, C. E., and Hickman, M. S. 2002. Projective curvature and integral invariants. *Acta Appl. Math.*, **74**(2), 177–193.

[43] Hasimoto, H. 1972. A soliton on a vortex filament. *J. Fluid Mech.*, **51**(2), 477–485.

[44] Hubert, E. 2007. *Generation properties of Maurer–Cartan invariants*. Tech. rept. inria-00194528. INRIA.

[45] Hubert, E. 2009. Differential invariants of a Lie group action: syzygies on a generating set. *J. Symbolic Comput.*, **44**(4), 382–416.

[46] Hubert, E., and Kogan, I. A. 2007a. Rational invariants of a group action. Construction and rewriting. *J. Symbolic Comput.*, **42**(1-2), 203–217.

[47] Hubert, E., and Kogan, I. A. 2007b. Smooth and algebraic invariants of a group action: local and global constructions. *Found. Comput. Math.*, **7**(4), 455–493.

[48] Hubert, E., and Olver, P. J. 2007. Differential invariants of conformal and projective surfaces. *SIGMA Symmetry Integrability Geom. Methods Appl.*, **3**, 097.

[49] Jensen, G. R. 1977. *Higher order contact of submanifolds of homogeneous spaces.* Lecture Notes in Math., vol. 610. Berlin: Springer.

[50] Kenney, J. P. 2009. *Evolution of Differential Invariant Signatures and Applications to Shape Recognition.* Ph.D. thesis, University of Minnesota.

[51] Kichenassamy, S., Kumar, A., Olver, P., Tannenbaum, A., and Yezzi, Jr., A. 1996. Conformal curvature flows: from phase transitions to active vision. *Arch. Rational Mech. Anal.*, **134**(3), 275–301.

[52] Kim, P. 2007. Invariantization of numerical schemes using moving frames. *BIT*, **47**(3), 525–546.

[53] Kim, P. 2008. Invariantization of the Crank-Nicolson method for Burgers' equation. *Phys. D*, **237**(2), 243–254.

[54] Kim, P., and Olver, P. J. 2004. Geometric integration via multi-space. *Regul. Chaotic Dyn.*, **9**(3), 213–226.

[55] Kogan, I. A. 2000. *Inductive approach to moving frames and applications in classical invariant theory.* Ph.D. thesis, University of Minnesota.

[56] Kogan, I. A., and Moreno Maza, M. 2002. Computation of canonical forms for ternary cubics. Pages 151–160 of: *Proceedings of the 2002 International Symposium on Symbolic and Algebraic Computation.* New York: ACM.

[57] Kogan, I. A., and Olver, P. J. 2003. Invariant Euler–Lagrange equations and the invariant variational bicomplex. *Acta Appl. Math.*, **76**(2), 137–193.

[58] Langer, J., and Perline, R. 1991. Poisson geometry of the filament equation. *J. Nonlinear Sci.*, **1**(1), 71–93.

[59] Lie, S. 1897. Über Integralinvarianten und ihre Verwertung für die Theorie der Differentialgleichungen. *Leipz. Berichte*, **49**, 369–410.

[60] Mansfield, E. L. 2001. Algorithms for symmetric differential systems. *Found. Comput. Math.*, **1**(4), 335–383.

[61] Marí Beffa, G. 2006. Poisson geometry of differential invariants of curves in some nonsemisimple homogeneous spaces. *Proc. Amer. Math. Soc.*, **134**(3), 779–791.

[62] Marí Beffa, G. 2008. Projective-type differential invariants and geometric curve evolutions of KdV-type in flat homogeneous manifolds. *Ann. Inst. Fourier (Grenoble)*, **58**(4), 1295–1335.

[63] Marí Beffa, G., Sanders, J. A., and Wang, J. P. 2002. Integrable systems in three-dimensional Riemannian geometry. *J. Nonlinear Sci.*, **12**(2), 143–167.

[64] Marí Beffa, Gloria. 2003. Relative and absolute differential invariants for conformal curves. *J. Lie Theory*, **13**(1), 213–245.

[65] McLachlan, R., and Quispel, R. 2001. Six lectures on the geometric integration of ODEs. Pages 155–210 of: *Foundations of Computational Mathematics*. London Math. Soc. Lecture Note Ser., vol. 284. Cambridge: Cambridge Univ. Press.

[66] McLenaghan, R. G., and Smirnov, R. G. 2010. *Hamilton–Jacobi theory via Cartan Geometry*. Singapore: World Sci. Publ. to appear.

[67] McLenaghan, R. G., Smirnov, R. G., and The, D. 2004. An extension of the classical theory of algebraic invariants to pseudo-Riemannian geometry and Hamiltonian mechanics. *J. Math. Phys.*, **45**(3), 1079–1120.

[68] Mikula, K., and Ševčovič, D. 2001. Evolution of plane curves driven by a nonlinear function of curvature and anisotropy. *SIAM J. Appl. Math.*, **61**(5), 1473–1501.

[69] Moons, T., Pauwels, E. J., van Gool, L. J., and Oosterlinck, A. 1995. Foundations of semi-differential invariants. *Int. J. Comput. Vision*, **14**(1), 25–47.

[70] Morozov, O. 2002. Moving coframes and symmetries of differential equations. *J. Phys. A*, **35**(12), 2965–2977.

[71] Morozov, O. I. 2005. Structure of symmetry groups via Cartan's method: survey of four approaches. *SIGMA Symmetry Integrability Geom. Methods Appl.*, **1**, 006.

[72] Musso, E., and Nicolodi, L. 2009. Invariant signature of closed planar curves. *J. Math. Imaging Vision*, **35**(1), 68–85.

[73] Olver, P. J. 1993. *Applications of Lie Groups to Differential Equations*. Second edn. Grad. Texts in Math., vol. 107. New York: Springer.

[74] Olver, P. J. 1995. *Equivalence, Invariants, and Symmetry*. Cambridge: Cambridge Univ. Press.

[75] Olver, P. J. 1999. *Classical Invariant Theory*. London Math. Soc. Stud. Texts, vol. 44. Cambridge: Cambridge Univ. Press.

[76] Olver, P. J. 2000. Moving frames and singularities of prolonged group actions. *Selecta Math. (N.S.)*, **6**(1), 41–77.

[77] Olver, P. J. 2001a. Geometric foundations of numerical algorithms and symmetry. *Appl. Algebra Engrg. Comm. Comput.*, **11**(5), 417–436.

[78] Olver, P. J. 2001b. Joint invariant signatures. *Found. Comput. Math.*, **1**(1), 3–67.

[79] Olver, P. J. 2007. Generating differential invariants. *J. Math. Anal. Appl.*, **333**(1), 450–471.

[80] Olver, P. J. 2008. Invariant submanifold flows. *J. Phys. A*, **41**(34), 344017.

[81] Olver, P. J. 2009a. Differential invariants of maximally symmetric submanifolds. *J. Lie Theory*, **19**(1), 79–99.

[82] Olver, P. J. 2009b. Differential invariants of surfaces. *Differential Geom. Appl.*, **27**(2), 230–239.

[83] Olver, P. J., and Pohjanpelto, J. 2005. Maurer–Cartan forms and the structure of Lie pseudo-groups. *Selecta Math. (N.S.)*, **11**(1), 99–126.

[84] Olver, P. J., and Pohjanpelto, J. 2008. Moving frames for Lie pseudo-groups. *Canad. J. Math.*, **60**(6), 1336–1386.

[85] Olver, P. J., and Pohjanpelto, J. 2009. Differential invariant algebras of Lie pseudo-groups. *Adv. Math.*, **222**(5), 1746–1792.

[86] Olver, P. J., Sapiro, G., and Tannenbaum, A. 1994. Differential invariant signatures and flows in computer vision: a symmetry group approach. Pages 255–306 of: *Geometry-Driven Diffusion in Computer Vision*. Comput. Imaging Vision, vol. 1. Dordrecht: Kluwer.

[87] Pauwels, E. J., Moons, T., van Gool, L. J., Kempenaers, P., and Oosterlinck, A. 1995. Recognition of planar shapes under affine distortion. *Int. J. Comput. Vision*, **14**(1), 49–65.

[88] Rund, H. 1966. *The Hamilton–Jacobi Theory in the Calculus of Variations: Its Role in Mathematics and Physics*. London–Toronto–New York: D. Van Nostrand Co., Ltd.

[89] Sapiro, G. 2001. *Geometric Partial Differential Equations and Image Analysis*. Cambridge: Cambridge University Press.

[90] Shakiban, C., and Lloyd, P. 2004. Signature curves statistics of DNA supercoils. Pages 203–210 of: *Geometry, Integrability and Quantization*. Softex, Sofia.

[91] Shakiban, C., and Lloyd, R. 2005. Classification of signature curves using latent semantic analysis. Pages 152–162 of: *Computer Algebra and Geometric Algebra with Applications*. Lecture Notes in Comput. Sci., vol. 3519. New York: Springer.

[92] Shemyakova, E., and Mansfield, E. L. 2008. Moving frames for Laplace invariants. Pages 295–302 of: *ISSAC '08: Proceedings of the Twenty-First International Symposium on Symbolic and Algebraic Computation*. New York: ACM.

[93] Starostin, E. L., and van der Heijden, G. H. M. 2007. The shape of a Möbius strip. *Nature Materials*, **6**, 563–567.

[94] Valiquette, F. 2009. *Applications of Moving Frames to Lie Pseudo-Groups*. Ph.D. thesis, University of Minnesota.

[95] Welk, M., Kim, P., and Olver, P. J. 2007. Numerical invariantization for morphological PDE schemes. Pages 508–519 of: *Scale Space and Variational Methods in Computer Vision*. Lecture Notes in Comput. Sci., vol. 4485. New York: Springer.

[96] Yezzi, A., Kichenassamy, S., Kumar, A., Olver, P., and Tannenbaum, A. 1997. A geometric snake model for segmentation of medical imagery. *IEEE Trans. Medical Imaging*, **16**(2), 199–209.

9

Lattices of Compact Semisimple Lie Groups

Jiří Patera

Abstract

An efficient construction is to be described of lattice points F_M of any density and any admissible symmetry in a finite region F of a real n-dimensional Euclidean space. The shape of F and the lattice symmetry of F_M is determined by a compact semisimple Lie group of rank n. The density of F_M is fixed by our choice of a positive integer M, where $1 \leq M < \infty$. The Lie group allows one to introduce systems of special functions discretely orthogonal on F_M.

9.1 Introduction

The goal of this chapter is to provide all of the details necessary for construction of an n-dimensional lattice L_M of any symmetry and density in the real Euclidean space $\mathbb{R}^n$. The motivation for such a construction might be the need to process digital data on L_M. This typically requires a system of orthogonal functions on a finite fragment $F_M \subset L_M$. Such functions are available although their description is outside of the scope of this chapter. Some of the functions are shown here. But for their properties one needs to go to the references provided.

The starting point is a compact simple Lie group G of rank n, or equivalently, the corresponding simple Lie algebra g. Symmetry of its weight lattice $P(g)$ is the symmetry of the lattice L_M we construct. Density of L_M is determined by our choice of natural number M, that is $L_M = P(g)/M$.

The specific result of the paper is an easy-to-use guide to finding the points of $F_M \subset L_M$ within a finite region $F(g) \subset \mathbb{R}^n$. The action of the affine Weyl group $W^{aff}(g)$ on F_M then extends the points to the entire infinite lattice L_M. In most cases the stage of our problem is $F_M \subset F$ and not the entire lattice L_M.

We start with an example which motivates the need for the lattice points on a square F_M of the lattice L_M, namely a Fourier expansion of digital data function $f(x)$ given of the points $x \in F_M$. It is transparent without much of the group theory because it involves the simplest of the simple Lie groups. The only uncommon aspect of the example is the central splitting (9.2) of $f(x)$ before its processing. In doing that we follow the last sectiom of [6].

9.2　Motivating example

Consider a square F with vertices $(0,0),(0,1),(1,0),(1,1)$ in the real Euclidean plane $\mathbb{R}^2$, and the square grid F_M of points $(x_i, y_j) \in F$,

$$F_M := \left\{ x_i, y_j \mid x_i, y_j \in \left\{ 0, \frac{1}{M}, \frac{2}{M}, \ldots, \frac{M-1}{M}, 1 \right\} \right\}, \qquad M \in \mathbb{N}. \qquad (9.1)$$

Thus there are $(M+1)^2$ points in F_M.

Suppose that a real function $f(x_i, y_j)$ is given by its values at the points $(x_i, y_j) \in F_M$. Our goal is to develop $f(x_i, y_j)$ into the finite series of up to $(M+1)^2$ terms, using a suitable set of special functions orthogonal on F_M.

First we split $f(x_i, y_j)$ into the sum of four component functions.

$$f(x_i, y_j) = f_{00}(x_i, y_j) + f_{10}(x_i, y_j) + f_{01}(x_i, y_j) + f_{11}(x_i, y_j), \qquad (9.2)$$

where

$$f_{00}(x_i, y_j) = \tfrac{1}{4}\{f(x_i, y_j) + f(1-x_i, y_j) + f(x_i, 1-y_j) + f(1-x_i, 1-y_j)\}$$

$$f_{10}(x_i, y_j) = \tfrac{1}{4}\{f(x_i, y_j) - f(1-x_i, y_j) + f(x_i, 1-y_j) - f(1-x_i, 1-y_j)\}$$

$$f_{01}(x_i, y_j) = \tfrac{1}{4}\{f(x_i, y_j) + f(1-x_i, y_j) - f(x_i, 1-y_j) - f(1-x_i, 1-y_j)\}$$

$$f_{11}(x_i, y_j) = \tfrac{1}{4}\{f(x_i, y_j) - f(1-x_i, y_j) - f(x_i, 1-y_j) + f(1-x_i, 1-y_j)\}$$

Clearly the four component functions add up to $f(x_i, y_j)$ and have the following symmetry properties

$$\begin{aligned}
f_{00}(x_i, y_j) &= f_{00}(1-x_i, y_j) = f_{00}(x_i, 1-y_j) = f_{00}(1-x_i, 1-y_j) \\
f_{10}(x_i, y_j) &= -f_{10}(1-x_i, y_j) = f_{10}(x_i, 1-y_j) = -f_{10}(1-x_i, 1-y_j) \\
f_{01}(x_i, y_j) &= f_{01}(1-x_i, y_j) = -f_{01}(x_i, 1-y_j) = -f_{01}(1-x_i, 1-y_j) \\
f_{11}(x_i, y_j) &= -f_{11}(1-x_i, y_j) = -f_{11}(x_i, 1-y_j) = f_{11}(1-x_i, 1-y_j)
\end{aligned} \qquad (9.3)$$

Therefore it suffices to know the component functions on $\tfrac{1}{4}$ of the points of F_M, that is on the points (x_i, y_j) with $0 \le i, j[M/2]$. By the symmetries (9.3), we know them on all F_M.

Our expansion functions are to be the products $C_r(x_i)C_s(y_j)$, where

$$C_p\left(\frac{k}{M}\right) = e^{\pi i pk/M} + e^{-\pi i pk/M}, \qquad k \in \{0, 1, \dots, M\}, \ p \in \mathbb{Z}^{>0}$$

$$C_0\left(\frac{k}{M}\right) = 1$$

with the orthogonality property

$$\sum_{k=0}^{M} \varepsilon_k C_p\left(\frac{k}{M}\right)C_q\left(\frac{k}{M}\right) = 2M\delta_{pq} \times \begin{cases} 1, & p = 0 \\ 4, & p = M \\ 2, & 1 \le p \le M-1 \end{cases}, \qquad 0 \le p, q \le M$$

$$\varepsilon_k = \begin{cases} 2, & 1 \le k \le M-1 \\ 1, & k = 0, M \end{cases}$$

and the obvious symmetry property

$$C_p(x) = (-1)^p C_p(1-x), \qquad p \in \mathbb{Z}, \quad 0 \le x \le 1.$$

Rather than computing the $(M+1)^2$ coefficients d_{rs} of the expansion

$$f(x_i, y_j) = \sum_{r,s=0}^{M} d_{rs} C_r(x_i)C_s(y_j), \qquad (x_i, y_j) \in F_M, \tag{9.4}$$

$$d_{rs} = \sum_{x_i, y_j \in F_M} \varepsilon_i \varepsilon_j f(x_i, y_j) C_r(x_i) C_s(y_j), \tag{9.5}$$

one can find the expansions of the four component functions, each containing only $\frac{1}{4}$ of the terms,

$$f_{00}(x_i, y_j) = \sum_{\substack{r,\,s \text{ even}}} d_{rs} C_r(x_i)C_s(y_j),$$

$$f_{10}(x_i, y_j) = \sum_{\substack{r \text{ odd} \\ s \text{ even}}} d_{rs} C_r(x_i)C_s(y_j),$$

$$0 \le r, s \le M.$$

$$f_{01}(x_i, y_j) = \sum_{\substack{r \text{ even} \\ s \text{ odd}}} d_{rs} C_r(x_i)C_s(y_j),$$

$$f_{11}(x_i, y_j) = \sum_{\substack{r,\,s \text{ odd}}} d_{rs} C_r(x_i)C_s(y_j),$$

Here due to the symmetries (9.3), the expansion coefficients are calculated by summing up over $\frac{1}{4}$ of the points of F_M:

$$d_{rs} = 4(-1)^{r+s} \sum_{i,j=0}^{[M/2]} \varepsilon_i \varepsilon_j f_{a,b}(x_i, y_j) C_r(x_i) C_s(y_j),$$

$$a = r \pmod 2, b = s \pmod 2$$

Two remarks about the example

There is a vast generalization of this example to any number of variables, to lattices of any symmetry and any density. It starts by recognizing that the quantities appearing here can be naturally defined as attributes of the semisimple compact Lie group SU(2) × SU(2). The generalization proceeds by replacing this group by any other compact semisimple Lie group and by identifying analogous attributes of the new group. It turns out that such a scenario is not only possible, but even not difficult in principle, although the number of technical details that need to be invoked is much greater. All that is a standard information found in the literature [1].

The only non-standard step in the example is the central splitting (9.2). It is an elementary property in the case of SU(2) and consequently not much more complicated for SU(2)×SU(2). For a general semisimple Lie group it has been noted only recently [6].

9.3 Simple Lie groups and simple Lie algebras

9.3.1 Simple roots

The Lie algebra L of the compact simple Lie group G of rank n, are recognized by their set of simple roots $\Pi = \{\alpha_1, \ldots, \alpha_n\}$, spanning the Euclidean space $\mathbb{R}^n$.

By uniform and standard methods for G of any type and rank, a number of related quantities and virtually all the properties of G and L are determined from Π. It is a non-orthonormal basis of $\mathbb{R}^n$. Geometric relation between the simple roots are concisely given by the conventions implied in the Dynkin diagrams of all compact simple Lie groups or simple Lie algebras:

A circle of the diagram stands for a simple root. Relative lengths of simple roots (open and black circles) are given by the convention:

$$\frac{\text{square length of open circle}}{\text{square length of black circle}} = \begin{cases} 2:1 & \text{all cases but } G_2 \\ 3:1 & G_2 \end{cases}$$

Relative angles of simple roots

$$\text{Relative angles} = \begin{cases} 90° & \text{no directly connecting line} \\ 120° & \text{single line} \\ 135° & \text{double line} \\ 150° & \text{triple line} \end{cases}$$

A standard additional convention is made about the square length of the longer roots, represented by open circles of the Dynkin diagram, namely

$$\langle \text{open circle}, \text{open circle} \rangle = 2.$$

Disconnected diagrams refer to root systems of semisimple but not simple Lie G.

9.3.2 Standard bases in $\mathbb{R}^n$

Various properties of G and L of any type are best described using up to four different bases, the set of simple roots being only one of them. The bases are given the α-basis,

α-basis	consists of simple roots, see Dynkin diagrams
ω-basis	$\langle \alpha_j \mid \omega_k \rangle = \frac{1}{2}\langle \alpha_j \mid \alpha_j \rangle \delta_{kj}$
$\check{\alpha}$-basis	$\alpha_j = \frac{1}{2}\langle \alpha_j \mid \alpha_j \rangle \check{\alpha}$
$\check{\omega}$-basis	$\omega_j = \frac{1}{2}\langle \alpha_j \mid \alpha_j \rangle \check{\omega}$

(9.6)

where $\langle \cdot \mid \cdot \rangle$ denotes scalar product in $\mathbb{R}^n$. In particular,

$$\langle \omega_k \mid \check{\alpha}_j \rangle = \langle \alpha_j \mid \check{\omega}_k \rangle = \delta_{jk}.$$

In the general matrix form, one has the relation between α- and ω-bases:

$$\alpha = C\omega, \qquad \omega = C^{-1}\alpha, \qquad C = \frac{2\langle \alpha_j \mid \alpha_k \rangle}{\langle \alpha_k \mid \alpha_k \rangle} \qquad \text{(Cartan matrix)}$$

Examples of Cartan matrices

$$C(A_1) = (2), \qquad C(A_2) = \begin{pmatrix} 2 & -1 \\ -1 & 2 \end{pmatrix}, \qquad C(A_3) = \begin{pmatrix} 2 & -1 & 0 \\ -1 & 2 & -1 \\ 0 & -1 & 2 \end{pmatrix},$$

$$C(C_2) = \begin{pmatrix} 2 & -1 \\ -2 & 2 \end{pmatrix}, \qquad C(G_2) = \begin{pmatrix} 2 & -3 \\ -1 & 2 \end{pmatrix}.$$

9.3.3 Reflections and affine reflections in $\mathbb{R}^n$

With each simple Lie group G of rank n there is associated a unique finite reflection group, acting in $\mathbb{R}^n$, and called the Weyl group of G. It is generated by reflections in $(n-1)$-dimensional mirrors orthogonal to simple roots and passing through the origin of $\mathbb{R}^n$.

First consider a reflection r_ζ in a single mirror, orthogonal to ζ and passing through the origin of $\mathbb{R}^n$. It is given by

$$r_\zeta x = x - \frac{2\langle \zeta, x \rangle}{\langle \zeta, \zeta \rangle} \zeta, \qquad x, \zeta \in \mathbb{R}^n$$

The affine reflection R_ζ in mirror orthogonal to ζ and passing through the point $\frac{1}{2}\zeta$ is given by

$$R_\zeta x = \zeta + r_\zeta x.$$

For example one has

$$r_\zeta \xi = -\zeta \qquad R_\zeta \zeta = 0 \qquad r_\zeta^2 x = x$$
$$r_\zeta 0 = 0 \qquad R_\zeta 0 = \zeta \qquad R_\zeta^2 x = x$$

Note that the mirrors of the reflections r_ζ and R_ζ are parallel at the distance $\frac{1}{2}\zeta$. Combining r_ζ and R_ζ one has a translation in $\mathbb{R}^n$:

$$r_\zeta R_\zeta x = x - \zeta, \qquad R_\zeta r_\zeta x = x + \zeta.$$

9.3.4 Weyl group and Affine Weyl group

The Weyl group W of G, or equivalently of L, is generated by n reflections in mirrors orthogonal to simple roots,

$$W := \langle r_{\alpha_1}, r_{\alpha_2}, \ldots, r_{\alpha_n} \rangle.$$

The set of all roots Δ of G consists of the set of distinct vectors obtained by the action of W on the set of simple roots.

$$\Delta = W(\text{simple roots})$$

Table 9.1 *Orders $|W|$ of the Weyl groups of series of classical simple Lie groups*

$A_n \ (n \geq 1)$	$B_n \ (n \geq 3)$	$C_n \ (n \geq 2)$	$D_n \ (n \geq 4)$
$(n+1)!$	$2^n n!$	$2^n n!$	$2^{n-1} n!$

Table 9.2 *Orders $|W|$ of the Weyl groups of the exceptional simple Lie groups*

E_6	E_7	E_8	F_4	G_2
$2^7 3^4 5$	$2^{10} 3^4 5 7$	$2^{14} 3^5 5^2 7$	$2^7 3^2$	12

The affine Weyl group of G is generated by the set of reflections that generate the finite Weyl group W augmented by one affine reflection R_ξ, where ξ is the highest root of Δ of G.

$$W^{\text{aff}} = \langle r_{\alpha_1}, r_{\alpha_2}, \ldots, r_{\alpha_n}, R_\xi \rangle.$$

Application of W^{aff} to the origin generates the root lattice Q od G. There are two lattices related to G:

The root lattice : $\qquad Q = \mathbb{Z}\alpha_1 + \cdots + \mathbb{Z}\alpha_n \subset \mathbb{R}^n$

The weight lattice : $\qquad P = \mathbb{Z}\omega_1 + \cdots + \mathbb{Z}\omega_n \subset \mathbb{R}^n,$

In general, $Q \subset P$ except for G of types E_8, F_4, and G_2 where $Q = P$. We introduce also the sectors,

Positive chamber : $\qquad P^+ = \mathbb{Z}^{\geq 0}\omega_1 + \cdots + \mathbb{Z}^{\geq 0}\omega_n$

The interior of P^+ : $\qquad P^{++} = \mathbb{Z}^{>0}\omega_1 + \cdots + \mathbb{Z}^{>0}\omega_n$

There is the unique highest root $\xi \in \Delta(G)$

$$\xi = \sum_{k=1}^{n} m_k \alpha_k,$$

where m_k are called the marks of G. They are used to define the fundamental region F, (simplex, $n+1$ vertices) where the orthogonality of our special functions is assured,

$$F = \left\{ 0, \frac{\check{\omega}_1}{m_1}, \ldots, \frac{\check{\omega}_n}{m_n} \right\}$$

Here is the list of the highest roots for all types of G together with their marks:

$$A_n : \qquad \xi = \sum_{k=1}^{n} \alpha_k$$

$$B_n : \qquad \xi = \alpha_1 + 2 \sum_{k=2}^{n} \alpha_k$$

$$C_n : \qquad \xi = 2 \sum_{k=1}^{n-1} \alpha_k + \alpha_n$$

$$D_n : \qquad \xi = \alpha_1 + 2 \sum_{k=2}^{n_2} \alpha_k + \alpha_{n-1} + \alpha_n$$

$$E_6 : \qquad \xi = \alpha_1 + 2\alpha_2 + 3\alpha_3 + 2\alpha_4 + \alpha_5 + 2\alpha_6$$

$$E_7 : \qquad \xi = 2\alpha_1 + 3\alpha_2 + 4\alpha_3 + 3\alpha_4 + 2\alpha_5 + \alpha_6 + 2\alpha_7$$

$$E_8 : \qquad \xi = 2\alpha_1 + 3\alpha_2 + 4\alpha_3 + 5\alpha_4 + 6\alpha_5 + 4\alpha_6 + 2\alpha_7 + 3\alpha_8$$

$$F_4 : \qquad \xi = 2\alpha_1 + 3\alpha_2 + 4\alpha_3 + 2\alpha_4$$

$$G_2 : \qquad \xi = 2\alpha_1 + 3\alpha_2$$

9.4 Lattice grids $F_M \subset F \subset \mathbb{R}^n$

Let $M \in \mathbb{N}$ be a fixed number. The refinement of the weight lattice P by M is the lattice

$$L_M = \frac{1}{M}P = \frac{\mathbb{Z}}{M}\omega_1 + \cdots + \frac{\mathbb{Z}}{M}\omega_n.$$

The lattice grid F_M is the intersection $F_M = F \cap L_M$.

For a fixed value of $M \in \mathbb{N}$, the points of F_M are found as all the sets $s = [s_0, s_1, \ldots, s_n]$ of non-negative integers satisfying the following sum rule:

$$M = s_0 + \sum_{k=1}^{n} m_k s_k, \qquad s_0, s_1, \ldots, s_n \in \mathbb{Z}^{\geq 0},$$

where m_k are the marks of G of rank n. There is a one-to-one correspondence between s and points of F_M. More precisely,

$$[s_0, s_1, \ldots, s_n] \iff \left(\frac{s_1}{M}\breve{\omega}_1, \frac{s_2}{M}\breve{\omega}_2, \ldots, \frac{s_n}{M}\breve{\omega}_n \right) \in F_M.$$

9.4.1 Examples of F_M

Drawings of other examples in 2D are found in [10, 11].

A_1: $F = \{0, \omega\}$, $M = s_0 + s_1 = 2$

$$[2,0] = 0, \qquad [1,1] = \tfrac{1}{2}\omega \qquad [02] = \omega$$

A_2: $F = \{0, \omega_1, \omega_2\}$, $M = s_0 + s_1 + s_2 = 2$

$$[2,0,0] = 0 \qquad [1,1,0] = \tfrac{1}{2}\omega_1$$
$$[0,2,0] = \omega_1 \qquad [1,0,1] - \tfrac{1}{2}\omega_2$$
$$[0,0,2] = \omega_2 \qquad [0,1,1] = \tfrac{1}{2}\omega_1 + \tfrac{1}{2}\omega_2$$

A_2: $F = \{0, \omega_1, \omega_2\}$, $M = s_0 + s_1 + s_2 = 3$

$$[3,0,0] = 0 \qquad [2,1,0] = \tfrac{1}{3}\omega_1 \qquad [1,2,0] = \tfrac{2}{3}\omega_1$$
$$[0,3,0] = \omega_1 \qquad [2,0,1] = \tfrac{1}{3}\omega_2 \qquad [1,0,2] = \tfrac{1}{3}\omega_2$$
$$[0,0,3] = \omega_2 \qquad [0,2,1] = \tfrac{2}{3}\omega_1 + \tfrac{1}{3}\omega_2 \qquad [0,1,2] = \tfrac{1}{3}\omega_1 + \tfrac{2}{3}\omega_2$$
$$[1,1,1] = \tfrac{1}{3}\omega_1 + \tfrac{1}{3}\omega_2$$

C_2: $F = \{0, \tfrac{1}{2}\check{\omega}_1, \check{\omega}_2\}$, $M = s_0 + 2s_1 + s_2 = 3$

$$[3,0,0] = 0 \qquad [1,1,0] = \tfrac{1}{3}\check{\omega}_1 \qquad [0,1,1] = \tfrac{1}{3}\check{\omega}_1 + \tfrac{1}{3}\check{\omega}_2$$
$$[0,0,3] = \check{\omega}_2 \qquad [2,0,1] - \tfrac{1}{3}\check{\omega}_2 \qquad [1,0,2] = \tfrac{2}{3}\check{\omega}_2$$

G_2: $F = \{0, \tfrac{1}{2}\check{\omega}_1, \tfrac{1}{3}\check{\omega}_2\}$, $M = s_0 + 2s_1 + 3s_2 = 3$

$$[3,0,0] = 0 \qquad [1,1,0] = \tfrac{1}{3}\check{\omega}_1 \qquad [0,0,1] = \tfrac{1}{3}\check{\omega}_2$$

D_4: $F = \{0, \omega_1, \tfrac{1}{2}\omega_2, \omega_3, \omega_4\}$, $M = s_0 + s_1 + 2s_2 + s_3 + s_4 = 3$

$$[3,0,0,0,0] = 0 \qquad [0,0,0,3,0] = \omega_3 \qquad [0,0,0,0,3] = \omega_4$$
$$[2,1,0,0,0] = \tfrac{1}{3}\omega_1 \qquad [2,0,0,1,0] = \tfrac{1}{3}\omega_3 \qquad [2,0,0,0,1] = \tfrac{1}{3}\omega_4$$
$$[1,2,0,0,0] = \tfrac{2}{3}\omega_1 \qquad [1,0,0,2,0] = \tfrac{2}{3}\omega_3 \qquad [1,0,0,0,2] = \tfrac{2}{3}\omega_4$$
$$[1,1,0,1,0] = \tfrac{1}{3}\omega_1 + \tfrac{1}{3}\omega_3 \qquad\qquad [1,0,0,1,1] = \tfrac{1}{3}\omega_3 + \tfrac{1}{3}\omega_4$$
$$[0,1,0,1,1] = \tfrac{1}{3}\omega_1 + \tfrac{1}{3}\omega_3 + \tfrac{1}{3}\omega_4 \qquad [1,0,1,0,0] = \tfrac{1}{3}\omega_2$$
$$[1,0,1,0,0] = \tfrac{1}{3}\omega_2 \qquad\qquad [0,1,1,0,0] = \tfrac{1}{3}\omega_1 + \tfrac{1}{3}\omega_2$$
$$[0,0,1,1,0] = \tfrac{1}{3}\omega_2 + \tfrac{1}{3}\omega_3 \qquad [0,0,1,0,1] = \tfrac{1}{3}\omega_2 + \tfrac{1}{3}\omega_4$$

B_3: $\xi = \alpha_1 + 2\alpha_2 + 2\alpha_3$

$M = 1$ $[1,0,0,0] = (0,0,0)$ $[0,1,0,0] = (1,0,0)$

$M = 2$ $[2,0,0,0] = (0,0,0)$ $[0,2,0,0] = (1,0,0)$

          $[1,1,0,0] = \frac{1}{2}(1,0,0)$ $[0,0,1,0] = \frac{1}{2}(0,1,0)$

$M = 4$ $[4,0,0,0] = (0,0,0)$ $[0,4,0,0] = (1,0,0)$

          $[2,2,0,0] = \frac{1}{2}(1,0,0)$ $[0,0,2,0] = \frac{1}{2}(0,1,0)$

          $[3,1,0,0] = \frac{1}{4}(1,0,0)$ $[2,0,1,0] = \frac{1}{4}(0,1,0)$

          $[0,2,0,1] = \frac{1}{4}(2,0,1)$ $[0,2,1,0] = \frac{1}{4}(2,1,0)$

          $[0,0,1,1] = \frac{1}{4}(0,1,1)$ $[1,3,0,0] = \frac{1}{4}(3,0,0)$

          $[1,1,0,1] = \frac{1}{4}(1,0,1)$ $[1,1,1,0] = \frac{1}{4}(1,1,0)$

          $[2,0,0,1] = \frac{1}{4}(0,0,1)$ $[0,0,0,2] = \frac{1}{2}(0,0,1)$

9.5 W-invariant functions orthogonal on F_M

Here are presented the functions of two families, called C- and S-family. They were identified in [9]. On a finite fragment $F_M \in \mathbb{R}^n$, containing $|F_M|$ points, at most $|F_M\|$ functions can be pairwise orthogonal. All information required for the exploitation of the orthogonality properties of the functions on F_M is provided in [1] for all compact simple Lie groups. Extensive reviews of the C- and S-functions are in [2, 3]. Transformation of the C- and S-functions into orthogonal polynomials of n variables is described in [7, 8].

Denote by W_λ the distinct points generated from $\lambda \in P^+$ by W, and by $|W_\lambda|$ the number of points in W_λ. We have $1 \le |W_\lambda| \le |W|$.

The functions are defined as

$$C\text{-function} \quad C_\lambda(x) = \sum_{\mu \in W_\lambda} e^{2\pi i \langle \mu, x \rangle}$$

$$S\text{-function} \quad S_\lambda(x) = \sum_{\mu \in W_\lambda} (-1)^{l(\mu)} e^{2\pi i \langle \mu, x \rangle}, \qquad x \in \mathbb{R}^n$$

Here $\langle \mu, x \rangle$ is the scalar product in $\mathbb{R}^n$, and $l(\mu)$ is the number of elementary reflections one needs to transform λ into μ.

Generic orbit of $W(A_2)$ has six points. Therefore generic C- and S-function of A_2 are the following,

$$C_\lambda(x) = e^{2\pi i \langle \lambda, x \rangle} + e^{2\pi i \langle r_1 \lambda, x \rangle} + e^{2\pi i \langle r_2 \lambda, x \rangle}$$

$$+ e^{2\pi i \langle r_2 r_1 \lambda, x \rangle} + e^{2\pi i \langle r_1 r_2 \lambda, x \rangle} + e^{2\pi i \langle r_1 r_2 r_1 \lambda, x \rangle}$$

Note that $r_1 r_2 r_1 = r_2 r_1 r_2$

$$S_\lambda(x) = e^{2\pi i \langle \lambda, x \rangle} - e^{2\pi i \langle r_1 \lambda, x \rangle} - e^{2\pi i \langle r_2 \lambda, x \rangle}$$
$$+ e^{2\pi i \langle r_2 r_1 \lambda, x \rangle} + e^{2\pi i \langle r_1 r_2 \lambda, x \rangle} - e^{2\pi i \langle r_1 r_2 r_1 \lambda, x \rangle}$$

Reviews of properties of C- and S-functions are in [2, 3] respectively. Two-dimensional examples of the functions are plotted in [10, 11] for the four semisimple Lie algebras of rank 2. In 1-dimension, the definitions give $C_\lambda(x) = 2\cos(2\pi\lambda x)$ and $S_\lambda(x) = 2i\sin(2\pi\lambda x)$.

9.6 Properties of elements of finite order

A conjugacy class of elements g of a compact simple Lie group G consists of all the elements hgh^{-1}, $\forall h \in G$. We say that g is of adjoint order $M \in \mathbb{N}$, provided $g^M = 1$, where M is the lowest possible integer. Elements of a conjugacy class are of the same order.

Any class $\underline{s}$ of elements of finite order (EFOs) in G is uniquely specified by the sets

$$[s_0, s_1, \ldots, s_n] := \{s_k \in \mathbb{Z}^{\geq 0}\} \mid \gcd\{s_k\} = 1, \ 0 \leq k \leq n\},$$

where n is the rank of G. The order M of the elements of $\underline{s}$ is given by

$$M = s_0 + \sum_{i=1}^{n} m_i s_i.$$

Elements of $\underline{s}$ of adjoint order M are represented by the point x relative to the $\check{\omega}$-basis

$$x = \frac{1}{M} \sum_{i=1}^{n} s_i \check{\omega}_i \in F \tag{9.7}$$

From (9.7) we see that the $n + 1$ vertices of the simplex F are the following points,

$$F = \left\{ 0, \frac{\check{\omega}_1}{m_1}, \frac{\check{\omega}_2}{m_2}, \ldots, \frac{\check{\omega}_n}{m_n} \right\}.$$

Theorem 9.1 (V. Kac, 1979) *Every* $[s_0, s_1, \ldots, s_n] \in F_M$ *such that* $\gcd\{s_0, s_1, \ldots, s_n\} = 1$, *corresponds to one conjugacy class of elements of order* M *in* G. *Every conjugacy class of elements of order* M *in* G *is specified in this way.*

The construction of matrices of the diagonal representatives of any conjugacy class and many other properties are described in [4, 5], see also [1] and the references therein.

References

[1] Hrivnák, J., and Patera, J. 2009. On discretization of tori of compact simple Lie groups. *J. Phys. A*, **42**(38), 385208.

[2] Klimyk, A., and Patera, J. 2006. Orbit functions. *SIGMA Symmetry Integrability Geom. Methods Appl.*, **2**, Paper 006.

[3] Klimyk, A., and Patera, J. 2007. Antisymmetric orbit functions. *SIGMA Symmetry Integrability Geom. Methods Appl.*, **3**, Paper 023.

[4] Moody, R. V., and Patera, J. 1984. Characters of elements of finite order in Lie groups. *SIAM J. Algebraic Discrete Methods*, **5**(3), 359–383.

[5] Moody, R. V., and Patera, J. 1987. Computation of character decompositions of class functions on compact semisimple Lie groups. *Math. Comp.*, **48**(178), 799–827.

[6] Moody, R. V., and Patera, J. 2006. Orthogonality within the families of *C*-, *S*-, and *E*-functions of any compact semisimple Lie group. *SIGMA Symmetry Integrability Geom. Methods Appl.*, **2**, Paper 076.

[7] Nesterenko, M., Patera, J., and Tereszkiewicz, A. 2010. *Orthogonal polynomials of compact simple Lie groups.* arXiv:1001.3683.

[8] Nesterenko, M., Patera, J., Szjewska, M., and Tereszkiewicz, A. 2010. *Orthogonal polynomials of compact simple Lie groups. The branching rules for polynomials,* J. Phys. A: Math. Theor. **43**, 495207 (20pp); arXiv:1007.4431v1

[9] Patera, J. 2005. Compact simple Lie groups and their *C*-, *S*-, and *E*-transforms. *SIGMA Symmetry Integrability Geom. Methods Appl.*, **1**, Paper 025.

[10] Patera, J., and Zaratsyan, A. 2005a. Discrete and continuous cosine transform generalized to Lie groups SU(2)×SU(2) and O(5). *J. Math. Phys.*, **46**(5), 053514.

[11] Patera, J., and Zaratsyan, A. 2005b. Discrete and continuous cosine transform generalized to Lie groups SU(3) and G(2). *J. Math. Phys.*, **46**(11), 113506.

10

Lectures on Discrete Differential Geometry

Yuri B. Suris

This series of lectures was based on the textbook with A. Bobenko:

A. I. Bobenko, Yu. B. Suris, *Discrete Differential Geometry: Integrable Structure.*
Grad. Studies in Math., vol. 98. Providence, RI: Amer. Math. Soc., 2008. This textbook
contains an extensive bibliography.

The preliminary version of the book is available online at arXiv:math.DG/
0504358.

The lectures aimed at giving an introduction to the main ideas of (an integrable part of) discrete differential geometry. I am grateful to organizers of
the Summer School on Symmetries and Integrability of Difference Systems
(Montréal, 2008) for inviting me to give these lectures and for the financial
support.

Discrete differential geometry (DDG) is a new field presently emerging on
the border between differential and discrete geometry. It develops discrete
analogues and equivalents of notions and methods of differential geometry of
smooth curves, surfaces etc. The smooth theory appears in a limit of the refinement of discretizations. Current interest in this field derives not only from its
importance in pure mathematics but also from its relevance for other fields, like
computer graphics, architectural design and numerics. *Integrable differential
geometry* deals with parametrized objects (surfaces and coordinate systems)
described by integrable differential equations. As integrability attributes one
counts traditionally: zero curvature representations, transformations with remarkable permutability properties, hierarchies of commuting flows etc. Development of DDG led, somewhat unexpectedly, among other things, to a better
understanding of the very notion of integrability. Since the above mentioned
textbook contains detailed bibliographical notes, we decided to omit them from
the present lectures and give only some brief hints. The reader is advised to
look in the book for more complete references.

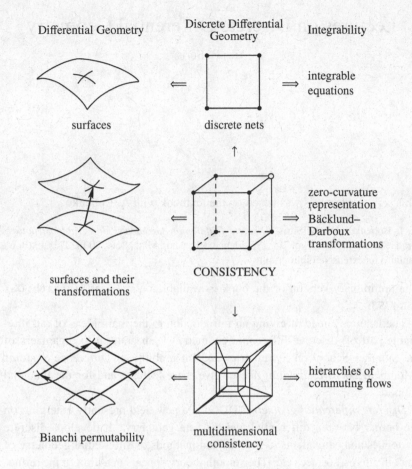

Figure 10.1 The consistency-principle of discrete differential geometry as conceptual basis of the differential geometry of special surfaces and of integrability.

The conceptual view of discrete differential geometry as the basis of the theory of surfaces and their transformations as well as of the theory of integrable systems is schematically represented in Figure 10.1. Some of the arrows on this figure will be explained in this text.

10.1 Basic notions

The basic notion of DDG is that of a *discrete net*, i.e., a map $f : \mathbb{Z}^m \to X$. Here X is some space; in the most straightforward examples, like geometric nets discussed below, $\mathcal{X}$ is just the ambient space of the underlying geometry, like $\mathcal{X} = \mathbb{RP}^N$, however more intricate mathematical models require for other spaces $\mathcal{X}$, and we will deal with these less trivial situations, as well. Notation for discrete nets: $f = f(u), f_i = f(u + e_i), f_{ij} = f(u + e_i + e_j)$, etc. Here e_i is the ith coordinate vector of $\mathbb{Z}^m$. The translation and difference operators are defined in a standard manner:

$$(\tau_i f)(u) = f(u + e_i), \qquad (\delta_i f)(u) = f(u + e_i) - f(u).$$

Thus, f_i, f_{ij} are synonymous to $\tau_i f, \tau_i \tau_j f$, etc. The symbol $C_{i_1 \ldots i_s}$ is for the elementary s-dimensional cube with the 2^s vertices $u + \epsilon_{i_1} e_{i_1} + \cdots + \epsilon_{i_s} e_{i_s}$, $c_i \subset \{0, 1\}$.

Definition 10.1 A (hyperbolic) 2D system (with fields assigned to vertices) is a geometric condition, equation, etc., which allows to determine the 4th point of an elementary square of $\mathbb{Z}^2$ if other three are arbitrarily prescribed.

These is schematically represented in Figure 10.2 (left). An example of a 2D system with complex-valued fields is given by the so called *Hirota equation* which reads

$$\frac{f_{12}}{f} = \frac{\alpha_2 f_2 - \alpha_1 f_1}{\alpha_2 f_1 - \alpha_1 f_2}, \tag{10.1}$$

Here α_1, α_2 are parameters, which are natural to think of as assigned to the edges parallel to the 1st, resp. 2nd coordinate direction. Putting formula (10.1) in a more memorizable form,

$$\alpha_1 f f_1 + \alpha_2 f_1 f_{12} - \alpha_1 f_{12} f_2 - \alpha_2 f_2 f = 0$$

(each term corresponding to an edge of an elementary square), makes it clear that Hirota equation is uniquely solvable for any of the four vertex variables which are actually on an equal footing.

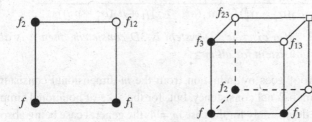

Figure 10.2 Left: an elementary square of a 2D system. Right: 3D consistency of a 2D system. Black circles mark the initial data; white circles mark the vertices uniquely determined by the initial data; the white square marks the vertex where the consistency condition appears.

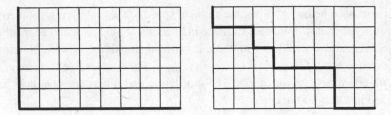

Figure 10.3 Initial data for a Goursat (left) and for a Cauchy (right) problem for a 2D system

Natural boundary value problems for hyperbolic 2D systems are *Goursat problems* (prescribing initial data along coordinate axes) and *Cauchy problems* (prescribing initial data along a noncharacteristic staircase line), see Figure 10.3.

One of the fundamental *organizing principles* of integrable DDG is the *multidimensional consistency principle*: discretizations of surfaces, coordinate systems and other parametrized objects should be extendable to multidimensionally consistent nets. The same should hold for (difference) equations describing these entities. Multidimensional consistency is, in our understanding, synonymous with integrability.

For a 2D system, the possibility to impose it everywhere on the *m*-dimensional lattice hinges on the case $m = 3$. The corresponding property of 3D *consistency* is schematically shown in Figure 10.2 (right) and should be understood as follows: suppose that four values f, f_1, f_2, f_3 are given. Then the 2D system defines f_{12}, f_{23} and f_{13}, and a further application of this system gives three a priori different values of f_{123}. The system is called 3D consistent if these three values automatically coincide for arbitrary initial data.

Exercise Show that Hirota equation (10.1) is 3D consistent, with

$$f_{123} = \frac{\alpha_3(\alpha_2^2 - \alpha_1^2)f_1f_2 + \alpha_1(\alpha_3^2 - \alpha_2^2)f_2f_3 + \alpha_2(\alpha_1^2 - \alpha_3^2)f_3f_1}{\alpha_3(\alpha_2^2 - \alpha_1^2)f_3 + \alpha_1(\alpha_3^2 - \alpha_2^2)f_1 + \alpha_2(\alpha_1^2 - \alpha_3^2)f_2}. \tag{10.2}$$

Theorem 10.2 *If a discrete 2D system is 3D consistent, then it is also m-dimensionally consistent for all $m \geq 3$.*

Proof The proof goes by induction from the *m*-dimensional consistency to the $(m + 1)$-dimensional consistency, but, for the sake of notational simplicity, we present the details only for the case $m = 4$, the general case being absolutely similar. Initial data for a 2D system on the 4D cube C_{1234} consist of the fields f and f_i for all $1 \leq i < j \leq 4$. From these data one first gets six fields f_{ij} for $1 \leq i < j \leq 4$, and then four fields f_{ijk} for $1 \leq i < j < k \leq 4$ (the fact that the latter are well defined is nothing but the assumed 3D consistency for the 3D cubes C_{ijk}). Now, one has six possibly different values for f_{1234}, coming from six elementary squares $\tau_i \tau_j C_{kl}$. To prove that these six values coincide, consider four 3D cubes $\tau_i C_{jkl}$. For instance, for the 3D cube $\tau_1 C_{234}$ the assumed 3D consistency assures that the three values for f_{1234} coming from three elementary squares

$$\tau_1 \tau_2 C_{34}, \qquad \tau_1 \tau_3 C_{24}, \qquad \tau_1 \tau_4 C_{23}$$

are all the same. Similarly, for the 3D cube $\tau_2 C_{134}$ the 3D consistency leads to the conclusion that the three values for f_{1234} coming from

$$\tau_1 \tau_2 C_{34}, \qquad \tau_2 \tau_3 C_{14}, \qquad \tau_2 \tau_4 C_{13}$$

coincide. Note that the elementary square $\tau_1 \tau_2 C_{34}$, the intersection of $\tau_1 C_{234}$ and $\tau_2 C_{134}$, is present in both lists, so that we now have five coinciding values for f_{1234}. Adding similar conclusions for the last two 3D cubes $\tau_3 C_{124}$ and $\tau_4 C_{123}$, we arrive at the desired result.   □

We postpone the discussion of the fundamental consequences of 3D consistency, like Bäcklund transformations and zero curvature representation, and of concrete examples, and turn now to the analogous notions in one dimension higher.

Definition 10.3   A (hyperbolic) 3D system (with fields assigned to vertices) is a geometric condition, equation, etc., which allows to determine the 8th point of an elementary cube of $\mathbb{Z}^3$ if other seven are arbitrarily prescribed.

A schematic presentation of a discrete 3D system is given in Figure 10.4 (left). A *Goursat problem* for a discrete 3D system consists of prescribing initial data along coordinate planes, while a *Cauchy problem* amounts to prescribing initial data along a noncharacteristic stepped surface.

The property of 4D consistency of a 3D system is schematically shown on Figure 10.4 (right) and is interpreted as follows. Suppose that the initial data f, f_i for $1 \le i \le 4$, and f_{ij} for $1 \le i < j \le 4$ are given. Then the 3D system defines f_{ijk} for $1 \le i < j < k \le 4$, and a further application of this equation gives four a priori different values of f_{1234}. The system is called 4D consistent if these three values automatically coincide for arbitrary initial data.

Theorem 10.4   *If a discrete 3D system is 4D consistent, then it is also m-dimensionally consistent for all $m \ge 4$.*

The proof is parallel to that of Theorem 10.2, and is left to the reader as an exercise.

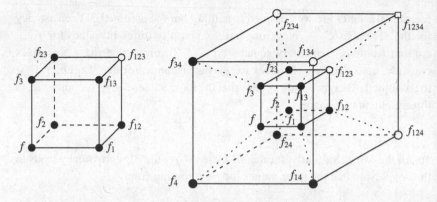

Figure 10.4 Left: an elementary cube of a 3D system. Right: 4D consistency of a 3D system. Black circles mark the initial data; white circles mark the vertices uniquely determined by the initial data; the white square marks the vertex where the consistency condition appears.

10.2 Bäcklund transformations

One of the major consequences of the multidimensional consistency is the existence of remarkable transformations with certain permutability properties, known under the collective name of *Bäcklund transformations*.

We start with 2D systems, for which the general construction can be described as follows. Let the system

$$Q(f, f_i, f_{ij}, f_j) = 0 \qquad (10.3)$$

be 3D consistent. Then for any solution $f \colon \mathbb{Z}^m \to \mathcal{X}$ of the system (10.3), one can construct a new solution of $f^+ \colon \mathbb{Z}^m \to \mathcal{X}$ of the same system, as soon as its value at one point, say $f^+(0)$, is prescribed. Indeed, the construction is performed pointwise, based on the defining relation

$$Q(f, f_i, f_i^+, f^+) = 0 \qquad (10.4)$$

which is required for all edges (f, f_i) of $\mathbb{Z}^m$. This is illustrated in Figure 10.5 (left). Such a solution f^+ is called a *Bäcklund transform* (BT) of f. Thus, performing a Bäcklund transformation is nothing but adding an additional lattice direction to the domain.

Due to the multidimensional consistency, Bäcklund transformations possess the following permutability property.

Theorem 10.5 *Let* $f^{(1)}, f^{(2)} \colon \mathbb{Z}^m \to \mathcal{X}$ *be two BT's of a solution* $f \colon \mathbb{Z}^m \to \mathcal{X}$ *of a 3D consistent 2D system. Then there exists a unique solution* $f^{(12)} \colon \mathbb{Z}^m \to \mathcal{X}$ *that is simultaneously a BT of* $f^{(1)}$ *and of* $f^{(2)}$, *see Figure 10.5 (right). It is*

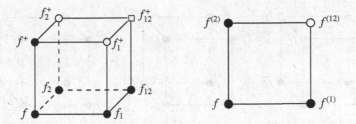

Figure 10.5 Left: construction of a BT for a 2D system. The transform is determined by the value f^+ marked by a black circle; values f_1^+ and f_2^+ marked by white circles are uniquely determined via the equation; value f_{12}^+ marked by the white square is well defined due to the consistency condition. Right: permutability of BTs. The common BT $f^{(12)}$ of $f^{(1)}$ and $f^{(2)}$ is uniquely defined via an algebraic relation.

given by the relation

$$Q(f, f^{(1)}, f^{(12)}, f^{(2)}) = 0. \tag{10.5}$$

Remark In the common case when the edges of the ith coordinate dimension of $\mathbb{Z}^m$ carry the parameters α_i, $1 \leq i \leq m$, so that equation (10.3) takes the form

$$Q(f, f_i, f_{ij}, f_j; \alpha_i, \alpha_j) = 0, \tag{10.6}$$

one can assume that the edges of the additional $(m + 1)$th direction carry the so called *Bäcklund parameter* λ, so that the defining equation (10.4) of the Bäcklund transformation takes the form

$$Q(f, f_i, f_i^+, f^+; \alpha_i, \lambda) = 0. \tag{10.7}$$

The superposition formula for two BTs with parameters λ, μ takes the form

$$Q(f, f^{(1)}, f^{(12)}, f^{(2)}; \lambda, \mu) = 0. \tag{10.8}$$

We now turn to 3D systems. As soon as such a system is 4D consistent, for any its solution $f: \mathbb{Z}^m \to X$ one can construct a new solution $f^+: \mathbb{Z}^m \to X$, as soon as its values $f^+(\mathbb{Z}e_i)$, $1 \leq i \leq m$, are prescribed. Thus, in the 3D case a Bäcklund transform is determined by m functions of one discrete variable. The construction is illustrated in Figure 10.6 (left).

Theorem 10.6 (a) *Let $f: \mathbb{Z}^m \to X$ be a solution of a 4D consistent 3D system, and let $f^{(1)}, f^{(2)}: \mathbb{Z}^m \to X$ be two of its BTs. Then there exists a family of further solutions $f^{(12)}: \mathbb{Z}^m \to X$ that are BTs of both $f^{(1)}$ and $f^{(2)}$. Such a solution is uniquely specified by an admissible choice of its value at one point, say $f^{(12)}(0)$. See Figure 10.6 (middle).*

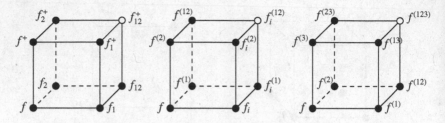

Figure 10.6 Left: continuation of a BT f^+ from the coordinate axes. Middle: continuation of a common BT $f^{(12)}$ of two BTs $f^{(1)}, f^{(2)}$ from the origin. Right: superposition principle of BT's for a 3D system.

(b) *Let $f\colon \mathbb{Z}^m \to X$ be a solution of a 4D consistent 3D system. Let $f^{(1)}, f^{(2)},$* *$f^{(3)}\colon \mathbb{Z}^m \to X$ be three of its BTs, and let three further solutions $f^{(12)}, f^{(23)},$* *$f^{(13)}\colon \mathbb{Z}^m \to X$ be given such that each $f^{(ij)}$ is a simultaneous BT of $f^{(i)}$* *and $f^{(j)}$. Then there exists a unique solution $f^{(123)}\colon \mathbb{Z}^m \to X$ that is a* *simultaneous BT of $f^{(12)}, f^{(23)}$ and $f^{(13)}$. See Figure 10.6 (right).*

10.3 Q-nets

Q-nets serve as a discretization of conjugate nets and belong to the most fundamental objects of DDG. They were introduced in the 2D setting by R. Sauer in mid-1930s and in the multidimensional setting by A. Doliwa and P. Santini in 1997.

Definition 10.7 A map $f\colon \mathbb{Z}^m \to \mathbb{R}^N$ is called an *m-dimensional Q-net* if all its elementary quadrilaterals are planar, i.e., if at every $u \in \mathbb{Z}^m$ and for every pair $1 \le i \ne j \le m$ the four points $f, \tau_i f, \tau_j f,$ and $\tau_i \tau_j f$ are coplanar:

$$\delta_i \delta_j f = c_{ji} \delta_i f + c_{ij} \delta_j f, \qquad i \ne j. \tag{10.9}$$

Here it is convenient to think that the real numbers c_{ij}, c_{ji}, as well as equation (10.9) itself, are assigned to elementary squares C_{ij} of $\mathbb{Z}^m$ parallel to the coordinate plane (ij).

m = **2: discrete surfaces parametrized by conjugate lines** To define a Q-surface $f\colon \mathbb{Z}^2 \to \mathbb{R}^N$, one can prescribe arbitrarily its two coordinate lines and then fill in the whole net inductively. The induction step consists in choosing f_{12} in the plane through f, f_1 and f_2, provided the latter three points are known (and are in general position). Thus, one has a freedom of two real parameters on each step.

m = 3: **the basic 3D system** Suppose that three coordinate surfaces of a three-dimensional Q-net f are given. Of course, each one of them is a Q-surface. To extend the net into the octant $\mathbb{Z}_+^3$, one proceeds by induction whose step consists in determining f_{123}, provided f, f_i and f_{ij} are known. The planarity condition determines the point f_{123} uniquely.

Theorem 10.8 *Given seven points f, f_i and f_{ij} ($1 \leq i < j \leq 3$) in $\mathbb{R}^N$, such that each of the three quadrilaterals (f, f_i, f_{ij}, f_j) is planar, define three planes $\tau_i \Pi_{jk}$ as those passing through the point triples (f_i, f_{ij}, f_{ik}), respectively. Then these three planes intersect at one point:*

$$f_{123} = \tau_1 \Pi_{23} \cap \tau_2 \Pi_{13} \cap \tau_3 \Pi_{12}. \tag{10.10}$$

Proof Planarity of the quadrilaterals (f, f_i, f_{ij}, f_j) assures that all seven initial points f, f_i and f_{ij} belong to the three-dimensional affine space Π_{123} through the four points f, f_i. Hence, the planes $\tau_i \Pi_{jk}$ lie in this three-dimensional space, and therefore generically they intersect at exactly one point. □

Remark In the spirit of local differential geometry, we always assume that all the data are in general position, without specifying this explicitly. In particular, in the previous theorem it was silently assumed that the four points f, f_i span a three-dimensional space, and that no three points f_i, f_{ij}, f_{ik} are collinear.

m = 4: **consistency** The content of Theorem 10.8 can be summarized as follows: Q-nets are described by a 3D system. This system turns out to possess a remarkable property.

Theorem 10.9 *The 3D system governing Q-nets is 4D consistent.*

Proof We have to show that the following four points automatically coincide:

$$f_{1234} = \tau_1 \tau_2 \Pi_{34} \cap \tau_1 \tau_3 \Pi_{24} \cap \tau_1 \tau_4 \Pi_{23},$$

and the three others obtained by cyclic shifts of indices. Thus, we have to prove that the six planes $\tau_i \tau_j \Pi_{kl}$ intersect in one point. Here we give a proof for the general position case only, when the ambient space $\mathbb{R}^N$ has dimension $N \geq 4$, and the affine space Π_{1234} through the five points f, f_i ($1 \leq i \leq 4$) is four-dimensional. It is easy to understand that the plane $\tau_i \tau_j \Pi_{kl}$ is the intersection of two three-dimensional subspaces $\tau_i \Pi_{jkl}$ and $\tau_j \Pi_{ikl}$. Indeed, the subspace $\tau_i \Pi_{jkl}$ through the four points $f_i, f_{ij}, f_{ik}, f_{il}$ contains also f_{ijk}, f_{ijl}, and f_{ikl}. Therefore, both $\tau_i \Pi_{jkl}$ and $\tau_j \Pi_{ikl}$ contain the three points f_{ij}, f_{ijk} and f_{ijl}, which determine the plane $\tau_i \tau_j \Pi_{kl}$. Now the intersection in question can be alternatively described as the intersection of the four three-dimensional subspaces $\tau_1 \Pi_{234}$,

$\tau_2\Pi_{134}$, $\tau_3\Pi_{124}$ and $\tau_4\Pi_{123}$ of one and the same four-dimensional space Π_{1234}. This intersection consists in the general case of exactly one point. $\square$

Exercise Find the way to prove the last theorem for the case $N = 3$.

Theorem 10.9 immediately yields Bäcklund transformations for Q-nets (known in this case as fundamental, or F-transformations). The characteristic property of F-transformations is the planarity of the "side quadrilaterals" (f, f_i, f_i^+, f^+). Thus, an m-dimensional Q-net and its F-transform build nothing but a two-layer slice of a $(m + 1)$-dimensional Q-net. F-transformations have permutability properties described in Theorem 10.6. Note that the admissible choice of $f^{(12)}(0)$ mentioned in part a) of this theorem means in the present case that this point has to be coplanar with $f(0)$, $f^{(1)}(0)$ and $f^{(2)}(0)$, so that there is a two-parameter family of nets $f^{(12)}$.

We finish this section by mentioning the continuous limit. Considering Q-nets with a small mesh size ϵ and sending $\epsilon \to 0$, one arrives at smooth nets described by the following analog of equations (10.9):

$$\partial_i\partial_j f = c_{ji}\partial_i f + c_{ij}\partial_j f, \qquad i \neq j. \tag{10.11}$$

In other words, $\partial_i\partial_j f \in \mathrm{span}(\partial_i f, \partial_j f)$, which is a characteristic property of the so called conjugate nets. Analogously, keeping one direction of a Q-net discrete, one arrives in the limit at a smooth conjugate net and its F-transform, characterized by the coplanarity of the vectors $\partial_i f^+, \partial_j f^+$, and $f^+ - f$. This analogy can be given a status of a mathematical theorem about approximation of smooth conjugate nets by discrete ones. On this way, permutability of F-transformations for conjugate nets, which is a difficult theorem of differential geometry, becomes an obvious consequence of the multidimensional consistency of Q-nets, combined with the approximation result mentioned above.

10.4 Circular nets

Circular nets serve as a discretization of orthogonal nets. They were introduced in the 3D setting by A. Bobenko in 1996 and in the multidimensional setting by J. Cieśliński, A. Doliwa and P. Santini in 1997.

Definition 10.10 A map $f: \mathbb{Z}^m \to \mathbb{R}^N$ is called an m-dimensional circular net if all its elementary quadrilaterals are circular, i.e., if at every $u \in \mathbb{Z}^m$ and for every pair $1 \leq i \neq j \leq m$ the four points $f, \tau_i f, \tau_j f$, and $\tau_i\tau_j f$ are concircular.

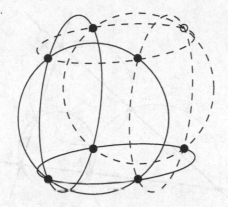

Figure 10.7 An elementary hexahedron of a circular net

We will show that this class of discrete nets is again described by a multidimensionally consistent 3D system.

$m = 2$: discrete curvature line parametrized surfaces To define a circular surface $f : \mathbb{Z}^2 \to \mathbb{R}^N$, one can prescribe arbitrarily its two coordinate lines and then fill in the whole net inductively. The induction step consists in choosing f_{12} on the circle through f, f_1 and f_2, provided the latter three points are known (and are in general position). Thus, one has a freedom of one real parameter on each step.

$m = 3$: the basic 3D system Suppose three coordinate surfaces of a three-dimensional circular net f are given. An inductive extension step is based on the following theorem:

Theorem 10.11 *Given seven points $f, f_i,$ and f_{ij} $(1 \le i < j \le 3)$ in $\mathbb{R}^N$, such that each of the three quadruples (f, f_i, f_j, f_{ij}) is inscribed in a circle C_{ij}, define three new circles $\tau_i C_{jk}$ as those passing through the triples (f_i, f_{ij}, f_{ik}), respectively. Then these new circles intersect at one point:*

$$f_{123} = \tau_1 C_{23} \cap \tau_2 C_{31} \cap \tau_3 C_{12};$$

see Figure 10.7.

Proof Under the conditions of the theorem, the seven points f, f_i, f_{ij} lie on some two-sphere S^2. Indeed, there is a unique sphere S^2 through the four points f, f_i. The circles C_{ij} through the triples (f, f_i, f_j) lie on S^2, and since $f_{ij} \in C_{ij}$, we find that $f_{ij} \in S^2$, as well. Perform a stereographic projection of

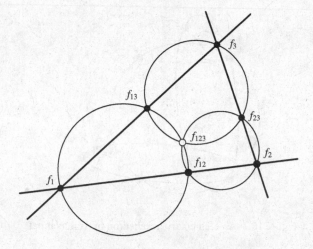

Figure 10.8 Miquel theorem

the sphere S^2 with the pole at the point f. Then the picture becomes planar, the image of f goes to infinity, the circles C_{ij} become the straight lines $(f_i f_j)$, and the claim of the theorem reduces to the classical Miquel theorem illustrated in Figure 10.8.   □

Exercise Formulate and prove the Miquel theorem.

Theorem 10.11 shows that finding the eightth point of a three-dimensional circular net from the seven known ones is a 3D system in the sense of Figure 10.4. Thus, one can say that the classical Miquel theorem just claims the existence of this 3D system. Exaggerating a bit, one can say that Miquel theorem lies at the heart of the classical theory of orthogonal systems and their transformations. This is an instance of a very remarkable fact: *classical incidence theorems lie at the basis of DDG and serve as its true roots and, through a continuous limit, as the ultimate explanation of classical differential geometry.*

$m \geq 4$: consistency One can easily extract from the proof of Theorem 10.11 the $m = 3$ case of the following statement.

Theorem 10.12 *If the coordinate surfaces of a Q-net $f: \mathbb{Z}^m \to \mathbb{R}^N$ are circular surfaces, then f is a circular net.*

Proof Consider an elementary hexahedron of a Q-net with three circular coordinate quadrilaterals (f, f_i, f_{ij}, f_j). Its eightth vertex is defined as the unique intersection point of the three planes $\tau_i \Pi_{jk}$ through the triples (f_i, f_{ij}, f_{ik}). Theorem 10.11 assures that the three circles $\tau_i C_{jk} = \tau_i \Pi_{jk} \cap S^2$ through the triples (f_i, f_{ij}, f_{ik}) intersect at a common point f_{123}. Therefore, the eighth vertex of a Q-net must coincide with f_{123} constructed as the intersection of three circles; in other words, it must be the eighth vertex of a circular net. Thus, the circularity constraint propagates in the construction of a Q-net from its coordinate surfaces. $\qquad\square$

As a corollary of Theorems 10.9 and 10.12 we deduce:

Theorem 10.13 *The 3D system governing circular nets is 4D consistent.*

Theorem 10.13 immediately yields Bäcklund transformations for circular nets (known in this case as Ribaucour transformations). The characteristic property of Ribaucour transformations is the circularity of the "side quadrilaterals" (f, f_i, f_i^+, f^+). Ribaucour transformations enjoy permutability properties described in Theorem 10.6. Note that the admissible choice of $f^{(12)}(0)$ mentioned in part a) of this theorem means in the present case that this point has to be concircular with $f(0)$, $f^{(1)}(0)$ and $f^{(2)}(0)$, so that there is a one-parameter family of nets $f^{(12)}$.

10.5 Q-nets in quadrics

Upon a stereographic projection, circular nets can be characterized as Q-nets in the sphere $\mathbb{S}^N \subset \mathbb{R}^{N+1}$. Also many other geometrically relevant nets turn out to be reductions of Q-nets to some quadrics or to intersections of quadrics. This is due to the following general fact: Q-nets can be consistently restricted to an arbitrary quadric in $\mathbb{R}^N$. A deep reason for this is, in turn, the following fundamental result, well known in the classical projective geometry.

Theorem 10.14 (on associated points) *Given eight distinct points which are the set of intersections of three quadrics in $\mathbb{RP}^3$, all quadrics through any subset of seven of the points must pass through the eighth point. Such sets of points are called associated.*

The relevance of this result for DDG is based on its corollary about the possibility to restrict Q-nets to an arbitrary quadric, observed by A. Doliwa in 1999.

Theorem 10.15 (Elementary hexahedron of a Q-net in a quadric) *If seven points f, f_i, and f_{ij} $(1 \le i < j \le 3)$ of an elementary hexahedron of a Q-net belong to a quadric $Q \subset \mathbb{RP}^N$, then so does the eightth point f_{123}.*

Proof The eight points f, f_i, f_{ij}, f_{123} lie in a three-dimensional space, and are known to form the intersection of three (degenerate) quadrics – the pairs of planes $\Pi_{jk} \cup \tau_i \Pi_{jk}$ for $(jk) = (12), (23), (31)$. Therefore, they are associated points. According to Theorem 10.14, any quadric Q through the seven points f, f_i, f_{ij} automatically goes through the eightth point.   □

Theorem 10.16 *If the coordinate surfaces of a Q-net $f: \mathbb{Z}^m \to \mathbb{RP}^N$ belong to a quadric Q, then so does the entire f.*

Proof For $m = 3$ this follows from Theorem 10.15. The claim for $m \geq 4$ follows from the m-dimensional consistency of Q-nets.   □

Another version of Theorem 10.15 can be formulated as follows. It is based on an obvious fact that for a nonisotropic line l with a nonempty intersection with Q this intersection consists generically of *two* points (because it is governed by a quadratic equation).

Theorem 10.17 *Let (f, f_1, f_{12}, f_2) be a planar quadrilateral in a quadric Q. Let l, l_1, l_2, l_{12} be nonisotropic lines in $\mathbb{RP}^N$ containing the corresponding points f, f_1, f_2, f_{12} and such that every two neighboring lines intersect. Denote the second intersection points of the lines with Q by $f^+, f_1^+, f_{12}^+, f_2^+$, respectively. Then the quadrilateral $(f^+, f_1^+, f_{12}^+, f_2^+)$ is also planar.*

Proof The eight points $f, f_1, f_2, f_{12}, f^+, f_1^+, f_{12}^+, f_2^+$ lie in a three-dimensional space, and are known to form the intersection of three quadrics, one of them being Q, and two others being the pairs of planes $\mathrm{span}(l, l_1) \cup \mathrm{span}(l_2, l_{12})$ and $\mathrm{span}(l, l_2) \cup \mathrm{span}(l_1, l_{12})$. Therefore, they are associated points. The pair of planes (f, f_1, f_{12}, f_2) and (f^+, f_1^+, f_2^+) contains seven of them; therefore the eightth point, f_{12}^+, must lie in the plane (f^+, f_1^+, f_2^+).   □

As a global corollary of this local statement, we immediately obtain the following:

Theorem 10.18 *Consider a quadric $Q \subset \mathbb{RP}^N$ and a Q-net $f: \mathbb{Z}^m \to Q$. Let a discrete congruence of nonisotropic lines $l: \mathbb{Z}^m \to \mathcal{L}^N$ be given such that $f(u) \in l(u)$ for all $u \in \mathbb{Z}^m$ and any two neighboring lines l, l_i intersect. Denote by $f^+(u)$ the second intersection point of $l(u)$ with Q, so that $l(u) \cap Q = \{f(u), f^+(u)\}$. Then $f^+: \mathbb{Z}^m \to Q$ is also a Q-net, called the Ribaucour transform of f.*

The importance of the reduction of Q-nets to quadrics hinges on the following circumstance. The transformation groups of various geometries, including those of Lie, Möbius and Laguerre, as well as Plücker line geometry, hyperbolic geometry and so on, can be modelled as subgroups of the projective transformation group of the suitable higher-dimensional space. Classically, such a subgroup is described as consisting of projective transformations which preserve some distinguished quadric called absolute. Since multidimensional Q-nets, being manifestly projectively invariant objects, can be restricted to an arbitrary quadric, one can require that *smooth geometric objects and their discretizations belong to the same geometry, i.e., are invariant with respect to the same transformation group.* This principle can be considered as a sort of the *discrete Erlangen program.*

10.6 T-nets

Definition 10.19 A map $y: \mathbb{Z}^m \to \mathbb{R}^N$ is called an m-dimensional T-net if for every $u \in \mathbb{Z}^m$ and for every pair of indices $i \neq j$, the discrete Moutard equation with minus signs,

$$\tau_i \tau_j y - y = a_{ij}(\tau_j y - \tau_i y), \tag{10.12}$$

holds with some $a_{ij}: \mathbb{Z}^m \to \mathbb{R}$; in other words, if all elementary quadrilaterals $(y, \tau_i y, \tau_i \tau_j y, \tau_j y)$ are planar and have parallel diagonals.

The definition of T-nets, considered irrespectively of the Koenigs nets, refers to the notion of parallelism and thus belongs to affine geometry. Of course, the coefficients a_{ij} must be skew-symmetric, $a_{ij} = -a_{ji}$.

***m* = 2: T-surfaces** To define a two-dimensional T-net $y: \mathbb{Z}^2 \to \mathbb{R}^N$, one can arbitrarily prescribe two coordinate curves. An induction step towards filling the whole net consists in choosing a point y_{12} on a line through y parallel to $y_2 - y_1$, i.e., in choosing a number a and setting $y_{12} = y + a(y_2 - y_1)$. Thus, one has a freedom of one real parameter on each step.

***m* = 3: basic 3D system** We show that three-dimensional T-nets are described by a well-defined 3D system. An inductive construction step of the net f is as follows.

Theorem 10.20 *Given seven points y, y_i, and y_{ij} ($1 \leq i \neq j \leq 3$) in $\mathbb{R}^N$ satisfying (10.12), there exists a unique point y_{123} such that all three quadrilaterals $(y_i, y_{ij}, y_{123}, y_{ik})$ have parallel diagonals.*

Proof Three equations (10.12) for the faces of an elementary cube adjacent to y_{123}, give:

$$\tau_i y_{jk} = (1 + (\tau_i a_{jk})(a_{ij} + a_{ki}))y_i - (\tau_i a_{jk})a_{ij}y_j - (\tau_i a_{jk})a_{ki}y_k.$$

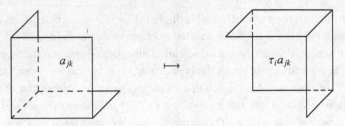

Figure 10.9 3D system on an elementary cube; fields on faces.

They lead to consistent results for y_{123} for arbitrary initial data if and only if the following conditions are satisfied:

$$1 + (\tau_1 a_{23})(a_{12} + a_{31}) = -(\tau_2 a_{31})a_{12} = -(\tau_3 a_{12})a_{31},$$
$$1 + (\tau_2 a_{31})(a_{23} + a_{12}) = -(\tau_3 a_{12})a_{23} = -(\tau_1 a_{23})a_{12},$$
$$1 + (\tau_3 a_{12})(a_{31} + a_{23}) = -(\tau_1 a_{23})a_{31} = -(\tau_2 a_{31})a_{23}.$$

These conditions constitute a system of six (linear) equations for three unknown variables $\tau_i a_{jk}$ in terms of the known a_{jk}. A direct computation shows that this system is not overdetermined but admits a unique solution:

$$\frac{\tau_1 a_{23}}{a_{23}} = \frac{\tau_2 a_{31}}{a_{31}} = \frac{\tau_3 a_{12}}{a_{12}} = -\frac{1}{a_{12}a_{23} + a_{23}a_{31} + a_{31}a_{12}}. \tag{10.13}$$

With $\tau_i a_{jk}$ so defined, equations (10.12) are fulfilled on all three quadrilaterals adjacent to y_{123}.   □

Equations (10.13) define a birational map $\{a_{jk}\} \mapsto \{\tau_i a_{jk}\}$, which can be considered as the fundamental 3D system related to T-nets. It is sometimes called the *star-triangle map*. Note that this is a 3D system with (scalar) fields on elementary squares rather than on vertices as in the previous examples. It can be symbolically illustrated as on Figure 10.9.

The local statement of Theorem 10.20 immediately yields the $m = 3$ case of the following theorem.

Theorem 10.21 *If the coordinate surfaces of a Q-net* $y: \mathbb{Z}^m \to \mathbb{R}^N$ *are T-surfaces, then* y *is a T-net.*

Proof Let the quadrilaterals (y, y_i, y_{ij}, y_j) be planar and have parallel diagonals. The planarity of the the quadrilaterals $(y_i, y_{ij}, y_{123}, y_{ik})$ defines the point y_{123} as the intersection point of three planes $\tau_i \Pi_{jk}$. Then these three quadrilaterals automatically have parallel diagonals. Indeed, by Theorem 10.20 there exists a point y_{123} with this property. It satisfies the planarity condition, and therefore it must coincide with the unique point defined by the planarity condition. □

$m \geq 4$: **consistency** The 4D consistency of T-nets is a consequence of the analogous property of Q-nets, since the T-constraint propagates in the construction of a Q-net from its coordinate surfaces. As a consequence, Theorem 10.21 holds for $m \geq 4$. On the level of formulas we have for T-nets with $m \geq 4$ the system (10.12), while the map $\{a_{jk}\} \mapsto \{\tau_i a_{jk}\}$ is given by

$$\frac{\tau_i a_{jk}}{a_{jk}} = -\frac{1}{a_{ij}a_{jk} + a_{jk}a_{ki} + a_{ki}a_{ij}}. \tag{10.14}$$

All indices i, j, k vary now between 1 and m, and for any triple of pairwise different indices (i, j, k), equations involving these indices solely, form a closed subset.

As usual, the 4D consistency yields the existence of Bäcklund transformations of T-nets with permutability properties, known in this case as *discrete Moutard transformations*. The admissible choice of $y^{(12)}(0)$ mentioned in part (a) of Theorem 10.6 means in the present case that this point has to lie on the line through $y(0)$ parallel to $y^{(2)}(0) - y^{(1)}(0)$, so that there is a one-parameter family of nets $y^{(12)}$.

Exercise Prove that the invariance of the left-hand side of equation (10.14) w.r.t. permutations of indices yields the existence of the function $z \colon \mathbb{Z}^m \to \mathbb{R}$ such that

$$a_{ij} = \frac{z_i z_j}{z z_{ij}}, \qquad i < j.$$

Plug this representation back in (10.14) to show that the function z solves the so called discrete BKP equation:

$$z_i z_{jk} - z_j z_{ik} + z_k z_{ij} - z z_{ijk} = 0, \qquad i < j < k. \tag{10.15}$$

We close this section by discussing the continuous limit. It is easy to realize that the discrete Moutard equation with minus signs, $z_{12} - z = a(z_2 - z_1)$, does not admit an (interesting) continuous limit. Indeed, setting

$$z_1 = z + \epsilon \partial_1 z + O(\epsilon^2), \qquad z_2 = z + \epsilon \partial_2 z + O(\epsilon^2), \qquad z_{12} = z + \epsilon(\partial_1 z + \partial_2 z) + O(\epsilon^2),$$

we find in the limit $\epsilon \to 0$ a first order equation

$$\partial_1 z + \partial_2 z = a(\partial_2 z - \partial_1 z).$$

However, upon changing the sign of z along every second row (or column), $z \mapsto (-1)^{\mu_2} z$, leads to the discrete Moutard equation with plus signs,

$$z_{12} + z = a(z_1 + z_2).$$

Exercise Check that the continuous limit $\epsilon \to 0$ of the latter equation (with $a = 1 + \epsilon^2 q/2$) is the classical second-order Moutard equation

$$\partial_1 \partial_2 z = qz.$$

10.7 A-nets

Definition 10.22 A map $f: \mathbb{Z}^m \to \mathbb{R}^3$ is called an m-dimensional A-net in $\mathbb{R}^3$ if for every $u \in \mathbb{Z}^m$ all the points $f(u \pm e_i)$, $i \in \{1, \ldots, m\}$, lie in some plane $\mathcal{P}(u)$ through $f(u)$.

Two-dimensional A-nets were introduced by R. Sauer in the 1930s as a discrete analog of surfaces parametrized along asymptotic lines; multidimensional A-nets are due to A. Doliwa (2001). Definition 10.22 belongs to the projective geometry and could equally well be formulated for the ambient space $\mathbb{RP}^3$. An A-net f is called *nondegenerate*, if all its elementary quadrilaterals $(f, \tau_i f, \tau_i \tau_j f, \tau_j f)$ are nonplanar.

Exercise In principle, it would be possible to consider discrete A-nets in $\mathbb{R}^N$ with $N > 3$, however this would not lead to an essential generalization. Show that a nondegenerate A-net in $\mathbb{R}^N$ actually lies in some three-dimensional subspace.

$m = 2$: **A-surfaces** serve as discrete counterparts of surfaces parametrized along asymptotic lines. For a construction of a discrete A-surface $f: \mathbb{Z}^2 \to \mathbb{R}^3$, one can start with two arbitrary discrete coordinate curves in general position (which means that no three neighboring points of such a curve are collinear). Of course, all four neighbors $f(\pm e_i)$ of the point $f(0)$ must lie in one plane $\mathcal{P}(0)$. Thus, the planes $\mathcal{P}(u)$ are well defined along the coordinate curves. Now these data can be recursively extended to an A-surface: one step of this procedure for the quadrant $\mathbb{Z}_+^2$, say, consists of choosing $f(u + e_1 + e_2)$ on the straight line through $f(u)$ which is the intersection of the planes $\mathcal{P}(u + e_1)$ and $\mathcal{P}(u + e_2)$. There is one free real parameter on each step of this extension procedure.

$m = 3$: **basic 3D system** Given any eight points f, f_i, f_{ij}, f_{123} in $\mathbb{R}^3$ ($1 \leq i < j \leq 3$), define the eight planes $\mathcal{P}, \mathcal{P}_i, \mathcal{P}_{ij}, \mathcal{P}_{123}$ as follows:

$$\mathcal{P}, \text{ the plane through } f_1, f_2, f_3,$$
$$\mathcal{P}_i, \text{ the plane through } f, f_{ij}, f_{ik},$$
$$\mathcal{P}_{ij}, \text{ the plane through } f_i, f_j, f_{123},$$
$$\mathcal{P}_{123}, \text{ the plane through } f_{12}, f_{13}, f_{23}.$$

Thus, one obtains two tetrahedra, one with the vertices $(f, f_{12}, f_{23}, f_{31})$ and with the face planes $(\mathcal{P}_1, \mathcal{P}_2, \mathcal{P}_3, \mathcal{P}_{123})$, and the second with the vertices (f_1, f_2, f_3, f_{123}) and with the face planes $(\mathcal{P}, \mathcal{P}_{12}, \mathcal{P}_{23}, \mathcal{P}_{31})$. Now the condition that the eight points are vertices of an elementary hexahedron of a three-dimensional A-net is expressed as the following eight incidence relations:

$$f \in \mathcal{P}, \qquad f_i \in \mathcal{P}_i, \qquad f_{ij} \in \mathcal{P}_{ij}, \qquad f_{123} \in \mathcal{P}_{123}. \qquad (10.16)$$

Thus, each vertex of each tetrahedron lies in the corresponding face plane of the other one. In other words, the two tetrahedra are inscribed in each other. Such pairs of tetrahedra are called *Möbius pairs*. We see that the geometry of an elementary hexahedron of a discrete three-dimensional A-net is exactly that of a Möbius pair of tetrahedra. Möbius demonstrated that eight conditions (10.16) are not independent: any one of them follows from the remaining seven.

Theorem 10.23 (Möbius pair of tetrahedra) *If the four vertices of one tetrahedron lie respectively in the four face planes of another, while three vertices of the second lie in three face planes of the first, then the remaining vertex of the second lies in the remaining face plane of the first.*

Proof A construction of a Möbius tetrahedra pair may be performed as follows. Let f be some point in $\mathbb{R}^3$, and let $\mathcal{P}$ and $\mathcal{P}_i$ ($i = 1, 2, 3$) be four planes through f such that any three of them are in general position. For $i = 1, 2, 3$, let f_i be an arbitrary point on the line $\mathcal{P} \cap \mathcal{P}_i$, and, for $(ij) = (12), (23), (31)$, let f_{ij} be an arbitrary point on the line $\mathcal{P}_i \cap \mathcal{P}_j$. Now, construct the planes $\mathcal{P}_{ij}$ through f_i, f_j, f_{ij}, and the plane $\mathcal{P}_{123}$ through f_{12}, f_{23}, f_{31}. Then Theorem 10.23 claims that the last point f_{123}, uniquely defined as the intersection of the planes $\mathcal{P}_{12}, \mathcal{P}_{23}, \mathcal{P}_{31}$, automatically lies in the plane $\mathcal{P}_{123}$.

The proof of this claim is based on the well-known incidence theorem of projective geometry, namely the theorem on quadrilateral sets of points, see Figure 10.10. The upper half of this figure represents the plane $\mathcal{P}$, and the lower half represents the plane $\mathcal{P}_{123}$. Each plane $\mathcal{P}_i$ is represented by a pair of lines $\mathcal{P}_i \cap \mathcal{P} = (ff_i)$ and $\mathcal{P}_i \cap \mathcal{P}_{123} = (f_{ij}f_{ik})$ meeting on $l = \mathcal{P} \cap \mathcal{P}_{123}$. Similarly, each plane $\mathcal{P}_{ij}$ is represented by a pair of lines $\mathcal{P}_{ij} \cap \mathcal{P} = (f_i f_j)$ and $\mathcal{P}_{ij} \cap \mathcal{P}_{123}$, which is a line passing through f_{ij} and meeting $(f_i f_j)$ on l. The claim of Theorem 10.23 can be now interpreted as follows: three lines in $\mathcal{P}_{123}$ which are traces of $\mathcal{P}_{12}, \mathcal{P}_{23}, \mathcal{P}_{31}$ in this plane, are concurrent (with the common point being f_{123}). But this is exactly what the theorem on quadrilateral sets says. $\square$

According to Theorem 10.23, we can think of A-nets as being governed by a 3D system in the sense of Figure 10.4. The fields attached to the vertices

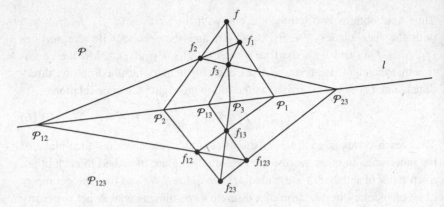

Figure 10.10 Möbius theorem

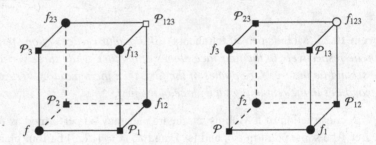

Figure 10.11 3D system with points assigned to the even sublattice and planes assigned to the odd sublattice.

of $\mathbb{Z}^3$ are pairs $(f, \mathcal{P})$ consisting of a point $f \in \mathbb{R}^3$ and a plane $\mathcal{P}$ through this point. The system is characterized by the defining property of A-nets, according to which all neighbors of a point f belong to the corresponding plane $\mathcal{P}$. Theorem 10.23 assures that the fields at any seven vertices of an elementary cube of $\mathbb{Z}^3$ determine the eighth one uniquely.

The construction of the Möbius tetrahedra performed in Theorem 10.23 can also be given a different, somewhat fancy combinatorial interpretation. In this interpretation, we assign points $f \in \mathbb{R}^3$ and planes $\mathcal{P} \subset \mathbb{R}^3$ to the even, resp. odd, sublattice of $\mathbb{Z}^4$, i.e., to $u \in \mathbb{Z}^4$ with the even (resp. odd) values of $|u| = u_1 + u_2 + u_3 + u_4$. An edge of $\mathbb{Z}^4$ connects a point f and a plane $\mathcal{P}$ if and only if $f \in \mathcal{P}$. It is not hard to realize that this condition determines a three-dimensional system (the fields at any seven vertices of a combinatorial 3D cube determine the eighth one uniquely); see Figure 10.11. For instance, the left diagram in Figure 10.11 corresponds to the construction with the following seven initial data: a point f, three planes $\mathcal{P}_i$ in general position through f, and three points $f_{ij} \in \mathcal{P}_i \cap \mathcal{P}_j$. Then the eighth field is the plane $\mathcal{P}_{123}$ through

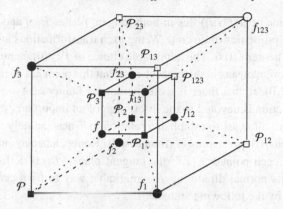

Figure 10.12 4D consistency of the 3D system with points on the even sub-lattice and planes on the odd sublattice.

f_{12}, f_{23} and f_{13}. The right diagram in Figure 10.11 corresponds to a projectively dual construction. Clearly, the fields at the eight vertices of any combinatorial 3D cube correspond to the four vertices and four face planes of a tetrahedron in $\mathbb{R}^3$. In particular, the cubes in Figure 10.11 encode both tetrahedra of a Möbius pair. Now, the defining properties of a Möbius tetrahedra pair and its construction are encoded in the hypercube in Figure 10.12, where the points f and the planes $\mathcal{P}$ with equal indices are connected by edges parallel to the fourth coordinate direction. In this interpretation Theorem 10.23 is nothing but the statement about the 4D consistency of the 3D system of Figure 10.11.

The previous, conventional combinatorial interpretation of Möbius tetrahedra pairs (or, what is the same, of A-nets), is reobtained upon a contraction of all edges of the fourth coordinate direction in Figure 10.12. As a result of this operation, the 4D hypercube turns into a 3D cube, whose vertices carry pairs $(f, \mathcal{P})$ with equal indices such that $f \in \mathcal{P}$.

$m \geq 4$: consistency A nonstandard interpretation of the Möbius theorem (Theorem 10.23) as a statement about 4D consistency of a certain 3D system, together with Theorem 10.4, yields the $(m+1)$-dimensional consistency of the latter for any $m \geq 4$. Returning back to the standard interpretation (by contraction of one of the coordinate directions), we see that the following statement holds.

Theorem 10.24 *A-nets are multidimensionally consistent.*

This justifies Definition 10.22. As a consequence, A-nets admit Bäcklund transformations, carrying the name of *discrete Weingarten transformations* in this case. Such a transformation can be characterized as follows: for every

$u \in \mathbb{Z}^m$, the line $(f(u)f^+(u))$ lies in both tangent planes $\mathcal{P}(u)$ and $\mathcal{P}^+(u)$. The permutability properties of discrete Weingarten transformations are described, as usual, by Theorem 10.6. The admissible choice of $f^{(12)}(0)$ mentioned in part (a) of this theorem means in the present case that this point has to lie on the line $\mathcal{P}^{(1)}(0) \cap \mathcal{P}^{(2)}(0)$, so that there is a one-parameter family of A-nets $f^{(12)}$. The rest of this section is devoted to the discussion of an important property of A-nets which will allow us to establish a relation to T-nets, namely the so called *discrete Lelieuvre representation* (found by B. Konopelchenko and U. Pinkall in 2000). At each point $u \in \mathbb{Z}^m$, the tangent plane $\mathcal{P}(u)$ of a discrete A-net defines also its normal direction. A remarkable way to fix a certain normal field is given by the following statement.

Theorem 10.25 *For a nondegenerate A-net $f: \mathbb{Z}^m \to \mathbb{R}^3$, there exists a normal field $n: \mathbb{Z}^m \to \mathbb{R}^3$ such that*

$$\delta_i f = \tau_i n \times n, \qquad i = 1, \ldots, m, \tag{10.17}$$

called a Lelieuvre normal field. It is uniquely defined by a value at one point $u_0 \in \mathbb{Z}^m$. All other Lelieuvre normal fields are obtained by $n(u) \mapsto \alpha n(u)$ for $|u| = u_1 + \cdots + u_m$ even, and $n(u) \mapsto \alpha^{-1} n(u)$ for $|u|$ odd, with some $\alpha \in \mathbb{R}$ (black-white rescaling).

Proof Let $v: \mathbb{Z}^m \to \mathbb{R}^3$ be some normal field to f. For every (nonplanar) elementary quadrilateral (f, f_i, f_{ij}, f_j) we have:

$$f_i - f = \alpha v_i \times v, \qquad\qquad f_j - f = \beta v_j \times v, \tag{10.18}$$

$$f_{ij} - f_j = \alpha_j v_{ij} \times v_j, \qquad\qquad f_{ij} - f_i = \beta_i v_{ij} \times v_i, \tag{10.19}$$

with some coefficients $\alpha, \beta, \alpha_j, \beta_i \in \mathbb{R}_*$ (associated with the edges of the quadrilateral). We will show that

$$\alpha \alpha_j = \beta \beta_i. \tag{10.20}$$

This will prove the theorem, because relation (10.20) is equivalent to the existence of the function $\rho: \mathbb{Z}^m \to \mathbb{R}_*$, associated with the vertices of $\mathbb{Z}^m$, such that

$$\alpha = \rho \rho_i, \qquad \beta = \rho \rho_j, \qquad \alpha_j = \rho_j \rho_{ij}, \qquad \beta_i = \rho_i \rho_{ij}. \tag{10.21}$$

A solution ρ of the latter system is completely determined by its value at one point, and any two solutions differ by a black-white rescaling. Comparison of (10.21) with (10.18), (10.19) shows that the vectors $n = \rho v$ satisfy (10.17).

To prove (10.20), build the scalar product of the first equation in (10.18) with v_j, and of the second with v_i. Taking into account that $\langle v_i \times v, v_j \rangle = -\langle v_j \times v, v_i \rangle$, and that $f_j - f$ is orthogonal to v_j, we find:

$$\frac{\alpha}{\beta} = -\frac{\langle f_i - f, v_j \rangle}{\langle f_j - f, v_i \rangle} = \frac{\langle f_i - f_j, v_j \rangle}{\langle f_i - f_j, v_i \rangle}.$$

Similarly, from (10.19) we derive:

$$\frac{\alpha_j}{\beta_i} = -\frac{\langle f_{ij} - f_j, v_i \rangle}{\langle f_{ij} - f_i, v_j \rangle} = \frac{\langle f_i - f_j, v_i \rangle}{\langle f_i - f_j, v_j \rangle}.$$

From the last two formulas, the relation (10.20) follows. $\qquad\square$

In particular, Lelieuvre normals n^+ of a discrete Weingarten transform f^+ can be uniquely fixed by the requirement that, together with the Lelieuvre normals n of the net f, they satisfy

$$f^+ - f = n^+ \times n. \tag{10.22}$$

Theorem 10.26 *A-nets in $\mathbb{R}^3$ (modulo parallel translations) are in a one-to-one correspondence, via the discrete Lelieuvre representation (10.17), with T-nets in $\mathbb{R}^3$ (modulo black-white rescalings).*

Proof It follows immediately from (10.17) that $(\tau_i \tau_j n - n) \times (\tau_i n - \tau_j n) = 0$, that is, the Lelieuvre normal field of an A-net satisfies the discrete Moutard equation

$$\tau_i \tau_j n - n = a_{ij}(\tau_j n - \tau_i n) \tag{10.23}$$

with some $a_{ij} \colon \mathbb{Z}^m \to \mathbb{R}$. Conversely, given a T-net $n \colon \mathbb{Z}^m \to \mathbb{R}^3$, formula (10.17) defines an exact form whose integration produces a discrete A-net $f \colon \mathbb{Z}^m \to \mathbb{R}^3$. $\qquad\square$

Theorem 10.26 gives another justification of the multidimensional consistency of A-nets.

10.8 T-nets in quadrics

In this section, we leave the world of integrable 3D systems and enter the world of integrable 2D systems. We will see how the latter appear as reductions of the former. Let $\mathbb{R}^N$ be equipped with a nondegenerate symmetric bilinear form $\langle \cdot, \cdot \rangle$ (which does not need to be positive definite), and let

$$Q = \{ f \in \mathbb{R}^N : \langle f, f \rangle = \kappa_0 \}$$

be a quadric in $\mathbb{R}^N$. We study T-nets $f \colon \mathbb{Z}^m \to Q$.

$m = 2$: basic 2D system Knowing two coordinate curves of a two-dimensional T-net $f: \mathbb{Z}^2 \to Q$ allows one to extend the net f to the whole of $\mathbb{Z}^2$. The induction step consists in computing $f_{12} = f + a(f_2 - f_1)$, where the coefficient a (attached to every elementary square of $\mathbb{Z}^2$) is determined by the condition $f_{12} \in Q$, provided $f, f_1, f_2 \in Q$. A simple computation using the formula

$$\langle f + a(f_2 - f_1), f + a(f_2 - f_1) \rangle = \kappa_0$$

shows that this condition is equivalent to

$$a = \frac{\langle f, f_1 - f_2 \rangle}{\kappa_0 - \langle f_1, f_2 \rangle}.$$

Thus, the elementary construction step consists of finding the fourth vertex of an elementary quadrilateral from the known three vertices, as symbolically represented in Figure 10.2.

$m \geq 3$: consistency In the multidimensional case, one can prescribe all coordinate lines of a T-net f in Q. Indeed, these data are independent, and one can, by induction, construct the whole net from them. The induction step is essentially two-dimensional and consists in determining f_{ij}, provided f, f_i and f_j are known. In order for this induction process to work without contradictions, equations

$$f_{ij} - f = a_{ij}(f_j - f_i), \qquad a_{ij} = \frac{\langle f, f_i - f_j \rangle}{\kappa_0 - \langle f_i, f_j \rangle} \qquad (10.24)$$

must be 3D consistent.

Theorem 10.27 *The 2D system* (10.24) *governing T-nets in Q is 3D consistent.*

Proof This can be checked by a tiresome computation, which, however, can be avoided by the following conceptual argument. T-nets in Q are the result of imposing two admissible reductions on Q-nets, namely the T-reduction and the restriction to a quadric Q. This reduces the effective dimension of the system by 1 (allows one to determine the fourth vertex of an elementary quadrilateral from the three known vertices), and transfers the original 3D equation into the 3D consistency of the reduced 2D equation. Indeed, after finding f_{12}, f_{23} and f_{13}, one can construct f_{123} according to the planarity condition (as intersection of three planes). Then both the T-condition and the Q-condition are fulfilled for all three quadrilaterals adjacent to f_{123}. Therefore, these quadrilaterals satisfy our 2D system. □

Exercise Prove that for a T-net $f: \mathbb{Z}^m \to Q$, the functions

$$\alpha_i = \langle f, f_i \rangle \qquad (10.25)$$

defined on the edges of $\mathbb{Z}^m$ of the ith coordinate direction, have the following property:

$$\tau_j \alpha_i = \alpha_i, \qquad i \neq j, \tag{10.26}$$

i.e., in every elementary square the opposite edges carry equal values of the corresponding α_i. If one assigns the value of α_i on the edge $(u, u + e_i)$ to the lattice point $u \in \mathbb{Z}^m$, then (10.26) is expressed as $\alpha_i = \alpha_i(u_i)$ for $i = 1, \ldots, m$.

With notation (10.25), the expression in (10.24) for the coefficients a_{ij} of the discrete Moutard equations takes the form

$$a_{ij} = \frac{\alpha_i - \alpha_j}{\kappa_0 - \langle f_i, f_j \rangle}. \tag{10.27}$$

10.9 K-nets

Two-dimensional K-nets serve as a discretization of surfaces with constant negative Gaussian curvature, which belong to the oldest and the most popular objects of the integrable differential geometry. These nets were introduced by R. Sauer in 1950 and by W. Wunderlich in 1951. The latter also found the discrete analogue of the classical Bäcklund transformations for such surfaces.

Definition 10.28 An A-net $f \colon \mathbb{Z}^m \to \mathbb{R}^3$ is called an m-dimensional K-net if for any elementary quadrilateral (f, f_i, f_{ij}, f_j),

$$|f_{ij} - f_j| = |f_i - f| \qquad and \qquad |f_{ij} - f_i| = |f_j - f|; \tag{10.28}$$

in other words, if the functions $\beta_i = |\delta_i f|$, defined on the edges parallel to the ith coordinate axes for $i = 1, \ldots, m$, have the labelling property (10.26) (depend on u_i only).

The property (10.28) of a net f is known as the *Chebyshev property*, so a quadrilateral (f, f_i, f_{ij}, f_j) satisfying (10.28) can be called a *Chebyshev quadrilateral*. Thus, a Chebyshev quadrilateral can be considered as a parallelogram bent in space along one of its diagonals.

Lemma 10.29 *A Chebyshev quadrilateral is symmetric under the* $180°$ *rotation about the line through the midpoints of its diagonals.*

Proof Let O_1 and O_2 denote the midpoints of the diagonals $[f, f_{ij}]$ and $[f_i, f_j]$, respectively (see Figure 10.13). It is enough to show that this line is orthogonal to both diagonals. But, as it follows from considering the congruent triangles $\triangle(f, f_i, f_{ij})$ and $\triangle(f, f_j, f_{ij})$, the point O_1 is equidistant from f_i and f_j, and

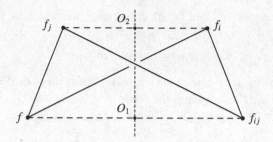

Figure 10.13 Geometry of a Chebyshev quadrilateral.

therefore belongs to the plane through O_2 orthogonal to $[f_i, f_j]$. Hence, the line $(O_1 O_2)$ is also orthogonal to $[f_i, f_j]$. For similar reasons, this line is orthogonal to the second diagonal as well. $\square$

Now a characterization of the Lelieuvre normals of a K-net follows immediately.

Theorem 10.30 *The Lelieuvre normal field* $n: \mathbb{Z}^m \to \mathbb{R}^3$ *of a K-net* $f: \mathbb{Z}^m \to \mathbb{R}^3$ *takes values, possibly upon a black-white rescaling, in some sphere* $S^2 \subset \mathbb{R}^3$, *thus being proportional to the Gauss map.*

Conversely, any T-net in the unit sphere $n: \mathbb{Z}^m \to \mathbb{S}^2$ *is the Gauss map and the Lelieuvre normal field of a K-net* $f: \mathbb{Z}^m \to \mathbb{R}^3$. *The functions*

$$\cos \alpha_i = \langle \tau_i n, n \rangle \tag{10.29}$$

have the labelling property (depend on u_i *only), which therefore holds also for*

$$\gamma_i = |\delta_i n| = 2 \left| \sin \frac{\alpha_i}{2} \right| \quad \text{and} \quad \beta_i = |\delta_i f| = |\sin \alpha_i|. \tag{10.30}$$

Proof The definition of K-nets is equivalent to the following conditions for the Lelieuvre normals:

$$|\tau_i \tau_j n \times \tau_j n| = |\tau_i n \times n|, \qquad |\tau_i \tau_j n \times \tau_i n| = |\tau_j n \times n|.$$

Because of the symmetry formulated in Lemma 10.29 (which clearly yields the rotational symmetry also for the directions of normal vectors), we derive:

$$|\tau_i \tau_j n| \cdot |\tau_j n| = |\tau_i n| \cdot |n|, \qquad |\tau_i \tau_j n| \cdot |\tau_i n| = |\tau_j n| \cdot |n|.$$

As a consequence,

$$|\tau_i \tau_j n| = |n|, \qquad |\tau_i n| = |\tau_j n|.$$

Thus, the Lelieuvre normal field of a discrete K-surface forms, possibly after a black-white rescaling, a T-net in a sphere, being an instance of the class considered in Section 10.8. This proves the first claim of the theorem.

Turning to the second claim, we start with a T-net n in the unit sphere $\mathbb{S}^2$, described by the equations

$$\tau_i \tau_j n - n = a_{ij}(\tau_j n - \tau_i n), \qquad a_{ij} = \frac{\langle n, \tau_i n - \tau_j n \rangle}{1 - \langle \tau_i n, \tau_j n \rangle}. \tag{10.31}$$

Due to exercise at the end of section 10.8, the edge functions $\cos \alpha_i = \langle \tau_i n, n \rangle$ depend on u_i only, and therefore $\gamma_i^2 = |\delta_i n|^2 = 2(1 - \cos \alpha_i) = 4 \sin^2(\alpha_i/2)$ also depend only on u_i. Define the discrete Λ-net $f \colon \mathbb{Z}^m \to \mathbb{R}^3$ by (10.17). Then

$$\beta_i^2 = |\delta_i f|^2 = 1 - \langle \tau_i n, n \rangle^2 = 1 - \cos^2 \alpha_i = \sin^2 \alpha_i,$$

which proves that (10.28) is fulfilled. $\qquad\qquad\qquad\qquad\qquad\qquad\qquad$ □

According to Theorem 10.30, K-nets f (modulo scalings and translations) are in a one-to-one correspondence with T-nets n in $\mathbb{S}^2$. Thus, one can determine a K-net by prescribing the values of n on the coordinate axes, i.e., m discrete curves in $\mathbb{S}^2$ through a common point $n(0)$.

Due to consistency of T-nets in a sphere (Theorem 10.27), we have:

Theorem 10.31 *K-nets are multidimensionally consistent.*

Therefore, as usual, one can define Bäcklund transformations for K-nets just by extending the domain by an additional lattice direction.

Definition 10.32 Two K-nets $f, f^+ \colon \mathbb{Z}^m \to \mathbb{R}^3$ with corresponding edges of equal length,

$$|f_i^+ - f^+| = |f_i - f|, \qquad i = 1, \dots, m,$$

are related by a Bäcklund transformation if they are related by a Weingarten transformation and the distance $|f^+ - f|$ is constant, i.e., does not depend on $u \in \mathbb{Z}^m$. The net f^+ is called a Bäcklund transform of f.

In particular, to specify a Bäcklund transform f^+ of a given m-dimensional K-net f, or, equivalently, a Moutard transform n^+ of the Gauss map n, one can prescribe just the value of $n^+(0) \in \mathbb{S}^2$.

Permutability of Bäcklund transformations for K-nets is a direct consequence of the 3D consistency of T-nets in $\mathbb{S}^2$, and can be formulated as in Theorem 10.5. Geometrically, the points of the fourth surface $f^{(12)}$ which is a simultaneous Bäcklund transform of $f^{(1)}$ and of $f^{(2)}$ lie in the intersection of the tangent planes to $f^{(1)}$ and to $f^{(2)}$ at the corresponding points, and are uniquely defined by the properties $|f^{(12)} - f^{(1)}| = |f^{(2)} - f|$ and $|f^{(12)} - f^{(2)}| = |f^{(1)} - f|$. In terms of the Gauss maps, the defining condition reads: $\langle n^{(1)}, n^{(12)} \rangle = \langle n, n^{(2)} \rangle$ and $\langle n^{(2)}, n^{(12)} \rangle = \langle n, n^{(1)} \rangle$.

10.10 Hirota equation for K-nets

For a convenient analytic description of K-nets and their Gauss maps, one can use the following matrix formalism. The space $\mathbb{R}^3$ can be identified with the Lie algebra su(2),

$$\begin{pmatrix} -ix_3 & -x_2 - ix_1 \\ x_2 - ix_1 & ix_3 \end{pmatrix} \in su(2) \quad \leftrightarrow \quad (x_1, x_2, x_3)^T \in \mathbb{R}^3. \tag{10.32}$$

The vector product in $\mathbb{R}^3$ and the matrix commutator in su(2) correspond as follows:

$$[x, y] = 2x \times y.$$

This isomorphism makes it unnecessary to distinguish between vectors in $\mathbb{R}^3$ and matrices in su(2). In other words, we use the following basis of the linear space su(2):

$$\mathbf{e}_1 = \begin{pmatrix} 0 & -i \\ -i & 0 \end{pmatrix} = -i\sigma_1, \quad \mathbf{e}_2 = \begin{pmatrix} 0 & -1 \\ 1 & 0 \end{pmatrix} = -i\sigma_2, \quad \mathbf{e}_3 = \begin{pmatrix} -i & 0 \\ 0 & i \end{pmatrix} = -i\sigma_3,$$

where σ_j are the Pauli matrices. We supply su(2) with the scalar product $\langle \cdot, \cdot \rangle$ induced from $\mathbb{R}^3$. It is easy to see that in the matrix form it may be represented as

$$\langle x, y \rangle = -\tfrac{1}{2} \operatorname{tr}(xy) = \tfrac{1}{2} \operatorname{tr}(xy^*), \tag{10.33}$$

where x^* stands for the Hermitian conjugate of x.

Rotations in $\mathbb{R}^3$ are conveniently described by the adjoint action of the Lie group SU(2), which consists of complex 2×2 matrices Φ satisfying the condition $\Phi\Phi^* = \Phi^*\Phi = 1$, where 1 is the 2×2 unit matrix (the group unit), so that $\Phi^{-1} = \Phi^*$. In terms of components:

$$\Phi = \begin{pmatrix} a & ib \\ i\bar{b} & \bar{a} \end{pmatrix}, \quad a, b \in \mathbb{C}, \ |a|^2 + |b|^2 = 1.$$

The property which makes SU(2) suitable for a description of rotations in $\mathbb{R}^3$ is the following: an arbitrary $\Phi \in$ SU(2) can be written as:

$$\Phi = \cos\theta \cdot 1 + \sin\theta \cdot x_0, \quad \text{with } x_0 \in su(2), \ \langle x_0, x_0 \rangle = 4. \tag{10.34}$$

In this notation, the action $\Phi^{-1} x \Phi$ on an arbitrary vector $x \in su(2)$ is nothing but the rotation of x around the vector x_0 by the angle 2θ.

Definition 10.33 A function $\Phi\colon \mathbb{Z}^m \to \mathrm{SU}(2)$ is called a frame of a given K-net $f\colon \mathbb{Z}^m \to \mathbb{R}^3$ with the Gauss map $n\colon \mathbb{Z}^m \to \mathbb{S}^2$ if it satisfies pointwise the formula

$$n = \Phi^{-1}\mathbf{e}_3\Phi. \tag{10.35}$$

Clearly, the frame is defined not uniquely but rather up to *gauge transformations* of the type

$$\Phi(u) \mapsto \exp(i\kappa(u)\sigma_3)\Phi(u) \tag{10.36}$$

with real-valued functions κ. We define also the *transition matrices*

$$U_j(u) = \Phi(u + e_j)\Phi^{-1}(u) \in \mathrm{SU}(2), \tag{10.37}$$

naturally assigned to the edges $(u, u + e_j)$ of $\mathbb{Z}^m$. The action of a gauge transformation (10.36) on the transition matrices is given by

$$U_j(u) \mapsto \exp(i\kappa_j\sigma_3)U_j(u)\exp(-i\kappa\sigma_3), \qquad \kappa = \kappa(u),\ \kappa_j = \kappa(u + e_j). \tag{10.38}$$

By definition, transition matrices satisfy the identity

$$(\tau_k U_j)U_k = (\tau_j U_k)U_j. \tag{10.39}$$

Theorem 10.34 *Let $n\colon \mathbb{Z}^m \to \mathbb{S}^2$ be a T-net in $\mathbb{S}^2$ with the labelling $\langle n_j, n \rangle = \cos\alpha_j$. Then the frame $\Phi\colon \mathbb{Z}^m \to \mathrm{SU}(2)$ can be gauged so that all transition matrices $U_j = U_j(u)$ have the form*

$$U_j = \begin{pmatrix} \cos(\alpha_j/2)e^{i\xi_j} & -i\sin(\alpha_j/2) \\ -i\sin(\alpha_j/2) & \cos(\alpha_j/2)e^{-i\xi_j} \end{pmatrix} \tag{10.40}$$

with some numbers $\xi_j = \xi_j(u) \in \mathbb{R}/(2\pi\mathbb{Z})$ assigned to the edges $(u, u + e_j)$. There exists a function $\phi\colon \mathbb{Z}^m \to \mathbb{R}/(4\pi\mathbb{Z})$ such that

$$\xi_j = \tfrac{1}{2}(\phi_j - \phi), \tag{10.41}$$

where, as usual, $\phi = \phi(u)$ and $\phi_j = \phi(u + e_j)$. This function satisfies the so-called Hirota equation

$$\sin\frac{1}{4}(\phi_{jk} + \phi_k - \phi_j - \phi) = \frac{\tan(\alpha_k/2)}{\tan(\alpha_j/2)}\sin\frac{1}{4}(\phi_{jk} + \phi_j - \phi_k - \phi). \tag{10.42}$$

In terms of variables $w = \exp(i\phi/2)\colon \mathbb{Z}^m \to \mathbb{S}^1$, this equation takes the form

$$\frac{w_{jk}}{w} = \frac{\tan(\alpha_k/2)\,w_k - \tan(\alpha_j/2)\,w_j}{\tan(\alpha_k/2)\,w_j - \tan(\alpha_j/2)\,w_k}, \tag{10.43}$$

cf. equation (10.1).

Proof For the transition matrices $U_j = \begin{pmatrix} a_j & ib_j \\ ib_j & \bar{a}_j \end{pmatrix}$ with $|a_j|^2 + |b_j|^2 = 1$ the relation

$$\cos(\alpha_j) = -\tfrac{1}{2}\operatorname{tr}(n_j n) = -\tfrac{1}{2}\operatorname{tr}(\Phi_j^{-1}\mathbf{e}_3\Phi_j\Phi^{-1}\mathbf{e}_3\Phi) = \tfrac{1}{2}\operatorname{tr}(U_j^{-1}\sigma_3 U_j\sigma_3)$$

is equivalent to $|a_j|^2 - |b_j|^2 = \cos(\alpha_j)$, so

$$|a_j|^2 = \cos^2(\alpha_j/2), \qquad |b_j|^2 = \sin^2(\alpha_j/2).$$

Thus, the transition matrices can be parametrized as follows:

$$U_j = \begin{pmatrix} \cos(\alpha_j/2)e^{i\xi_j} & -i\sin(\alpha_j/2)e^{i\eta_j} \\ -i\sin(\alpha_j/2)e^{-i\eta_j} & \cos(\alpha_j/2)!ee^{-i\xi_j} \end{pmatrix}, \tag{10.44}$$

with $\xi_j, \eta_j \in \mathbb{R}/(2\pi\mathbb{Z})$. Plugging (10.44) into equation (10.39), one can prove that there exists a gauge transformation (10.38) leading to all $\eta_j = 0$, so that all transition matrices are as in (10.40). The coefficients ξ_j satisfy then the relations

$$\tau_k\xi_j + \xi_k \equiv \tau_j\xi_k + \xi_j \tag{10.45}$$

and

$$\cos\frac{\alpha_j}{2}\sin\frac{\alpha_k}{2}\sin\frac{1}{2}(\tau_k\xi_j + \xi_j) = \sin\frac{\alpha_j}{2}\cos\frac{\alpha_k}{2}\sin\frac{1}{2}(\tau_j\xi_k + \xi_k). \tag{10.46}$$

Equation (10.45) can be interpreted as the exactness of the discrete one-form ξ, so there exists a function ϕ defined on the vertices of the lattice such that (10.41) holds for all edges $(u, u + e_j)$. Plugging this into (10.46), we arrive at the Hirota equation (10.42). $\square$

Exercise Fill in the omitted details of the proof of Theorem 10.34.

For discrete K-surfaces, that is, for K-nets with $m = 2$, one can, as usual, modify the notion of Lelieuvre normals so that the passage to the continuous limit can be performed in a straightforward way. Indeed, after the usual change of variables $n(u) \mapsto (-1)^{u_2}n(u)$, the Gauss map $n: \mathbb{Z}^2 \to \mathbb{S}^2$ will turn into a net in $\mathbb{S}^2$ satisfying the discrete Moutard equation with plus signs:

$$\tau_1\tau_2 n + n = a_{12}(\tau_1 n + \tau_2 n), \qquad a_{12} = \frac{\langle n, \tau_1 n + \tau_2 n\rangle}{1 + \langle\tau_1 n, \tau_2 n\rangle}. \tag{10.47}$$

Alternatively, such nets are called discrete Lorentz-harmonic nets. Of course, this change of variables must be supplied with the corresponding change of the Lelieuvre formulas (10.17):

$$\delta_1 f = \tau_1 n \times n = \delta_1 n \times n, \qquad \delta_2 f = n \times \tau_2 n = n \times \delta_2 n. \tag{10.48}$$

This will also yield the change of sign $\cos \alpha_2 \mapsto -\cos \alpha_2$, that is, $\alpha_2 \mapsto \pi - \alpha_2$ for the labelling of edges of the second coordinate direction.

Modifications in Theorem 10.34 for the case of discrete Lorentz-harmonic nets in a sphere, necessary for a well-behaved continuous limit, are given in the following statement.

Theorem 10.35 *Let $n: \mathbb{Z}^2 \to \mathbb{S}^2$ be a Lorentz-harmonic net in $\mathbb{S}^2$ with the labelling $\langle n_j, n \rangle = \cos \alpha_j$. Then the frame $\Phi: \mathbb{Z}^2 \to \mathrm{SU}(2)$ can be gauged so that the transition matrices $U_j = U_j(u)$ have the form*

$$U_1 = \begin{pmatrix} \cos(\alpha_1/2)e^{i\xi_1} & -i\sin(\alpha_1/2) \\ -i\sin(\alpha_1/2) & \cos(\alpha_1/2)e^{-i\xi_1} \end{pmatrix}, \tag{10.49}$$

$$U_2 = \begin{pmatrix} \cos(\alpha_2/2) & i\sin(\alpha_2/2)e^{i\eta_2} \\ i\sin(\alpha_2/2)e^{-i\eta_2} & \cos(\alpha_2/2) \end{pmatrix}, \tag{10.50}$$

with some $\xi_1 = \xi_1(u), \eta_2 = \eta_2(u) \in \mathbb{R}/(2\pi\mathbb{Z})$ assigned to the edges of the first, resp. of the second, coordinate direction. There exists a function $\phi: \mathbb{Z}^2 \to \mathbb{R}/(4\pi\mathbb{Z})$ such that

$$\xi_1 = \tfrac{1}{2}(\phi_1 - \phi), \qquad \eta_2 = \tfrac{1}{2}(\phi_2 + \phi),$$

This function satisfies Hirota equation in the following form:

$$\sin \tfrac{1}{4}(\phi_{12} - \phi_1 - \phi_2 + \phi) = \tan(\alpha_1/2)\tan(\alpha_2/2)\sin \tfrac{1}{4}(\phi_{12} + \phi_1 + \phi_2 + \phi). \tag{10.51}$$

The transition from the multidimensionally consistent version (10.42) of the Hirota equation to the essentially two-dimensional version (10.51), which has the advantage of a well-behaved continuous limit, is achieved by a transformation analogous to the one already familiar to us from the consideration of the discrete Moutard equation, namely $\phi(u) \mapsto (-1)^{u_2}\phi(u)$. As discussed above, this has to be accompanied by $\alpha_2 \mapsto \pi - \alpha_2$, and so $\tan(\alpha_2/2)$ must be replaced by $\cot(\alpha_2/2)$. The formulas for a Bäcklund transformation for the Hirota equation change into

$$\sin\frac{1}{4}(\tau_1\phi^+ - \phi^+ + \tau_1\phi - \phi)$$

$$= \frac{\tan(\alpha_1/2)}{\tan(\gamma/2)} \sin\frac{1}{4}(\tau_1\phi^+ + \phi^+ - \tau_1\phi - \phi), \tag{10.52}$$

$$\sin\frac{1}{4}(\tau_2\phi^+ - \phi^+ - \tau_2\phi + \phi)$$

$$= \tan(\alpha_2/2)\tan(\gamma/2)\sin\frac{1}{4}(\tau_2\phi^+ + \phi^+ + \tau_2\phi + \phi). \tag{10.53}$$

Here the Bäcklund parameter γ comes from $\cos\gamma = \langle n^+, n \rangle$. A superposition formula for Bäcklund transformations reads:

$$\sin\frac{1}{4}(\phi^{(12)} + \phi^{(2)} - \phi^{(1)} - \phi) = \frac{\tan(\gamma_2/2)}{\tan(\gamma_1/2)} \sin\frac{1}{4}(\phi^{(12)} - \phi^{(2)} + \phi^{(1)} - \phi). \quad (10.54)$$

Equation (10.51) depends only on the product $\tan(\alpha_1/2)\tan(\alpha_2/2)$, and therefore remains invariant under the change of parameters $\alpha_j \mapsto \alpha_j(\lambda)$ such that

$$\tan\frac{\alpha_1(\lambda)}{2} = \lambda\tan\frac{\alpha_1}{2}, \qquad \tan\frac{\alpha_2(\lambda)}{2} = \lambda^{-1}\tan\frac{\alpha_2}{2}. \quad (10.55)$$

Geometrically, this means that each solution of the Hirota equation describes a whole one-parameter family of discrete K-surfaces (the so called *associated family*). Making the change of parameters (10.55) in the transition matrices (10.49), (10.50), one comes to spectral parameter dependent transition matrices. Up to scalar normalizing factors $\cos(\alpha_j(\lambda)/2)$, they are given in the following statement.

Theorem 10.36 *Hirota equation* (10.51) *for discrete Lorentz-harmonic nets in a sphere admits a discrete zero curvature representation* (10.39) *with the following spectral parameter dependent matrices:*

$$U_1^0(\lambda) = \begin{pmatrix} e^{i(\phi_1-\phi)/2} & -i\lambda\tan(\alpha_1/2) \\ -i\lambda\tan(\alpha_1/2) & e^{-i(\phi_1-\phi)/2} \end{pmatrix}, \quad (10.56)$$

$$U_2^0(\lambda) = \begin{pmatrix} 1 & i\lambda^{-1}\tan(\alpha_2/2)e^{i(\phi_2+\phi)/2} \\ i\lambda^{-1}\tan(\alpha_2/2)e^{-i(\phi_2+\phi)/2} & 1 \end{pmatrix}. \quad (10.57)$$

These formulas admit a rather straightforward continuous limit. For this aim, one can set $\tan(\alpha_1/2) = \tan(\alpha_2/2) = \epsilon/2$. Then in the limit $\epsilon \to 0$ the Hirota equation (10.51) turns into the sine-Gordon equation

$$\partial_1\partial_2\phi = \sin\phi \quad (10.58)$$

for a function $\phi\colon \mathbb{R}^2 \to \mathbb{R}$. Formulas (10.52), (10.53) defining a Bäcklund transformation turn in the limit $\epsilon \to 0$ into

$$\partial_1\phi^+ + \partial_1\phi = \frac{2}{c}\sin\frac{\phi^+ - \phi}{2}, \qquad \partial_2\phi^+ - \partial_2\phi = 2c\sin\frac{\phi^+ + \phi}{2}, \quad (10.59)$$

where the Bäcklund parameter is denoted by $c = \tan(\gamma/2)$. The superposition formula (10.54) for Bäcklund transformations does not depend on ϵ at all, so that it remains true for the sine-Gordon equation, as well:

$$\sin\frac{1}{4}(\phi^{(12)} + \phi^{(2)} - \phi^{(1)} - \phi) = \frac{c_2}{c_1}\sin\frac{1}{4}(\phi^{(12)} - \phi^{(2)} + \phi^{(1)} - \phi). \quad (10.60)$$

Finally, Theorem 10.36 yields in the limit $\epsilon \to 0$ the zero curvature representation

$$\partial_2 U_1 - \partial_1 U_2 + [U_1, U_2] = 0 \tag{10.61}$$

for the sine-Gordon equation with the matrices

$$U_1 = \frac{i}{2} \begin{pmatrix} \partial_1 \phi & -\lambda \\ -\lambda & -\partial_1 \phi \end{pmatrix}, \tag{10.62}$$

$$U_2 = \frac{i}{2} \begin{pmatrix} 0 & \lambda^{-1} e^{i\phi} \\ \lambda^{-1} e^{-i\phi} & 0 \end{pmatrix}. \tag{10.63}$$

Thus, the main integrability attributes of the sine-Gordon equation are recovered from the DDG framework.

11

Symmetry Preserving Discretization of Differential Equations and Lie Point Symmetries of Differential-Difference Equations

Pavel Winternitz

Abstract

In the first four sections of this chapter we consider an ordinary differential equation of any order invariant under some nontrivial group G of local point transformations. We show how such an ODE can be approximated by a difference scheme invariant under the same group G. Some advantages of such invariant schemes are pointed out. The schemes are exact for first-order equations. They can be solved analytically for some second-order equations. Used for numerical calculations the invariant schemes provide better qualitative descriptions of solutions than standard methods, specially close to singularities. The last two sections are devoted to methods of determining the Lie point symmetries of differential difference equations on fixed nontransforming lattices.

Introduction

Lie group theory started out as a theory of continuous transformations in the space of independent and dependent variables figuring in a system of differential equations. These point transformations were so constructed as to leave the space of solutions invariant, i.e., transform solutions into solutions. After Sophus Lie's seminal work in the end of the 19th and beginning of the 20th century. Lie theory developed in several directions, one being abstract group theory, another applications. In particular Lie group theory has evolved into a very general and powerful tool for obtaining exact (analytic) solutions of large classes of ordinary and partial differential equations. The symmetry theory of differential equations has been reviewed in modern books and review articles [5, 6, 25, 35, 36, 69, 82].

The purpose of these lectures is to provide a partial review of a different application of Lie group theory that is much more recent, namely, the application of these continuous groups to discrete equations.

We will consider two different types of applications. One can be called *Symmetry preserving discretization of differential equations* and its main application is to numerical methods of solving ordinary and partial differential equations (ODEs and PDEs). The other could be called *Continuous symmetries of discrete equations*. In this case the difference or differential-difference equations are considered as independent objects, not as approximations of differential equations. The aim is to use group theory to solve them.

A vigorous development of this field started in the 1990s and previous reviews include [22, 47, 83, 84]. For many original contributions and partial reviews we refer to proceedings of a series of conferences on *Symmetries and integrability of difference equations* (SIDE1–SIDE9) and to special issues of journals devoted to this topic [12, 13, 33, 43, 49, 56, 65, 81].

Early work on continuous symmetries of difference equations is due to Maeda [59–61]. The approach followed in the present chapter started with the papers [20, 21, 44, 45].

In this chapter we concentrate on just some of the recent results obtained in collaboration with colleagues and students. Thus Sections 11.1–11.4 are devoted to the discretization of ordinary differential equations (ODE) in a manner that preserves their Lie point symmetries. The ODE is replaced by an ordinary difference system (OΔS), i.e., two equations representing both the original ODE and the lattice. The group transformations act on the equation and the lattice. The results presented here mainly follow the articles [7, 8, 18, 19, 72]. In Sections 11.5, 11.6 we treat symmetries of differential-difference equations on fixed non-transforming lattices and follow mainly [57].

A number of well-developed topics has been omitted in this review and will be overviewed in a book in preparation. Among them we mention the following.

1. The symmetry classification of discrete dynamical systems on fixed lattices, in particular differential-difference equations describing the dynamics of molecular chains [26, 41, 46]
2. Nonlinear difference equations allowing (nonlinear) superposition formulas [27, 70, 74, 79]
3. Methods of determining symmetries of given difference equations on given lattices [29, 50–52]
4. Symmetry classification and solutions of first- and second-order differential systems [18, 19, 73]

5. Umbral calculus and quantum physics on lattices [53, 54]

6. Generalized symmetries of difference equations that in the continuous limit "contract" to point symmetries [30–32]. See also the chapter in this volume by D. Levi and R. Yamilov.

7. Point symmetries of specific difference or differential-difference equations, like the discrete Burgers equations [29] or generalized Toda field theories [40, 62].

8. Lie point symmetries of partial difference equations [3, 9, 14, 16, 17, 52, 80].

11.1 Symmetry preserving discretization of ODEs

11.1.1 Formulation of the problem

Let us consider an ODE in the form

$$y^{(n)} = f(x, y, y', \ldots, y^{(n-1)}), \qquad y^{(k)} = \frac{\mathrm{d}^k y}{\mathrm{d}x^k} \tag{11.1}$$

that allows a (local) Lie point symmetry group G of (local) point transformations

$$\tilde{x} = \Lambda(x, y), \qquad \tilde{y} = \Omega(x, y). \tag{11.2}$$

The Lie algebra $\mathfrak{g}$ of the group G is realized by vector fields of the form

$$X = \xi(x, y)\partial_x + \phi(x, y)\partial_y. \tag{11.3}$$

The symmetry group can be used to lower the order of the equation and if it is large enough, reduce the solution to quadratures. This may produce an explicit general solution of (11.1). On the other hand, the obtained solution may be implicit. This essentially means that the ODE (11.1) is replaced by an algebraic or transcendental equation of the form

$$H(x, y, C_1, \ldots, C_n) = 0 \tag{11.4}$$

where C_i are integration constants and H may be quite complicated. If we need to know the behavior of y as a function of x for given initial or boundary conditions it may be easier to obtain this information from (11.1) directly, rather than from (11.4).

If the group G is not large enough, or does not have the appropriate structure it will still be necessary to resort to numerical solutions of (11.1).

In standard numerical methods the ODE (11.1) is approximated by a difference equation on an a priori chosen lattice, usually a uniform one given, e.g., by the equation

$$x_n = hn + x_0 \tag{11.5}$$

where h and x_0 are fixed constants (e.g., $h = 1$, $x_0 = 0$). The derivatives in (11.1) are approximated by discrete derivatives, e.g.,

$$y_n^{(1)} = \frac{y(x+h) - y(x)}{h}, \qquad y_n^{(2)} = \frac{y(x+2h) - 2y(x+h) + y(x)}{2h}, \tag{11.6}$$

etc. The discrete derivatives are then substituted into (11.1) and the obtained difference equation is then solved (numerically).

Following such a procedure we loose most or all of the symmetry properties of the ODE (11.1).

Since the symmetry group G of (11.1) determines many properties of the solution space, it also determines much of the phenomena described by the ODE. In any case, when discretizing a differential equation it is desirable to preserve as much of the original symmetry as possible. The hope is that such a symmetry adapted approach will provide better numerical solutions. It should improve qualitative features like stability and behavior near singularities and also quantitative ones like providing better accuracy for the same cost.

The idea of including Lie point symmetries in numerical methods is an example of geometric integration [10, 11, 37]. Essentially this means that one gives up the idea of an universal computational method for all ODEs and instead identifies specific features of a differential equation and then incorporates them in the numerics. Such features can be integrability, symplecticity, Hamiltonian structure, conservation laws, point symmetries, etc.

Our aim is to preserve Lie point symmetries and we shall follow an approach using symmetry adapted lattices that themselves are transformed by symmetry groups [18–22, 47, 83].

A different approach is due to Yanenko and Shokin [76, 87] but we will not be concerned with it in these lectures. For further developments see also [34].

11.1.2 Lie point symmetries of ordinary difference schemes

We shall consider "difference schemes" rather than just difference equations. Let us restrict ourselves here to the case when we have just one independent variable x and one dependent one y. Both are continuous variables, however the independent variable x is "sampled" at the discrete points x_i and the dependent variable y is evaluated at these points. We thus have a lattice of points on a line with

$$x_i, y_i = y(x_i), \qquad M \le i \le N \tag{11.7}$$

where the range of values of i may be finite, infinite or semiinfinite.

An ordinary difference scheme (OΔS) will be a set of two relations between K points

$$E_a(\{x_k\}_{k=n+M}^{n+N}, \{y_k\}_{k=n+M}^{n+N}) = 0, \qquad a = 1, 2,$$
$$K = N - M + 1, \qquad n, M, N \in \mathbb{Z}, \ N > M. \tag{11.8}$$

If a continuous limit of (11.8) exists then one of the two equations (11.8) should go into an ordinary differential equation (ODE) of order $K' \le K - 1$, the other into an identity (like $0 = 0$). Thus (11.8) represents both a difference equation and a lattice. If the values of $\{x_k, y_k\}$ are given at $K - 1$ neighboring points we must be able to calculate them at the next point. That imposes some independence conditions on E_1 and E_2, e.g., the corresponding Jacobian must satisfy

$$\frac{\partial(E_1, E_2)}{\partial(x_{n+N}, y_{n+N})} \ne 0, \qquad \forall n. \tag{11.9}$$

The general solution of the OΔS (11.8) will have the form

$$y = f(x, C_1, C_2, \dots, C_{2(K-1)})$$
$$x_n = x(n, C_1, C_2, \dots, C_{2(K-1)}) \tag{11.10}$$

where $C_1, \dots, C_{2(K-1)}$ are integration constants. A simple example of an OΔS is

$$E_1 = y_{n+1} - 2y_n + y_{n-1} = \tfrac{1}{2}(x_{n+1}^2 - 2x_n^2 + x_{n-1}^2) \tag{11.11}$$
$$E_2 = x_{n+1} - 2x_n + x_{n-1} = 0. \tag{11.12}$$

The general solution is

$$y(x) = \tfrac{1}{2}x^2 + a_1 x + a_2, \qquad x_n = x_0 + nh \tag{11.13}$$

where x_0, h, a_1, a_2 are integration constants. The lattice defined by (11.12) is a uniform one with x_0 as its origin and h its scale. To take the continuous limit (for any OΔS) we expand y_{n+k} into a Taylor series, putting

$$x_{n+j} = x_n + h_{n+1} + h_{n+2} + \cdots + h_{n+j}, \qquad h_{n+k} = x_{n+k} - x_{n+k-1},$$
$$\tilde{h}_j = h_{n+1} + h_{n+2} + \cdots + h_{n+j},$$
$$y_{n+j} = y_n + \tilde{h}_j y_n' + \frac{1}{2!}\tilde{h}_j^2 y_n'' + \frac{1}{3!}\tilde{h}_j^3 y_n''' + \cdots, \qquad h_k = \alpha_k \varepsilon, \ \alpha_k \sim 1 \tag{11.14}$$

and take the limit $\varepsilon \to 0$. For instance in the case (11.11), (11.12) we obtain

$$\lim_{\varepsilon \to 0} E_1 = y'' - 1 = 0, \qquad \lim_{\varepsilon \to 0} E_2 = 0 = 0. \tag{11.15}$$

We see that the OΔS (11.11), (11.12) is an exact discretization of the ODE (11.15), i.e., the function $y(x)$ given in (11.13) is the exact general solution of (11.15) and passes exactly through all the points x_n involved in (11.11), (11.12) (for any values of the constants a_0, a_1, x_0 and h).

The OΔS (11.8) may be of interest in its own right and its solutions may describe some discrete physical phenomena. Alternatively it may be obtained as a discretization of an ODE and used to obtain numerical solutions of the ODE.

Lie point symmetries of the OΔS (11.8) are defined in the same way as for ODEs. Thus the transformations will have the form (11.2). Restricting to infinitesimal one-parameter transformations we have

$$\tilde{x} = x + \varepsilon \xi(x, y), \qquad \tilde{y} = y + \varepsilon \phi(x, y), \qquad \varepsilon \ll 1. \tag{11.16}$$

The functions $\xi(x, y)$ and $\phi(x, y)$ are the same ones as figure in the vector field (11.3).

We shall write the vector field acting at the point (x_n, y_n) as

$$X_n = \xi_n \partial_{x_n} + \phi_n \partial_{y_n}, \qquad \xi_n = \xi(x_n, y_n), \qquad \phi_n = \phi(x_n, y_n). \tag{11.17}$$

The prolongation acting at all points of the lattice figuring in the OΔS (11.8) is

$$\mathrm{pr}^D X = \sum_{k=n+M}^{n+N} (\xi_k \partial_{x_k} + \phi_k \partial_{y_k}) \tag{11.18}$$

(the superscript D stands for "discrete").

If the OΔS (11.8) is given then the algorithm for determining the vector field (11.17), i.e., the functions ξ_n and ϕ_n is

$$\mathrm{pr}^D X E_a \big|_{E_1 = E_2 = 0} = 0, \qquad a = 1, 2 \tag{11.19}$$

Equation (11.19) splits into a set of determining equations since ξ_k and ϕ_k depend on x_k and y_k only and the prolongation and the equation (11.8) introduce ξ_{n+k}, ϕ_{n+k} evaluated at other points.

If on the other hand we are approximating a given ODE (11.1) then the symmetry algebra of the ODE is determined by standard methods [69]. We can then consider the vector fields (11.17) to be known and they will be used to construct an invariant OΔS.

In the continuous case the vector fields are prolonged to the usual differential jet space: the space of independent and dependent variables $\{x, u\}$ and all derivatives $u_x, u_{xx}, \ldots, u_{kx}$ (up to the order of the equation). We have

$$\mathrm{pr}\, X = \xi \partial_x + \phi \partial_u + \phi^x \partial_{u_x} + \phi^{xx} \partial_{u_{xx}} + \cdots + \phi^{nx} \partial_{u_{nx}} \tag{11.20}$$

where ϕ^{nx} is defined recursively [69]

$$\phi^x = \mathrm{D}_x \phi - (\mathrm{D}_x \xi) u_x, \qquad \ldots, \qquad \phi^{nx} = \mathrm{D}_x \phi^{(n-1)x} - (\mathrm{D}_x \xi) u_{nx} \tag{11.21}$$

and D_x is the total derivative operator.

The invariant ODE (11.1) can be rewritten as

$$E(I_1, I_2, \ldots, I_{J_1}) = 0, \tag{11.22}$$

where $I_\mu(x, u, u_x, \ldots, u_{nx})$ for $\mu = 1, \ldots, J_1$ form a basis of the differential invariants of order n of the group G in the differential jet space:

$$\mathrm{pr}^{(n)} X_i I_\mu = 0, \qquad \mu = 1, \ldots, J_1. \tag{11.23}$$

Here $\{X_i\}$ is a basis of the symmetry algebra of the ODE (11.1).

The symmetry preserving discretization of the ODE (11.1) is obtained as follows.

1 Take the basis elements X_i of the symmetry algebra of the ODE (11.1) and prolong them to the *discrete* jet space as in (11.18).
2 Find the invariants $I^D(\{x_k, y_k\}_{k=n+M}^{n+N})$ of the action of G in the discrete jet space, i.e., find a basis of solutions of the equations

$$\mathrm{pr}^D X_i I_\nu^D = 0, \qquad \nu = 1, \ldots, J. \tag{11.24}$$

3 Separate solutions of (11.24) into two sets

$$\{I_1^D, \ldots, I_{J_1}^D\}, \quad \{I_{J_1+1}^D, \ldots, I_{J_1+J_2}^D\}, \qquad J_1 + J_2 = J \tag{11.25}$$

such that in the continuous limit we have

$$\lim_{x_{k+1} \to x_k} I_a^D = \begin{cases} I_a & 1 \le a \le J_1 \\ 0 & J_1 + 1 \le a \le J_1 + J_2. \end{cases} \tag{11.26}$$

4 Write the invariant OΔS (11.8) in the form

$$E_a(I_1^D, \ldots, I_{J_1}^D, I_{J_1+1}^D, \ldots, I_{J_1+J_2}^D) = 0, \qquad a = 1, 2 \tag{11.27}$$

where we have

$$\lim_{x_{k+1} \to x_k} E_1 = E, \qquad \lim_{x_{k+1} \to x_k} E_2 \equiv 0. \tag{11.28}$$

In (11.28) E is the same as in (11.22) and the second equation reduces to $0 = 0$.

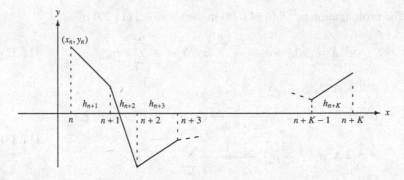

Figure 11.1 Stencil for OΔS with $K + 1$ points

11.1.3 The continuous limit

In order to consider limits from the discrete jet space to the continuous one, in particular limits of the prolongations of vector fields and of the invariants, we introduce a different set of local coordinates in the discrete jet space. The transformation is between the sets of variables $\{x_k, y_k : n \le k \le n + N\}$ and

$$\{x_n, y_n, h_{n+1}, \ldots, h_{n+N}, p_{n+1}^{(1)}, p_{n+2}^{(2)}, \ldots, p_{n+N}^{(N)}\} \tag{11.29}$$

with

$$h_{n+k} = x_{n+k} - x_{n+k-1}$$

$$p_{n+1}^{(1)} = \frac{y_{n+1} - y_n}{x_{n+1} - x_n}, \quad p_{n+2}^{(2)} = 2\frac{p_{n+2}^{(1)} - p_{n+1}^{(1)}}{x_{n+2} - x_n}, \quad \ldots,$$

$$p_{n+N}^{(N)} = N\frac{p_{n+N}^{(N-1)} - p_{n+N-1}^{(N-1)}}{x_{n+N} - x_n} \tag{11.30}$$

The continuous limit corresponds to

$$h_{n+k} \to 0, \quad p_{n+1}^{(1)} \to y', \quad p_{n+2}^{(2)} \to y'', \quad \ldots, \quad p_{n+N}^{(N)} \to y^{(N)}. \tag{11.31}$$

Thus, e.g., $p_m^{(j)}$ is the jth discrete derivative evaluated on the $j + 1$ points (m, $m - 1, m - 2, \ldots, m - j$) of the stencil on Figure 11.1. Equally well we could use a shifted stencil with points ($n + M, n + M + 1, \ldots, n + N - 1, n + N$) with $N - M = K$. The transformation from $\{x_{n+k}, y_{n+k}\}$ to the set (11.29) amounts to a change of variable for the stencil of Fig. 11.1.

The prolongation $\mathrm{pr}^D X$ of (11.18) in the variables (11.29) is

$$\mathrm{pr}^D X = \xi_n \partial_{x_n} + \phi_n \partial_{y_n} + \sum_{k=1}^{K} \phi^{(k)} \partial_{p_{n+k}^{(k)}} + \sum_{k=1}^{K} \kappa^k \partial_{h_{n+k}} \tag{11.32}$$

$$\phi^{(1)} = \Delta^{\mathrm{T}} \phi_n - p_{n+1}^{(1)} \Delta^{\mathrm{T}} \xi_n$$

$$\phi^{(2)} = \Delta^{\mathrm{T}} \phi^{(1)} - p_{n+2}^{(2)} \frac{1}{h_{n+1} + h_{n+2}} (h_{n+2} \Delta^{\mathrm{T}} \xi_{n+1} + h_{n+1} \Delta^{\mathrm{T}} \xi_n)$$

$$\phi^{(N)} = \Delta^{\mathrm{T}} \phi^{(N-1)} - p_{n+N}^{(N)} \frac{1}{\sum_{j=1}^{N} h_{n+j}} \sum_{j=1}^{N} h_{n+j} \Delta^{\mathrm{T}} \xi_{n+j-1} \tag{11.33}$$

$$\kappa^k = \xi_{n+k} - \xi_{n+k-1} \quad k = 1, \ldots, N$$

where Δ^{T} is the total discrete derivative defined as

$$\Delta^{\mathrm{T}} F(x_n, y_n, p_{n+1}^{(1)}, p_{n+2}^{(2)}, \ldots)$$

$$= \frac{1}{h_{n+1}} \{ F(x_{n+1}, y_{n+1}, p_{n+2}^{(1)}, p_{n+3}^{(2)} \ldots) - F(x_n, y_n, p_{n+1}^{(1)}, p_{n+2}^{(2)} \ldots) \}. \tag{11.34}$$

In the continuous limit $h_j \to 0 \ \forall j$ we have

$$\lim_{h_j \to 0} \mathrm{pr}^D X = \mathrm{pr}\, X, \tag{11.35}$$

i.e., (11.32) (and hence also (11.18)) reduces to the standard prolongation formula (11.20) for an ODE.

11.2 Examples of symmetry preserving discretizations

11.2.1 Equations invariant under $\mathrm{SL}_1(2, \mathbb{R})$

As a first example let us consider one of the four inequivalent realizations of $\mathrm{sl}(2, \mathbb{R})$ by vector fields in two variables called $\mathrm{sl}_1(2, \mathbb{R})$ in [7, 8, 72]

$$\mathrm{sl}(2, \mathbb{R})_1 \qquad X_1 = \partial_y \qquad X_2 = y \partial_y \qquad X_3 = y^2 \partial_y. \tag{11.36}$$

This is really a one-dimensional realization and corresponds to the projective action of $\mathrm{SL}(2, \mathbb{R})$ on a real line. The group action is

$$\tilde{x} = x, \qquad \tilde{y} = \frac{\alpha y + \beta}{\gamma y + \delta}, \qquad \alpha, \beta, \gamma, \delta \in \mathbb{R}, \quad \alpha \delta - \beta \gamma = 1. \tag{11.37}$$

One invariant is manifest, namely $I_1 = x$. Let us now calculate the lowest order y-dependent differential and difference invariants of the $\mathrm{SL}(2, \mathbb{R})$ action (11.37). There are two standard ways of doing this. One is an infinitesimal method using the appropriate prolongations of the vectors fields. The other is the method of moving frames, using the group action (that can be obtained by integrating the vector fields). Here we shall use the method of moving frames, explained in P. Olver's lectures in this volume. Let us first consider the differential case. Differentiating (11.37) we obtain the transformation formulas for the first three derivatives

$$\tilde{y}' = \frac{y'}{(\gamma y + \delta)^2}, \qquad \tilde{y}'' = \frac{1}{(\gamma y + \delta)^3}\{(\gamma y + \delta)y'' - 2\gamma y'^2\}, \tag{11.38}$$

$$\tilde{y}''' = \frac{1}{(\gamma y + \delta)^4}\{(\gamma y + \delta)^2 y''' - 6\gamma y' y''(\gamma y + \delta) + 6\gamma^2 y'^3\}. \tag{11.39}$$

Following the general method we now choose a section in the prolonged space. We choose

$$\tilde{y} = 0, \qquad \tilde{y}' = 1, \qquad \tilde{y}'' = 0, \tag{11.40}$$

substitute into (11.37), (11.38) and solve for the group parameters β, γ, δ in terms of y, y', y'' and α. The parameter α is obtained from the condition $\alpha\delta - \beta\gamma = 1$. Finally

$$\alpha = \varepsilon y'^{-1/2}, \qquad \beta = -\varepsilon y(y')^{-1/2}$$

$$\gamma = \frac{\varepsilon}{2}y'^{-3/2}y'', \qquad \delta = -\frac{\varepsilon}{2}y'^{-3/2}(yy'' - 2y'^2). \tag{11.41}$$

Substituting the group parameters (11.41) into (11.39) we obtain the lowest-order differential invariant I_2. We have:

$$I_1 = x, \qquad I_2 = \frac{1}{y'^2}\left(y'y''' - \frac{3}{2}y''^2\right). \tag{11.42}$$

Not surprisingly, the lowest-order invariant I_2 under the $SL(2, \mathbb{R})$ action (11.37) is the Schwarzian derivative (11.42) and the invariant ODE is $I_2 = F(I_1)$, i.e.,

$$\frac{1}{y'^2}\left(y'y''' - \frac{3}{2}y''^2\right) = F(x) \tag{11.43}$$

where $F(x)$ is an arbitrary function.

In order to construct a difference scheme approximating a third-order ODE we must use at least 4 points. In the discrete jet space we can use the co-ordinates (11.7) or (11.29). The set (11.29) is better adapted to the continuous limit. Using the method of moving frames one can construct invariants that directly have the proper continuous limit. The expressions for the invariant may however be quite complicated and the calculations cumbersome.

We shall again use the method of moving frames but stick to the coordinates (x_{n+k}, y_{n+k}), $k = 0, 1, 2, 3$. The transformation formulas in these coordinates are quite simple

$$\tilde{y}_{n+k} = \frac{\alpha y_{n+k} + \beta}{\gamma y_{n+k} + \delta}, \qquad \tilde{x}_{n+k} = x_{n+k}, \qquad k = 0, 1, 2, 3. \tag{11.44}$$

We choose the section

$$\tilde{y}_n = 0, \qquad \tilde{y}_{n+1} = 1, \qquad \tilde{y}_{n+2} \to \infty \tag{11.45}$$

and then calculate the parameters β, γ, δ in terms of y_n, y_{n+1}, y_{n+2} and α using (11.44) for $k = 0, 1, 2$. We find

$$\beta = -\alpha y_n, \qquad \gamma = -\alpha \frac{y_{n+1} - y_n}{y_{n+2} - y_{n+1}}, \qquad \delta = \alpha \frac{(y_{n+1} - y_n)y_{n+2}}{y_{n+2} - y_{n+1}} \tag{11.46}$$

Substituting (11.46) into (11.44) with $k = 3$, we obtain the invariants

$$I_1^{n+k} = x_{n+k}, \qquad I_2^{n+3} = \frac{(y_{n+3} - y_{n+1})(y_{n+2} - y_n)}{(y_{n+3} - y_{n+2})(y_{n+1} - y_n)}. \tag{11.47}$$

Again, not surprisingly, the $\mathrm{SL}_1(2, \mathbb{R})$ invariant is the anharmonic ratio of the 4 values $y_n, y_{n+1}, y_{n+2}, y_{n+3}$.

Thus, for each set of 4 points (x_{n+k}, y_{n+k}) we have 5 independent invariants that we can choose as $\{I_2^{n+3} \equiv R, x_n, h_{n+1}, h_{n+2}, h_{n+3}\}$. To obtain the continuous limit we put

$$y_{n+k} = y(x_{n+k}), \qquad x_{n+k} = x_n + \sum_{j=1}^{k} h_{n+j}, \qquad k = 1, 2, 3$$

$$h_{n+j} = \alpha_{n+j}\varepsilon, \qquad \varepsilon \to 0 \tag{11.48}$$

where α_{n+j} are constants of the order of 1. We expand y_{n+k} into a Taylor series about x_n, keeping terms up to order ε^3. The invariant formed out of (11.47) that has the correct limit is

$$J^{n+3} = \frac{6h_{n+3}h_{n+1}}{h_{n+2}(h_{n+2} + h_{n+3})(h_{n+1} + h_{n+2})(h_{n+1} + h_{n+2} + h_{n+3})}$$
$$\times \left[\frac{(h_{n+1} + h_{n+2})(h_{n+2} + h_{n+3})}{h_{n+1}h_{n+2}} - I_2^{n+3} \right]. \tag{11.49}$$

We have

$$\lim_{\varepsilon \to 0} J^{n+3} = \frac{1}{y'^2} \left[y'y''' - \frac{3}{2}y''^2 \right] + O(\varepsilon). \tag{11.50}$$

Thus the invariant OΔS that approximates the EDO (11.43) is

$$J^{n+3} = F(x_n, h_{n+1}, h_{n+2}, h_{n+3}), \qquad \phi(x_n, h_{n+1}, h_{n+2}, h_{n+3}) = 0. \tag{11.51}$$

11.2.2 Equations invariant under $SL_2(2, \mathbb{R})$

A second realization of $sl(2, R)$ (and its extension to $gl(2, \mathbb{R})$) is represented by the differential operators

$$X_1 = \partial_y, \qquad X_2 = x\partial_x + y\partial_y, \qquad X_3 = 2xy\partial_x + y^2\partial_y, \qquad X_4 = x\partial_x. \quad (11.52)$$

This realization $\{X_1, X_2, X_3\}$ of $sl(2, \mathbb{R})$ allows two independent differential invariants of order 2 and 3 respectively:

$$I_1 = \frac{2xy'' + y'}{y'^3}, \qquad I_2 = \frac{x^2(y'y''' - 3y''^2)}{y'^5}. \quad (11.53)$$

Adding invariance under X_4 we obtain just one invariant: $J = I_2/I_1^{3/2}$. The $SL_2(2, \mathbb{R})$ invariant ODEs that we can construct are

$$\frac{2xy'' + y'}{y'^3} = \gamma \quad (11.54)$$

$$\frac{x^2(y'y''' - 3y''^2)}{y'^5} = F(I_1) \quad (11.55)$$

where γ is a constant and F is an arbitrary function.

We can add the requirement that (11.55) be invariant under $GL_,(2, \mathbb{R})$, i.e., $F(z) = \alpha z^{3/2}$

$$x^2(y'y''' - 3y''^2) = \alpha(2xy'' + y')^{3/2}y'^{1/2}. \quad (11.56)$$

The ODE (11.56) can in principle be integrated using the solvable subalgebra $\{X_1, X_2, X_4\}$ of (11.52). Indeed X_1 allows us to lower the order by one putting $y'(x) = u(x)$

$$x'(uu'' - 3u'^2) = \alpha(2xu' + u)^{3/2}u^{1/2}, \qquad y(x) = \int_0^x u(s)\,ds + y_0 \quad (11.57)$$

Equation (11.57) inherits the symmetry algebra $Y_1 = u\partial_u$, $Y_2 = x\partial_x$ from (11.56) We use Y_1 to put

$$u = e^w, \qquad w_x = v \quad (11.58)$$

and obtain the first-order ODE

$$x^2(v_x - 2v^2) = \alpha(2xv + 1), \qquad w(x) = \int_0^x v(s)\,ds + w_0. \quad (11.59)$$

This ODE inherits the symmetry algebra $Y = x\partial_x - v\partial_v$ and this allows us to reduce the problem to quadratures by the transformation:

$$v = e^{-f}, \qquad x = ze^f. \quad (11.60)$$

We obtain

$$\frac{df}{dz} = -\frac{2z^2 + 2\alpha z + \alpha}{(2z+1)(z+\alpha)z}.$$ (11.61)

Integrating (11.61) we obtain f as a function of z and in principle we can return to the variables y et x. However this leads to a transcendental equation that cannot be solved explicitly for z in terms of x, so it is necessary to resort to a numerical solution in any case.

Five functionally independent $SL_2(2, \mathbb{R})$ difference invariants exist on a four-point stencil. We choose them as

$$I_1^{n+k} = \frac{y_{n+k} - y_{n+k-1}}{\sqrt{x_{n+k} x_{n+k-1}}} \qquad k = 1, 2, 3$$

$$I_2^{n+k+1} = \frac{y_{n+k+1} - y_{n+k-1}}{\sqrt{x_{n+k+1} x_{n+k-1}}} \qquad k = 1, 2.$$ (11.62)

The invariants that have the correct limits are

$$J_1^{n+2} = 8 \frac{I_2^{n+2} - (I_1^{n+1} + I_1^{n+2})}{I_1^{n+1} I_1^{n+2}(I_1^{n+1} + I_1^{n+2})} = \frac{2xy'' + y'}{y'^3} + O(\varepsilon)$$ (11.63)

$$J_2^{n+3} = \frac{3}{2} \frac{J_1^{n+3} - J_1^{n+2}}{I_1^{n+1} + I_1^{n+2} + I_1^{n+3}} = \frac{x^2}{y'^5}(y'y''' - 3y''^2) + O(\varepsilon)$$ (11.64)

An OΔS approximating (11.54) and (11.55) is given by

$$J_1^{n+2} = \gamma$$ (11.65)

$$J_2^{n+3} = F(J_1^{n+2}, I_1^{n+1}, I_1^{n+2}, I_1^{n+3})$$ (11.66)

respectively, together with the lattice equation

$$\phi(I_1^{n+1}, I_1^{n+2}, I_1^{n+3}) = 0$$ (11.67)

with

$$F(J_1^{n+2}, 0, 0, 0,) = F(J_1), \qquad \phi(0, 0, 0) \equiv 0.$$ (11.68)

11.2.3 Equations invariant under the similitude group of the Euclidean plane

Let us consider the Lie algebra of the similitude group Sim(2), namely, sim(2) realized as

$$P_1 = \partial_x, \qquad P_2 = \partial_y, \qquad L = y\partial_x - x\partial_y, \qquad D = x\partial_x + y\partial_y$$ (11.69)

and construct the corresponding third-order differential and difference invariants. This time we shall use the infinitesimal method.

The differential invariants are constructed by solving the 4 differential equations

$$\text{pr}\,X\,F(x, y, y', y'', y''') = 0 \tag{11.70}$$

for $X = P_1, P_2, L$ and D. The result for the Euclidean algebra e(2) $\{P_1, P_2, L\}$ is

$$I_1 = \frac{y''}{(1 + y'^2)^{3/2}}, \qquad I_2 = \frac{y'''(1 + y'^2) - 3y''^2 y'}{(1 + y'^2)^3}. \tag{11.71}$$

Adding the dilation operator D we obtain just one Sim(2) invariant, namely,

$$J = \frac{I_2}{I_1^2}. \tag{11.72}$$

Thus the second- and third-order E(2) invariant equations are

$$y'' = K(1 + y'^2)^{3/2} \tag{11.73}$$

$$y'''(1 + y'^2) - 3y''^2 y' = (1 + y'^2)^3 F(I_1) \tag{11.74}$$

where K is a constant and F an arbitrary function. The lowest-order Sim(2) invariant equation is obtained by putting $F(z) = Kz^2$ in (11.74), i.e.

$$y'''(1 + y'^2) - 3y'y''^2 = Ky''^2. \tag{11.75}$$

The corresponding difference invariants are obtained by solving the equations

$$\text{pr}^D\,X\,F(x_{n+k}, y_{n+k}) = 0, \qquad k = 0, 1, 2, 3 \tag{11.76}$$

for $X = \{P_1, P_2, L, D\}$. Taking $X = \{P_1, P_2, L\}$ we obtain 5 difference invariants

$$
\begin{aligned}
\xi_{n+k} &= h_{n+k}\left[1 + \left(\frac{y_{n+k} - y_{n+k-1}}{h_{n+k}}\right)^2\right]^{1/2} & k &= 1, 2, 3 \\
\eta_{n+k} &= (y_{n+k+1} - y_{n+k})h_{n+k} - (y_{n+k} - y_{n+k-1})h_{n+k+1} & k &= 1, 2.
\end{aligned}
\tag{11.77}
$$

Expanding into power series we find the combinations of the invariants that have the correct continuous limits

$$
\begin{aligned}
J_1 &= \frac{2\alpha\eta_{n+2}}{\xi_{n+3}\xi_{n+2}(\xi_{n+2} + \xi_{n+3})} + \frac{2\beta\eta_{n+1}}{\xi_{n+1}\xi_{n+2}(\xi_{n+1} + \xi_{n+2})} \\
&= I_1 + O(\varepsilon) \qquad \alpha + \beta = 1
\end{aligned}
\tag{11.78}
$$

$$
\begin{aligned}
J_2 &= \frac{6}{\xi_{n+1} + \xi_{n+2} + \xi_{n+3}}\left\{-\frac{\eta_{n+2}}{\xi_{n+3}\xi_{n+2}(\xi_{n+2} + \xi_{n+3})} + \frac{\eta_{n+1}}{\xi_{n+1}\xi_{n+2}(\xi_{n+1} + \xi_{n+2})}\right\} \\
&= I_2 + O(\varepsilon).
\end{aligned}
\tag{11.79}
$$

An invariant OΔS approximating, e.g., (11.74) is

$$J_2 = F(J_1, \xi_{n+1}, \xi_{n+2}, \xi_{n+3}) \qquad \phi(\xi_{n+1}, \xi_{n+2}, \xi_{n+3}) = 0 \tag{11.80}$$

with

$$F(J_1, 0, 0, 0) = F(J_1) \qquad \phi(0, 0, 0) \equiv 0. \qquad (11.81)$$

11.3 Applications to numerical solutions of ODEs

11.3.1 General procedure for testing the numerical schemes

This section reports on the numerical experiments performed using the schemes described in the previous sections. The schemes are used to compute the solution for initial value problems on a given interval. Before describing the results for each class of symmetries analyzed above in Section 11.2, we first describe some general procedures to implement and test the various methods.

Reference solution

For test-problems for which an analytical solution is not available, a very accurate and reliable reference solution is computed numerically and used to assess the performance of the point symmetry preserving scheme. This is done using MATLAB's standard adaptive Runge–Kutta scheme ODE45, with a very strict tolerance on the error set at tol $= 10^{-9}$. The first step is to convert the nth order equation (11.1) for $y(x)$ into a system of n first-order ODEs for $u_1(x) = y(x)$, $u_2(x) = y'(x), \ldots, u_n(x) = y^{(n-1)}(x)$. Then (11.1) becomes the system

$$u_1' = u_2, \qquad u_2' = u_3, \qquad \ldots, \qquad u_n' = f(x, u_1, u_2, \ldots, u_n). \qquad (11.82)$$

Given initial conditions $u_1(x_0), u_2(x_0), \ldots, u_n(x_0)$, one then proceeds to compute the solution on the interval $[x_0, x_F]$, where the scheme adaptively selects the local integration step so that its local error estimate satisfies the imposed tolerance. Those very high order, very accurate (and very costly numerically) solutions are used to generate start-up values as well as error estimations for the point symmetry preserving schemes as described next.

Start-up values

The symmetry preserving schemes require a number of start-up values ($y_0 = y(x_0)$, $y_1 = y(x_1)$ for the second order case; also $y_2 = y(x_2)$ for the third order cases). For given initial values $y(x_0)$, $y'(x_0)$, (and $y''(x_0)$ for the third order case), the start-up value $y_0 = y(x_0)$ is directly available, while the values for y_1 and y_2 are obtained as the the numerical reference solution (obtained as described above) at the nodes x_1 and x_2.

Error analysis

Given the discrete mesh x_n, $n = 0, 1, 2, \ldots, N$ and corresponding solution y_n generated by the point-symmetry preserving scheme, the corresponding errors are obtained by comparing y_n with $y_{\text{ref}}(x_n)$. Although the user has no direct input on the actual mesh used by the MATLAB's solver, it is possible for the user to request specific output points for the discrete solutions, so that given x_n, one can obtain a very reliable numerical approximation $y_{\text{ref}}(x_n)$, accurate with the prescribed tolerance.

Equivalent standard schemes

To better assess the new schemes proposed here, their performance for various test-cases is compared with that of the standard finite difference schemes that uses the same number of grid points $x_{s,n}$, $n = 0, 1, 2, \ldots, N$. Although the symmetry preserving scheme finite mesh is typically non-uniform, for simplicity, the standard mesh is assumed to be uniform, so that $x_{s,n} = x_0 + nh$ with $h = (x_F - x_0)/N$. The discrete standard scheme is obtained using the following standard procedure (given here for the third order case, easily adapted for the second order case). Given the four points $(x_{s,n-1}, y_{s,n-1})$, $(x_{s,n}, y_{s,n})$, $(x_{s,n+1}, y_{s,n+1})$, $(x_{s,n+2}, y_{s,n+2})$:

1 Obtain the interpolating polynomial $P_3(x)$ through the four given points
2 Evaluate analytically $P_3'(x_{s,n+1/2})$, $P_3''(x_{s,n+1/2})$, $P_3'''(x_{s,n+1/2})$, which gives:

$$P_3'(x_{s,n+1/2}) = \frac{1}{24h}(27(y_{s,n+1} - y_{s,n}) - (y_{s,n+2} - y_{s,n-1}))$$

$$P_3''(x_{s,n+1/2}) = \frac{1}{2h^2}(y_{s,n+2} - (y_{s,n+1} + y_{s,n}) + y_{s,n-1})$$

(11.83)

$$P_3'''(x_{s,n+1/2}) = \frac{1}{h^3}(y_{s,n+2} - 3y_{s,n+1} + 3y_{s,n} - y_{s,n-1})$$

3 Substitute those expressions in the equation being discretized, evaluated at $x = x_{s,n+1/2}$.

11.3.2 Numerical experiments for a third-order ODE invariant under $\text{SL}_1(2, \mathbb{R})$

Numerical experiments are conducted for (11.43) for the case $F(x) = \sin(x)$:

$$\frac{1}{y'^2}\left(y'y''' - \frac{3}{2}y''^2\right) = \sin(x).$$

(11.84)

The solution is sought for x in the interval $[0, 2]$ (also $[0, 6]$) with initial conditions $y(1) = 0$, $y'(1) = -10$, $y''(1) = 1$. A uniform mesh is used here, it is compatible with the difference invariants in (11.47). With $h_n = h_{n+1} = h_{n+2} = h$ corresponding to a uniform mesh, the other invariant difference equation in (11.47) becomes:

$$I_2^{n+3} = R = 4\left(1 - \frac{h^2}{2}F(x_{n+1}, h)\right) \tag{11.85}$$

with R defined as $R = (y_{n+3} - y_{n+1})(y_{n+2} - y_n)/((y_{n+3} - y_{n+2})(y_{n+1} - y_n))$ and we select $F(x_{n+1}, h) = F(x_{n+1} + h/2)$ to achieve second order accuracy. This leads to the following explicit expression for y_{n+3}:

$$y_{n+3} = \frac{(y_{n+2} - y_n)y_{n+1} - K(y_{n+1} - y_n)y_{n+2}}{(y_{n+2} - y_n) - K(y_{n+1} - y_n)} \tag{11.86}$$

where $K = 4(1 - (h^2/2)F(x_{n+1} + h/2))$. This explicit expression for y_{n+3} is remarkably simple.

On the other hand, the standard scheme for the same problem is nonlinear. Substituting the finite difference approximations for y', y'', y''' in (11.83) into the ODE (11.84) leads to a nonlinear equation for y_{n+3} to be solved iteratively.

First, we compare the discretization errors using the invariant scheme and the standard scheme on the interval $[0, 2]$ on which the solution is smooth. Both schemes display a second order convergence rate. The standard scheme has errors which are smaller by a factor of 6, but in terms of computational efforts, the invariant scheme is much more efficient, as it gives an explicit formula for y_{n+3} unlike the standard scheme that requires a nonlinear iterative solver at each step. However, if the integration interval is $[0, 6]$, remarkably different conclusions are obtained. The solution develops a singularity around $x = 3$. At that point, both the standard scheme and the adaptive Runge–Kutta solver from MATLAB fail to converge. On the other hand, the invariant scheme integrates right through the singularity. The solution obtained with the three schemes (reference, standard, invariant) is displayed in Figure 11.2 for the coarse resolution $h = 0.1$. In Figure 11.3, the solution is shown with the invariant scheme for three resolutions $h = 0.1, 0.01, 0.001$. To better observe the behavior of the solution near the singularity, the plot uses a log scale, and the absolute value of the solution is shown. Excellent numerical convergence is observed, with the solutions corresponding to the three resolutions matching very closely each other (of course, the singularity is better captured by the finest mesh).

The most striking feature shown on Figures 11.2 and 11.3 is that the symmetry preserving difference scheme provides a numerical solution $u(x)$ for the entire region $0 \le x \le 6$, $x \ne x_0$, even though the solution has a pole at x_0 close to 3. A similar phenomenon was observed in a previous study of a specific type of first order systems of ODEs, namely matrix Riccati equations [2, 28].

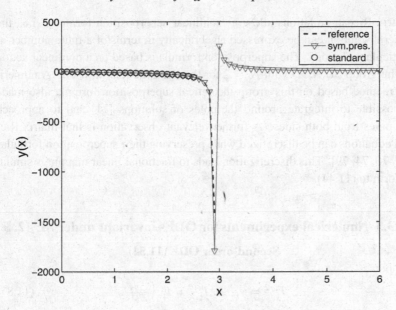

Figure 11.2 Solution of (11.84) for the symmetry preserving scheme and the standard scheme, $h = 0.1$ on the interval $[0, 6]$.

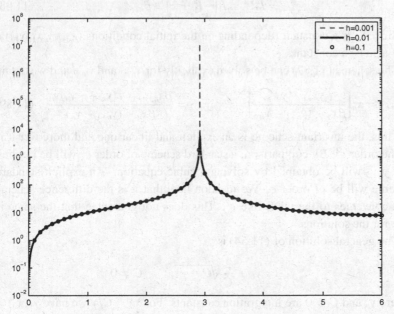

Figure 11.3 Solution for the symmetry preserving scheme, $h = 0.1, 0.01,$ 0.001. Example (11.84) on $[0, 6]$. Absolute value of solution on y axis, logarithmic scale.

Matrix Riccati equations allow a "nonlinear superposition formula," i.e., the general solution can be expressed algebraically in terms of a finite number of particular solutions. The superposition formula is based on a nonlinear action of the group $SL(N, \mathbb{R})$ with $N = 2$ for the Riccati equation itself. A numerical method based on this group theoretical superposition formula also made it possible to integrate around the poles of solutions [71] and to approach the poles from both sides. A further relevant observation is that matrix Riccati equations can be discretized while preserving their superposition formulas [27, 70, 74, 79]. This discretization leads to fractional linear mappings similar in form to (11.44).

11.3.3 Numerical experiments for ODEs invariant under $SL_2(2, \mathbb{R})$

Second-order ODE (11.54)

We put

$$J_1^{n+2} = \gamma, \qquad I_1^{n+2} = I_1^{n+1} \equiv I_1. \tag{11.87}$$

From (11.87) we obtain

$$I_2^{n+2} = I_1\left(\frac{\gamma}{4}I_1^2 + 2\right) \equiv \beta. \tag{11.88}$$

Since I_1 is a constant (depending on the initial conditions (x_0, y_0, x_1, y_1)) β will also be a constant.

The scheme (11.87) can be solved explicitly for x_{n+2} and y_{n+2} and we obtain

$$x_{n+2} = x_n\left[\frac{y_{n+1} - y_n}{\beta x_n - (y_{n+1} - y_n)}\right]^2, \qquad y_{n+2} = \frac{\beta x_n y_{n+1} - (y_{n+1} - y_n)y_n}{\beta x_n - (y_{n+1} - y_n)}. \tag{11.89}$$

Thus, the invariant scheme is an explicit and linear one and moreover it is of the order ϵ^2. By comparison, a standard scheme of order ϵ^2 will be implicit and y_{n+2} will be obtained by solving a cubic equation. An explicit standard scheme will be of order ϵ. We mention here that it is the difference scheme that converges to the ODE like ϵ^2. This does not guarantee that the same is true for the solutions.

The general solution of (11.54) is

$$y_{1,2} = y_b \pm \frac{2}{C}\sqrt{C - \gamma x}, \qquad C \neq 0$$

where y_b and $C \neq 0$ are integration constants. For $x_0 = C/\gamma$ we have $y_1(x_0) = y_2(x_0) = y_b$, i.e., the two solutions intersect.

On Figure 11.4 we show the exact solutions y_1 (increasing branch) and y_2 (decreasing branch) for $\gamma = 150$, $y_b = 5$, $C = e^2$. The step for the exact

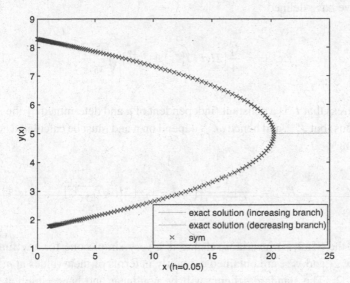

Figure 11.4 Behavior of the symmetry preserving scheme near the singularity for (11.54)

solution was $h = 0.05$. The symmetry preserving method integrates y_1 up to the singularity at $x = x_0 = C/\gamma \sim 20.3$ then continues along the second branch y_2 to its 'initial' value. The standard method fails to converge close to the singularity (where the solution becomes complex).

Third-order ODE (11.55)

From (11.55) we see that an invariant difference scheme for the third-order equation (11.55) is obtained by putting

$$J_2^{n+3} = F(J_1^{n+2}), \qquad I_1^{n+3} = I_1^{n+2} = I_1^{n+1} \equiv I. \qquad (11.90)$$

Alternatively, we can put

$$J_2^{n+3} = F\left(\frac{J_1^{n+2} + J_1^{n+3}}{2}\right), \qquad I_1^{n+2} = I_1^n \equiv I. \qquad (11.91)$$

Both schemes converge to the ODE like ϵ^2. The scheme (11.90) can be solved explicitly for x_{n+3}, y_{n+3} and we obtain

$$x_{n+3} = \frac{x_{n+1}}{(1 - \omega_{n+2})^2} \qquad y_{n+3} = \frac{y_{n+1} - \omega_{n+2} y_{n+2}}{1 - \omega_{n+1}} \qquad (11.92)$$

where we have defined

$$\omega_{n+2} = \frac{I_1^2}{4}[2I_1 F(J_1^{n+2}) + J_1^{n+2}]\sqrt{\frac{x_{n+1}}{x_{n+2}}}. \tag{11.93}$$

We stress that I_1 is a constant (independent of n and determined by the initial conditions) but J_1^{n+2} and hence ω_{n+2} depend on n and must be calculated at each step using

$$J_1^{n+1} = \frac{4}{\sqrt{x_{n+2}x_n}I_1^3}[y_{n+2} - y_n - 2I_1\sqrt{x_{n+2}x_n}] \tag{11.94}$$

Thus the $SL(2,\mathbb{R})$ invariant scheme is a very simple one for any function $F(J_1)$: x_{n+3} and y_{n+3} are obtained explicitly in terms of their values at $n, n+1$ and $n+2$. The standard scheme will be nonlinear and hence implicit. The condition $I_1^n = I_1^{n+1}$ in (11.90) is actually quite restrictive for third (and higher) order ODEs. The constant I is determined by the initial conditions and (11.90) imposes a relation between $y(0)$, $y'(0)$ and $y''(0)$. For general initial conditions a relation of the type $I_1^{n+1} = \gamma I_1^n$ is more suitable (γ is determined in terms of the initial conditions).

Let us now specialize to (11.56) with $\alpha = -1$. We put $I_1^{n+2} = \gamma I_1^{n+1}$ where γ is determined by the initial conditions. As the step h tends to zero, γ will tend to $\gamma = 1$.

In Figure 11.5 we compare the accuracies of the standard and symmetry preserving scheme. Since in this case no exact analytic solution is available we compare with a reference solution using a MATLAB Runge–Kutta scheme with a tolerance on the error set at tol $= 10^{-9}$. The initial conditions were set at $y_0 = 1$, $y_0' = 10$, $y_0'' = -4$ and this corresponds to a solution with no singularity on the real axis for $0 \le x \le 16$. We see that the accuracy is better for the symmetry preserving scheme by a factor of 10.

A singular solution is shown in Figure 11.6. The initial conditions were set at $y_0 = 1$, $y_0' = 1$, $y_0'' = 3$ and a singularity occurs for $x \sim 1.7$. MATLAB solvers and standard schemes stop providing solutions close to the singularity. The symmetry preserving method approaches the singularity closely and then continues along the second branch of the singular solution towards an appropriate initial condition. Qualitatively we have the same features as for the second order equation (11.54). The solution itself stays finite but its derivative becomes infinite at the singularity.

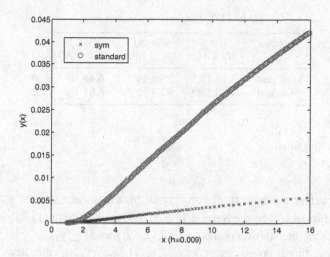

Figure 11.5 Discretization errors for standard and symmetry preserving schemes for (11.56), $\alpha = -1$ for a regular solution

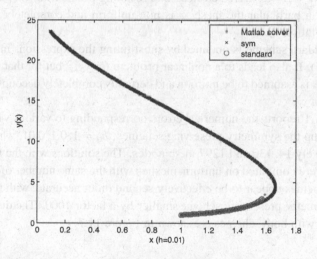

Figure 11.6 Behavior of the symmetry preserving scheme near a singularity for (11.56)

11.3.4 Numerical experiments for third-order ODE invariant under Sim(2)

The test-case consists in solving (11.75) for $K = 1$, with x in the interval $[0, 10]$ and with initial values $y(0) = 0$, $y'(0) = -10$, $y''(0) = 1$. The lattice equation

Table 11.1 *Discretization errors*

Scheme	$h = 1$ ($N = 14$)	$h = 0.1$ ($N = 130$)	$h = 0.01$ ($N = 1297$)
Sym. pres.	$2.14 \ 10^{-5}$	$2.98 \ 10^{-7}$	$6.45 \ 10^{-9}$
Standard	$4.20 \ 10^{-2}$	$5.83 \ 10^{-4}$	$6.01 \ 10^{-6}$

is chosen in the form

$$\frac{\xi_{n+1}}{\xi_{n+2}} = \frac{\xi_{n+2}}{\xi_{n+3}} = \gamma. \tag{11.95}$$

Start-up values for $x_0 = 0$, $x_1 = h_0$, $x_2 = 2h_0$ are generated for a given h_0 using the MATLAB solver, see Section 11.3.1. The constant γ in (11.95) is then computed using the start-up points $(x_0, y_0), (x_1, y_1)$, and (x_2, y_2).

Given the three points $(x_n, y_n), (x_{n+1}, y_{n+1}), (x_{n+2}, y_{n+2})$, the new point (x_{n+3}, y_{n+3}) is obtained as the solution of the nonlinear system consisting of equation $J_2 = K J_1^2$ (with $K = 1$) and (11.95). In the present set of experiments, the values $\alpha = \beta = \frac{1}{2}$ were selected. The resulting problem for (x_{n+3}, y_{n+3}) is nonlinear, in particular the mesh x_n is non-uniform and completely coupled with the solution y_n.

The standard scheme is obtained by substituting the expressions in (11.83) into (11.75). It also leads to a nonlinear problem for y_{n+2}, but for that scheme, the mesh x_n is assumed to be uniform and certainly completely decoupled from the solution y_n.

Table 11.1 reports the numerical errors corresponding to various values for h_0 to start up the symmetry preserving schemes: $h_0 = 1, 0.1, 0.01$, which lead to respectively $14, 130$ and 1297 mesh nodes. The solutions with the standard schemes were computed on uniform meshes with the same number of nodes.

Both schemes appear to be effectively second order accurate, with the error in the symmetry preserving scheme smaller by a factor 1000. The discretization errors with both schemes are shown in Figure 11.7.

11.4 Exact solutions of invariant difference schemes

Let us now consider a different aspect of invariant difference schemes. Namely, if the ODE that we are solving happens to be an Euler–Lagrange equation for a first-order Lagrangian, then it may be possible to solve the corresponding invariant difference scheme exactly. This problem was discussed in [19] and also in Chapter 1 of the present volume (written by Dorodnitsyn and Kozlov).

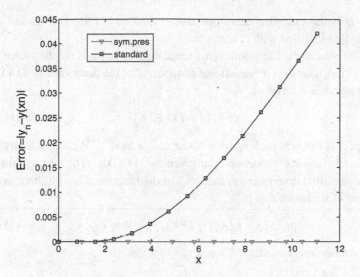

Figure 11.7 Discretization errors for the symmetry preserving scheme and the standard scheme, $h = 1$.

Here we shall just briefly present some results on symmetries of continuous and discrete Lagrangians and consider one example. We restrict to second-order ODEs

$$y'' = f(x, y, y') \tag{11.96}$$

and hence to three-point ordinary difference schemes (OΔS)

$$E_a(x_{n-1}, x_n, x_{n+1}, y_{n-1}, y_n, y_{n+1}) = 0. \tag{11.97}$$

11.4.1 Lagrangian formulation for second-order ODEs

It has been known since E. Noether's fundamental work that conservation laws for differential equations are connected with their symmetry properties [4, 36, 66, 69]. For convenience we present here some well-known results adapted to the case of second order ODEs.

Let us consider the functional

$$\mathbb{L}(y) = \int_I L(x, y, y') \, dx, \qquad I \subset \mathbb{R}, \tag{11.98}$$

where $L(x, y, y')$ is called a first-order Lagrangian. The functional (11.98) achieves its extremal values when $y(x)$ satisfies the Euler–Lagrange equation

$$\frac{\delta L}{\delta y} = \frac{\partial L}{\partial y} - D\left(\frac{\partial L}{\partial y'}\right) = 0, \qquad D = \frac{\partial}{\partial x} + y'\frac{\partial}{\partial y} + y''\frac{\partial}{\partial y'} + \cdots, \tag{11.99}$$

where D is the total derivative operator. Equation (11.99) is a second-order ODE to be identified with (11.96).

Let us consider a Lie point transformation G generated by the vector field (11.3). The group G is a "variational symmetry" of the functional $\mathbb{L}$ if and only if the Lagrangian satisfies

$$\mathrm{pr}\, X(L) + L D(\xi) = 0, \tag{11.100}$$

when $\mathrm{pr}\, X$ is the first prolongation of the vector field X. We will actually need a weaker invariance condition than given by (11.100). The vector field X is an "infinitesimal divergence symmetry" of the functional $\mathbb{L}(y)$ if there exists a function $V(x, y)$ such that [4]

$$\mathrm{pr}\, X(L) + L D(\xi) = D(V), \qquad V = V(x, y). \tag{11.101}$$

Two important statements for us are:

1. If X is an infinitesimal divergence symmetry of the functional $\mathcal{L}$, it generates a symmetry group of the corresponding Euler–Lagrange equation. The symmetry group of (11.99) can of course be larger than the one generated by symmetries of the Lagrangian.
2. Noether's theorem [4, 36, 66, 69] can be based on the following Noether-type identity which holds for any vector field and any function L:

$$\mathrm{pr}\, X(L) + L D(\xi) = (\phi - \xi y')\frac{\delta L}{\delta y} + D\Big(\xi L + (\phi - \xi y')\frac{\partial L}{\partial y'}\Big). \tag{11.102}$$

It follows that if X is a divergence symmetry of L, i.e. (11.100) or (11.101) is satisfied, then there exists a first integral

$$\xi L + (\phi - \xi y')\frac{\partial L}{\partial y'} - V = K = \text{const} \tag{11.103}$$

of the corresponding Euler–Lagrange equation.

The above considerations tell us how to obtain invariant ODEs and conservation laws from divergence invariant Lagrangians. They do not tell us how to obtain invariant Lagrangians for invariant equations. This amounts to "variational integration," as opposed to variational differentiation.

A procedure that we shall use below to find invariant Lagrangians for differential equations can be summed up as follows.

Step 1. Start from a given ODE (11.96) and its symmetry algebra with basis

$$X_\alpha = \xi_\alpha(x, y)\frac{\partial}{\partial x} + \phi_\alpha(x, y)\frac{\partial}{\partial y}, \qquad \alpha = 1, \ldots, k.$$

Find the invariants of X_α in the space $\{x, y, y', \Lambda\}$, where Λ is the Lagrangian. The appropriate prolongation in this case is

$$\text{pr } X = \xi \frac{\partial}{\partial x} + \phi \frac{\partial}{\partial y} + \phi^x \frac{\partial}{\partial y'} - (D\xi)\Lambda \frac{\partial}{\partial \Lambda}, \qquad \phi^x = D(\phi) - y'D(\xi) \quad (11.104)$$

and we require that $L(x, y, y')$ should satisfy

$$\text{pr } X(\Lambda - L)|_{\Lambda = L} = 0. \tag{11.105}$$

Each basis element X_α provides us with an equation of the form

$$\xi_\alpha \frac{\partial L}{\partial x} + \phi_\alpha \frac{\partial L}{\partial y} + \phi_\alpha^x \frac{\partial L}{\partial y'} - LD(\xi_\alpha) = 0. \tag{11.106}$$

Step 2. Solve the partial differential equations (11.106). This will give us the general form of an invariant Lagrangian. It may involve arbitrary functions of the invariants of X.

Step 3. Request that the Euler–Lagrange equation (11.99) should coincide with the equation we started from. This will further restrict the invariant Lagrangian and determine whether one exists.

If this procedure does not yield a suitable Lagrangian, then Step 1 can be weakened. We can request that the Lagrangian be invariant under some subgroup of the symmetry group of the given ODE, rather then the entire group. We then go through Step 2, verify whether the obtained Lagrangian is divergence invariant under the entire group, or at least a larger subgroup. In any case, each divergence symmetry of the Lagrangian will provide a first integral of the ODE.

For ODEs the Lagrangian formalism is not the only integration method. The existence of one-parameter symmetry group provides a reduction to a first-order ODE directly. The existence of a two-parameter symmetry group makes it possible to integrate in quadratures. An invariant Lagrangian provides an alternative. Indeed, assume that we know two first integrals

$$f_1(x, y, y') = A, \qquad f_2(x, y, y') = B, \tag{11.107}$$

then we eliminate y' from these two equations and obtain the general solution

$$y = F(x, A, B), \tag{11.108}$$

of the corresponding ODE by purely *algebraic* manipulations. It is this method of invariant Lagrangians that generalizes to difference equations and is particularly useful when direct methods fail.

11.4.2 Lagrangian formulation for second order difference equations

The variational formulation of discrete equations and a discrete analog of Noether's theorem are much more recent (see [15, 21] and the Chapter 1 by V. Dorodnitsyn and R. Kozlov in this volume). Here we briefly overview the results that we shall need below.

Let us consider a finite difference functional

$$\mathbb{L}_h = \sum_{\Omega} \mathcal{L}_{n+1}(x_n, x_{n+1}, y_n, y_{n+1}) h_{n+1}, \tag{11.109}$$

defined on some one-dimensional lattice Ω determined by the equation

$$\Psi(x_n, x_{n+1}, y_n, y_{n+1}) = 0. \tag{11.110}$$

In the continuous case, a Lagrangian L provides an equation (the Euler–Lagrange equation) that inherits all the symmetries of L. In the discrete case we require that the Lagrangian (11.109) should provide two equations: the entire difference scheme. Moreover, the three-point difference system should inherit the symmetries of the two-point Lagrangian.

Let us again consider a Lie group of point transformations, generated by a Lie algebra of vector fields X_α of the form (11.3). The infinitesimal invariance condition of the functional (11.109) on the lattice (11.110) is given by two equations:

$$\xi_n \frac{\partial \mathcal{L}_{n+1}}{\partial x_n} + \xi_{n+1} \frac{\partial \mathcal{L}_{n+1}}{\partial x_{n+1}} + \phi_n \frac{\partial \mathcal{L}_{n+1}}{\partial y_n} + \phi_{n+1} \frac{\partial \mathcal{L}_{n+1}}{\partial y_{n+1}} + \mathcal{L}_{n+1} \Delta^T(\xi) = 0, \tag{11.111}$$

$$X(\Psi) = 0. \tag{11.112}$$

Let us consider a variation of the difference functional (11.109) along some curve $y = y(x)$ at some point (x, y). The variation will effect only two terms in the sum (11.109):

$$\mathbb{L}_h = \cdots + \mathcal{L}_n h_n + \mathcal{L}_{n+1} h_{n+1} + \cdots, \tag{11.113}$$

so we get the following expression for the variation of the difference functional

$$\delta \mathbb{L}_h = \left[\frac{\partial \mathcal{L}_{n+1}}{\partial x_n} h_{n+1} + \frac{\partial \mathcal{L}_n}{\partial x_n} h_n + \mathcal{L}_n - \mathcal{L}_{n+1} \right] \delta x$$
$$+ \left[\frac{\partial \mathcal{L}_{n+1}}{\partial y_{n+1}} h_{n+1} + \frac{\partial \mathcal{L}_n}{\partial y_n} h_n \right] \delta y \tag{11.114}$$

where $\delta y = \phi' \delta x$.

Thus, for an arbitrary curve the stationary value of difference functional is given by any solution of the *two* equations, called *quasiextremal equations*

$$\frac{\partial \mathcal{L}_{n+1}}{\partial x_n} h_{n+1} + \frac{\partial \mathcal{L}_n}{\partial x_n} h_n + \mathcal{L}_n - \mathcal{L}_{n+1} = 0,$$

$$\frac{\partial \mathcal{L}_{n+1}}{\partial y_{n+1}} h_{n+1} + \frac{\partial \mathcal{L}_n}{\partial y_n} h_n = 0.$$

(11.115)

Both of them tend to the differential Euler-Lagrange equation in the continuous limit. Together they represent the entire difference scheme and could be called "the discrete Euler–Lagrange system." The difference between these two equations, or some other function of them that vanishes in the continuous limit will represent the lattice.

Now let us consider a vector field (11.3) with given coefficients $\xi(x, y)$ and $\phi(x, y)$. Variations along the integral curves of this vector field are given by $\delta x = \xi \, da$ and $\delta y = \phi \, da$, where da is a variation of a group parameter. A stationary value of the difference functional (11.109) along the flow generated by this vector field is given by the equation (11.114)

$$\delta \mathbb{L}_n = 0, \qquad \delta x = \xi_n \, da, \qquad \delta y = \phi_n \, da \qquad (11.116)$$

which depends explicitly on the coefficients of the generator.

If we have a Lie algebra of vector fields of dimension 2 or more, then a stationary value of the difference functional (11.109) along the entire flow will be achieved on the intersection of the solutions of all equations of the type (11.116), i.e., on the quasiextremals (11.115).

On the other hand, (11.115) can be interpreted as a three-point difference scheme. For instance, given two points (x_n, y_n) and (x_{n-1}, y_{n-1}), we can calculate (x_{n+1}, y_{n+1}). In the continuous limit both of these equations will provide the same second-order differential equation. Thus, one of the quasiextremal equations can be identified with E_1 of (11.8) and the difference between the two of them with the lattice equation $E_2 = 0$.

It has been shown elsewhere [15, 21, 22] that if the functional (11.109) is invariant under some group G, then the quasiextremal equations (11.115) are also invariant with respect to G. As in the continuous case, the quasiextremal equations can be invariant with respect to a larger group than the corresponding Lagrangian.

A useful operator identity, valid for any Lagrangian $\mathcal{L}(x_n, x_{n+1}, y_n, y_{n+1})$ and any vector field X is:

$$\xi_n \frac{\partial \mathcal{L}_{n+1}}{\partial x_n} + \xi_{n+1} \frac{\partial \mathcal{L}_{n+1}}{\partial x_{n+1}} + \phi_n \frac{\partial \mathcal{L}_{n+1}}{\partial y_n} + \phi_{n+1} \frac{\partial \mathcal{L}_{n+1}}{\partial y_{n+1}} + \mathcal{L}_{n+1} \Delta^T (\xi_n)$$

$$= \xi_n \left[\frac{\partial \mathcal{L}_{n+1}}{\partial x_n} + \frac{h_n}{h_{n+1}} \frac{\partial \mathcal{L}_n}{\partial x_n} - \Delta^T \mathcal{L}_n \right] + \phi_n \left[\frac{\partial \mathcal{L}_{n+1}}{\partial y_n} + \frac{h_n}{h_{n+1}} \frac{\partial \mathcal{L}_n}{\partial y_n} \right]$$

$$+ \Delta^T \left[h_n \phi_n \frac{\partial \mathcal{L}_n}{\partial y_n} + h_n \xi_n \frac{\partial \mathcal{L}_n}{\partial x_n} + \xi \mathcal{L}_n \right]. \quad (11.117)$$

From (11.117) we obtain the following discrete analog of Noether's theorem.

Theorem 11.1   *Let the Lagrangian density $\mathcal{L}_{n+1}$ be divergence invariant under a Lie group G of local point transformations generated by vector fields X of the form (11.3), i.e., let us have*

$$\mathrm{pr}\, X(\mathcal{L}_{n+1}) + \mathcal{L}_{n+1} \Delta^T \xi_n = \Delta^T V \quad (11.118)$$

for some function $V(x_n, y_n)$. Then each element X of the Lie algebra corresponding to G provides us with a first integral of the quasiextremal equations (11.115), namely

$$K = h_n \phi_n \frac{\partial \mathcal{L}_n}{\partial y_n} + h_n \xi_n \frac{\partial \mathcal{L}_n}{\partial x_n} + \xi_n \mathcal{L}_n - V. \quad (11.119)$$

Proof [15, 21]   On solutions of the quasiextremal equations (11.115), equation (11.117) reduces to

$$\Delta^T \left[h_n \phi_n \frac{\partial \mathcal{L}_n}{\partial y_n} + h_n \xi_n \frac{\partial \mathcal{L}_n}{\partial x_n} + \xi_n \mathcal{L}_n \right] = \Delta^T V \quad (11.120)$$

(we have used (11.117)). The result (11.119) follows immediately.   □

The fundamental equation (11.119) is the discrete analog of (11.103) for ODEs.

Let us compare the situation for second order ODEs and for three-point difference schemes. For a second order ODE a Lagrangian that is divergence invariant under a two-dimensional symmetry group provides two integrals of motion. From them we can eliminate the remaining first derivative and obtain the general solution, depending on two arbitrary constants (the two first integrals). Moreover, we do not really need a Lagrangian. Once we have a two dimensional symmetry group of the ODE, we can integrate in quadratures.

For three-point difference schemes we have two equations to solve, namely the system (11.8). Equivalently, we have a set of points (x_n, y_n), labeled by an integer n. Any 3 neighboring points are related by two equations that we can write, e.g., as

$$y_{n+1} = F_1(x_n, y_n, x_{n-1}, y_{n-1}), \qquad x_{n+1} = \Omega_1(x_n, y_n, x_{n-1}, y_{n-1}). \qquad (11.121)$$

Given some starting values $(x_0, y_0, x_{-1}, y_{-1})$, we can solve (11.121) for (x_n, y_n) with $n \geq 1$, and $n \leq -2$. The solution will depend on four constants K_i, $i = 1, \ldots, 4$, and can be written as

$$y_n = y_n(x_n, K_1, K_2, K_3, K_4), \qquad (11.122)$$
$$x_n = x_n(K_1, K_2, K_3, K_4). \qquad (11.123)$$

The two quasiextremal equations (11.115) correspond to the system (11.121).

A one-parameter symmetry group of the Lagrangian $\mathcal{L}$ will provide us with a first integral (11.119), i.e., an equation of the form

$$f(x_n, y_n, x_{n+1}, y_{n+1}) = K_1. \qquad (11.124)$$

compatible with the system (11.115). We can solve (11.124) for, e.g., y_{n+1}, substitute into (11.121) and thus simplify this system.

A two-dimensional symmetry group will provide two first integrals of the form (11.119). We can solve for x_{n+1} and y_{n+1}. Then system (11.121) is reduced to a two-point difference scheme. Quite often it is possible to solve it by integration methods that allow one to integrate a two-point difference scheme explicitly.

A three-dimensional symmetry group provides three first integrals of the type (11.119). From them we can express x_{n-1}, y_{n-1} and y_n in terms of x_n. This provides us with the solution (11.122) and a two-point difference equation relating x_{n+1} and x_n. If this equation can be solved, we have a complete solution of the problem. Finally, if we have four first integrals, then we get the general solution of the system by purely algebraic manipulations.

11.4.3 Example: Second-order ODE with three-dimensional solvable symmetry algebra

Let us consider a family of solvable Lie algebras depending on one constant k:

$$X_1 = \frac{\partial}{\partial x}, \quad X_2 = \frac{\partial}{\partial y}, \quad X_3 = x\frac{\partial}{\partial x} + ky\frac{\partial}{\partial y}, \qquad k \neq 0, \frac{1}{2}, 1, 2. \qquad (11.125)$$

The invariant differential equation has the form

$$y'' = y'^{(k-2)/(k-1)}. \qquad (11.126)$$

This equation can be obtained by the usual variational procedure from the Lagrangian

$$L = \frac{(k-1)^2}{k}(y')^{k/(k-1)} + y, \tag{11.127}$$

which admits operators X_1 and X_2 for any parameter k and X_3 for $k = -1$:

$$\text{pr}\, X_1 L + LD(\xi_1) = 0;$$
$$\text{pr}\, X_2 L + LD(\xi_2) = 1 = D(x);$$
$$\text{pr}\, X_3 L + LD(\xi_3) = (k+1)L.$$

It is possible to show that there is no Lagrangian function $L(x, y, y')$ which gives (11.126) with $k \neq -1$ as its Euler's equation and is divergence invariant for all three symmetries (11.125).

For arbitrary k there are two first integrals

$$J_1 = \frac{1-k}{k}(y')^{k/(k-1)} + y = A^0, \qquad J_2 = (k-1)(y')^{1/(k-1)} - x = B^0.$$

Eliminating y' we find the general solution:

$$y = \frac{1}{k}\left(\frac{1}{k-1}\right)^{(k-1)}(x + B^0)^k + A^0. \tag{11.128}$$

In the case $k = -1$ we have a further first integral corresponding to the symmetry X_3:

$$J_3 = \frac{2}{\sqrt{y'}}(y - xy') + xy = C^0.$$

It is functionally dependent on J_1 and J_2 since a second-order ODE can possess only two functionally independent first integrals. In this case we have the following relation:

$$4 - J_1 J_2 - J_3 = 0. \tag{11.129}$$

Thus, the integral J_3 is not independent and is of no use in the present context.

Now let us turn to the discrete case and first consider the case $k = -1$. Let us choose the Lagrangian to be

$$\mathcal{L} = -4\sqrt{p_{n+1}^{(1)}} + \frac{y_n + y_{n+1}}{2}, \qquad p_{n+1}^{(1)} = \frac{y_{n+1} - y_n}{x_{n+1} - x_n} \tag{11.130}$$

as a discrete Lagrangian, which is invariant for X_1 and X_3 and divergence invariant for X_2:

$$\text{pr}\, X_1 \mathcal{L} + \mathcal{L}\Delta^T(\xi_1) = 0;$$
$$\text{pr}\, X_2 \mathcal{L} + \mathcal{L}\Delta^T(\xi_2) = 1 = \Delta^T(x); \tag{11.131}$$
$$\text{pr}\, X_3 \mathcal{L} + \mathcal{L}\Delta^T(\xi_3) = 0.$$

From the Lagrangian we obtain the quasiextremal equations:

$$\frac{\delta \mathcal{L}}{\delta y} : -\frac{4}{h_n + h_{n+1}}\left(\frac{1}{\sqrt{P_{n+1}^{(1)}}} - \frac{1}{\sqrt{P_n^{(1)}}}\right) = 1;$$

$$\frac{\delta \mathcal{L}}{\delta x} : 4\left(\sqrt{P_{n+1}^{(1)}} - \sqrt{P_n^{(1)}}\right) - \frac{y_n + y_{n+1}}{2} + \frac{y_{n-1} + y_n}{2} = 0.$$

(11.132)

This system of equations is invariant with respect to all three operators (11.125). The application of the difference analog of the Noether theorem gives us three first integrals:

$$I_1 = -2\sqrt{P_{n+1}^{(1)}} + \frac{y_n + y_{n+1}}{2} = A, \qquad I_2 = -\frac{2}{\sqrt{P_{n+1}^{(1)}}} - \frac{x_n + x_{n+1}}{2} = B,$$

$$I_3 = \frac{2(x_{n+1}y_n - y_{n+1}x_n)}{h_{n+1}\sqrt{P_{n+1}^{(1)}}} + \frac{x_{n+1}y_n + y_{n+1}x_n}{2} = C.$$

(11.133)

In contrast to the continuous case the three difference first integrals I_1, I_2 and I_3 are functionally independent and instead of (11.129) we have the following relation:

$$4 - I_1 I_2 - I_3 = \frac{1}{4}h_{n+1}^2 P_{n+1}^{(1)} = \frac{4\varepsilon^2}{(\varepsilon + 2)^2}.$$

(11.134)

This coincides with (11.129) in the continuous limit $\varepsilon \to 0$. We see that the expression $h_{n+1}^2 P_{n+1}^{(1)}$ is also a first integral of (11.132). This allows to introduce a convenient lattice, namely:

$$\frac{1}{4}h_n^2 P_n^{(1)} = \frac{1}{4}h_{n+1}^2 P_{n+1}^{(1)} = \frac{4\varepsilon^2}{(\varepsilon + 2)^2}, \qquad \varepsilon = \text{const}, \ 0 < \varepsilon \ll 1.$$

(11.135)

Substituting $P_{n+1}^{(1)}$ from (11.135) into I_2, we obtain a two-term recursion relation for x, namely

$$x_{n+1} - (1 + \varepsilon)x_n - \varepsilon B = 0,$$

(11.136)

or

$$-(1 + \varepsilon)x_{n+1} + x_n - \varepsilon B = 0,$$

(11.137)

depending on the sign choice for $\sqrt{P_{n+1}^{(1)}}$. These equations can be solved and we obtain a lattice satisfying

$$x_n = (x_0 + B)(1 + \varepsilon)^n - B, \qquad x_0 > -B$$

(11.138)

for the first equation and a lattice satisfying

$$x_n = (x_0 + B)(1 + \varepsilon)^{-n} - B, \qquad x_0 < -B$$

(11.139)

for the second equation. Using the expressions for I_1, we get the general solution for y (the same for both lattices (11.138) and (11.139)) as

$$y_n = A - \frac{4}{x_n + B} \frac{1 + \varepsilon}{(1 + \varepsilon/2)^2}. \qquad (11.140)$$

This agrees with the continuous case up to order ε^2.

We have used the three integrals I_1, I_2 and I_3 to obtain the general solution of the difference scheme (11.132). Indeed, the solution (11.138), (11.140) for x_n, y_n depends on 4 constants (A, B, x_0, ε), as it should.

The difference scheme is not compatible with a regular lattice, but requires an exponential one, as in (11.138). The only non-algebraic step in the integration was the solution of (11.136): a linear two point equation with constant coefficients.

Let us now consider the ODE (11.126) for $k \neq -1$. The equation still has the three-dimensional symmetry algebra (11.125) but only X_1 and X_2 are symmetries of the Lagrangian. They make it possible to reduce the OΔS (11.97) to just one three-point ordinary difference equation for x_n.

Let us first consider the more general case of the algebra

$$X_1 = \frac{\partial}{\partial x}, \qquad X_2 = \frac{\partial}{\partial y} \qquad (11.141)$$

corresponding to the invariant ODE

$$y'' = F(y'), \qquad (11.142)$$

where F is an arbitrary function.

The equation can be obtained from the first-order Lagrangian

$$L = y + G(y'), \qquad F(y') = \frac{1}{G''(y')}. \qquad (11.143)$$

The Lagrangian admits symmetries X_1 and X_2:

$$\mathrm{pr}\, X_1 L + L D(\xi_1) = 0;$$
$$\mathrm{pr}\, X_2 L + L D(\xi_2) = 1 = D(x).$$

With the help of Noether's theorem we obtain the following first integrals:

$$J_1 = y + G(y') - y' G'(y'), \qquad J_2 = G'(y') - x. \qquad (11.144)$$

It is sufficient to have two first integrals to write out the general solution of a second order ODE without quadratures. More explicitly, we can solve the second equation (11.144) for y' in terms of x and obtain

$$y' = H(J_2 + x), \qquad H(J_2 + x) = [G']^{-1}(J_2 + x) \qquad (11.145)$$

(i.e. H is the inverse function of G^{-1}).

Substituting into the first equation in (11.144), we obtain

$$y(x) = J_1 - G[H(J_2 + x)] + (J_2 + x)H(J_2 + x). \tag{11.146}$$

Now we are in a position to show how one can find a variational discrete model and its conservation laws by means of Lagrange-type technique. Let us choose a difference Lagrangian in the form

$$\mathcal{L}_{n+1} = \frac{y_n + y_{n+1}}{2} + G(p_{n+1}^{(1)}), \tag{11.147}$$

then

$$\begin{aligned}
&\text{pr}\, X_1 \mathcal{L}_{n+1} + \mathcal{L}_{n+1} \Delta^T(\xi_1) = 0; \\
&\text{pr}\, X_2 \mathcal{L} + \mathcal{L} \Delta^T(\xi_2) = 1 = \Delta^T(x).
\end{aligned} \tag{11.148}$$

The variations of $\mathcal{L}$ yield the following quasiextremal equations:

$$\frac{\delta \mathcal{L}}{\delta y} : G'(p_{n+1}^{(1)}) - G'(p_n^{(1)}) = \frac{h_{n+1} + h_n}{2}; \tag{11.149}$$

$$\frac{\delta \mathcal{L}}{\delta x} : -\frac{y_n + y_{n+1}}{2} - G(p_{n+1}^{(1)}) + p_{n+1}^{(1)} G'(p_{n+1}^{(1)}) + \frac{y_n + y_{n-1}}{2}$$
$$+ G(p_n^{(1)}) - p_n^{(1)} G'(p_n^{(1)}) = 0. \tag{11.150}$$

Due to the invariance of the Lagrangian with respect to the operators X_1 and X_2, the difference analog of Noether's theorem yields two first integrals

$$I_1 = y_n + G(p_{n+1}^{(1)}) - p_{n+1}^{(1)} G'(p_{n+1}^{(1)}) + \frac{x_{n+1} - x_n}{2} p_{n+1}^{(1)}, \tag{11.151}$$

$$I_2 = G'(P_{n+1}^{(1)}) - \frac{x_n + x_{n+1}}{2}. \tag{11.152}$$

We can solve (11.152) for $p_{n+1}^{(1)}$ to obtain

$$p_{n+1}^{(1)} = \Phi_1(I_2, x_n + x_{n+1}). \tag{11.153}$$

Substituting into the equation for I_1 we obtain

$$y = \Phi_2(I_1, I_2, x_n, x_{n+1}). \tag{11.154}$$

Calculating $p_{n+1}^{(1)}$ from (11.154) and setting it equal to (11.153), we obtain a three point recursion relation for x. Solving it (if we can), we turn (11.154) into an explicit general solution of the difference scheme (11.149), (11.150).

Now let us return to the ODE under consideration, namely (11.126). The Lagrangian in the continuous case is given in (11.127). We choose its discrete analogue to be

$$\mathcal{L}_{n+1} = \frac{(k-1)^2}{k} \left(\frac{y_{n+1} - y_n}{x_{n+1} - x_n} \right)^{k/(k-1)} + \frac{y_n + y_{n+1}}{2} \tag{11.155}$$

and it satisfies the invariance conditions (11.148). The quasiextremal equations (11.149) (11.150) in this case are

$$(k-1)[(p_{n+1}^{(1)})^{1/(k-1)} - (p_n^{(1)})^{1/(k-1)}] = \frac{h_n + h_{n+1}}{2} \qquad (11.156)$$

$$\frac{y_{n+1} - y_{n-1}}{2} - \frac{k-1}{k}[(p_{n+1}^{(1)})^{k/(k-1)} - (p_n^{(1)})^{k/(k-1)}] = 0. \qquad (11.157)$$

The two first integrals provided by the symmetry subalgebra (11.141) are

$$I_1 = \frac{1}{2}(y_n + y_{n+1}) - \frac{k-1}{k}(p_{n+1}^{(1)})^{k/(k-1)} \qquad (11.158)$$

$$I_2 = -\frac{1}{2}(x_n + x_{n+1}) + (k-1)(p_{n+1}^{(1)})^{1/(k-1)}. \qquad (11.159)$$

From (11.159) following the general method we obtain

$$p_{n+1}^{(1)} = \left[\frac{2I_2 + x_n + x_{n+1}}{2(k-1)}\right]^{k-1}. \qquad (11.160)$$

We replace y_{n+1} in (11.158) by $y_{n+1} = y_n + h_{n+1}p_{n+1}^{(1)}$ and substitute (11.160) for $p_{n+1}^{(1)}$ to obtain

$$y_n = I_1 - \frac{1}{2}h_{n+1}\left[\frac{2I_2 + x_n + x_{n+1}}{2(k-1)}\right]^{k-1} + \frac{k-1}{k}\left[\frac{2I_2 + x_n + x_{n+1}}{2(k-1)}\right]^k \qquad (11.161)$$

Using (11.161) and its upshift for y_{n+1} we can calculate $p_{n+1}^{(1)} = (y_{n+1} - y_n)/(x_{n+1} - x_n)$, substitute into (11.160) and obtain a three-term difference equation for x_n. The solution y_n is then given explicitly by (11.161).

11.5 Lie point symmetries of differential-difference equations

11.5.1 Formulation of the problem

In this section we take a different point of view than in the previous ones. We shall consider a differential-difference equation (DΔE) on a given fixed (non transforming) lattice. Such an equation involves continuous dependent variables u_α, $\alpha = 1, \ldots, q$, and both continuous x_j ($j = 1, \ldots, p$) and discrete n_a ($1 \le a \le N$) independent variables. For simplicity (mainly of notation) we take $q = 1$, $p = 1$ or 2, and $N = 1$.

The DΔE to be considered will thus have one of the forms

$$E_n(t, u_j, \dot{u}_j, \ddot{u}_j, \dddot{u}_j, \ldots) = 0 \qquad (11.162)$$

or

$$E_n(x, y, u_j, u_{j,x}, u_{j,y}, u_{j,xx}, u_{j,xy}, u_{j,yy}, \dots) \tag{11.163}$$

$u, t, x, y \in \mathbb{R}, n + L \le j \le n + M, n, L, M \in \mathbb{Z}, L < M$.

The aim of this section is to present an algorithm for calculating the Lie point symmetries of such equations. As usual the main tool will be vector fields and their prolongations. The vector fields will be taken in the form

$$\widehat{X} = \tau_n(t, u_n)\partial_t + \phi_n(t, u_n)\partial_{u_n} \tag{11.164}$$

or

$$\widehat{X} = \xi_n(x, y, u_n)\partial_x + \eta_n(x, y, u_n)\partial_y + \phi_n(x, y, u_n)\partial_{u_n}, \tag{11.165}$$

respectively.

Once the formalism is established, we shall treat several examples.

11.5.2 The evolutionary formalism and commuting flows for differential equations

Two different but equivalent infinitesimal formalisms exist for calculating Lie point symmetries of *differential equations* [69]. One is that of 'standard' vector fields

$$\widehat{X} = \sum_{i=1}^{p} \xi_i(\vec{x}, \vec{u})\partial_{x_i} + \sum_{\alpha=1}^{q} \phi_\alpha(\vec{x}, \vec{u})\partial_{u_\alpha} \tag{11.166}$$

acting on the independent variables x_i and dependent ones u_α in the considered differential equation.

The other is that of the *evolutionary* vector fields

$$\widehat{X}^E = \sum_{\alpha=1}^{q} Q_\alpha(\vec{x}, \vec{u}, \vec{u}_{\vec{x}})\partial_{u_\alpha}, \tag{11.167}$$

acting only on the dependent variables.

The equivalence of the two formalisms is due to the fact that the total derivatives D_{x_i} are themselves 'generalized' symmetry operators. Thus for any differential equation

$$\mathcal{E}(x_i, u_\alpha, u_{\alpha,x_i}, \dots) = 0 \tag{11.168}$$

we have

$$\text{pr }\widehat{X}^E \mathcal{E}|_{\mathcal{E}=0} = \left(\text{pr }\widehat{X} - \sum_{i=1}^{p} \xi_i D_{x_i}\right)\mathcal{E}|_{\mathcal{E}=0} = 0. \tag{11.169}$$

Here pr $\widehat{X}^E$ and pr $\widehat{X}$ are the appropriate differential prolongations of $\widehat{X}^E$ and $\widehat{X}$. Relation (11.169) implies that for point transformations we have

$$Q_\alpha = \phi_\alpha - \sum_{i=1}^{p} \xi_i u_{\alpha,x_i}. \tag{11.170}$$

For all details we refer to, e.g., P. Olver's textbook [69].

An advantage of the standard formalism is its direct relation to the group transformations obtained by integrating the equations

$$\frac{d\tilde{x}_i}{d\lambda} = \xi_i(\tilde{\vec{x}}, \tilde{\vec{u}}), \qquad \frac{d\tilde{u}_\alpha}{d\lambda} = \phi_\alpha(\tilde{\vec{x}}, \tilde{\vec{u}}),$$
$$\tilde{x}_i|_{\lambda=0} = x_i, \quad \tilde{u}_\alpha|_{\lambda=0} = u_\alpha, \qquad i = 1, \ldots, p, \ \alpha = 1, \ldots, q. \tag{11.171}$$

One advantage of the evolutionary formalism is its direct relation to the existence of flows commuting with the studied equation (11.168)

$$\frac{d\tilde{u}_\alpha}{d\lambda} = Q_\alpha, \tag{11.172}$$

where Q_α is the characteristic of the vector field (as in (11.170) for point transformations).

Another advantage is that the evolutionary formalism can easily be adapted to the case of higher symmetries.

11.5.3 The evolutionary formalism and commuting flows for differential-difference equations

An alternative method of calculating symmetries of DΔE on a fixed lattice is to construct commuting flows in two variables. Let us consider the equation

$$\dot{u}_n \equiv u_{n,t} = \mathcal{F}_n(t, u_n, u_{n+1}). \tag{11.173}$$

We introduce an additional variable λ, the group parameter and consider the flow of $u_n(t, \lambda)$ in this variable

$$u_{n,\lambda} = Q_n(t, u_n, \dot{u}_n). \tag{11.174}$$

Let us now require that the flows (11.173) and (11.174) be compatible, i.e., commute. Thus we impose

$$u_{n,t\lambda} = u_{n,\lambda t}. \tag{11.175}$$

We replace $u_{n,\lambda}$ using (11.174), $\dot{u}_n$ and $\ddot{u}_n$ using (11.173)) and its differential consequences and obtain

$$Q_{n,t} + Q_{n,u_n}\mathcal{F}_n + Q_{n,\dot{u}_n}(\mathcal{F}_{n,t} + \mathcal{F}_{n,u_n}\mathcal{F}_n + \mathcal{F}_{n,u_{n+1}}\mathcal{F}_{n+1})$$
$$= \mathcal{F}_{n,u_n}Q_n + \mathcal{F}_{n,u_{n+1}}Q_{n+1}. \quad (11.176)$$

This derivation of (11.176) is completely equivalent to the following procedure. We first introduce an evolutionary vector field

$$\widehat{X}_{\mathrm{E}}^{\mathrm{SD}} = Q_n(t, u_n, \dot{u}_n, \dots)\partial_{u_n}, \quad (11.177)$$

and its prolongation

$$\mathrm{pr}\,\widehat{X}_{\mathrm{E}}^{\mathrm{SD}} = Q_n\partial_{u_n} + Q_{n+1}\partial_{u_{n+1}} + (D_tQ_n)\partial_{\dot{u}_n} + \cdots. \quad (11.178)$$

We then apply this prolonged field to (11.173), require

$$\mathrm{pr}\,\widehat{X}_{\mathrm{E}}^{\mathrm{SD}}[\dot{u}_n - \mathcal{F}_n(t, u_n, u_{n+1})]\big|_{(\dot{u}_n=\mathcal{F}_n,\ddot{u}_n=D_t\mathcal{F}_n)} = 0, \quad (11.179)$$

and reobtain (11.176). Above the superscript SD stands for "semidiscrete," the subscript E for "evolutionary."

Let us now specialize to the case of point symmetries. The quantity Q_n in (11.174) and (11.177) is the *characteristic* of the vector field $\widehat{X}_{\mathrm{E}}^{\mathrm{SD}}$. For point symmetries it has the form

$$Q_n(t, u_n, \dot{u}_n, \cdots) = \phi_n(t, u_n) - \tau_n(t, u_n)\dot{u}_n. \quad (11.180)$$

The total derivative D_t is itself a (generalized) symmetry of the DΔE (11.162) and in particular (11.173). This provides us with a relation between ordinary and evolutionary vector fields and their prolongations, namely

$$\mathrm{pr}\,\widehat{X}^{\mathrm{SD}} = \mathrm{pr}\,\widehat{X}_{\mathrm{E}}^{\mathrm{SD}} + \tau_n(t, u_n)D_t. \quad (11.181)$$

Putting (11.180) and (11.178) into (11.181) we obtain

$$\mathrm{pr}\,\widehat{X}^{\mathrm{SD}} = \tau_n\partial_t + \phi_n\partial_{u_n} + \phi_n^t\partial_{u_{n,t}} + \phi_n^{[n+1]}\partial_{u_{n+1}} \quad (11.182)$$

$$\phi_n^{[n]} = D_t\phi_n - (D_t\tau_n)\dot{u}_n, \qquad \phi_n^{[n+1]} = \phi_{n+1} + (\tau_n - \tau_{n+1})\dot{u}_{n+1}. \quad (11.183)$$

We see that the "obvious" prolongation (11.178) of the evolutionary vector field (11.177) provides, via (11.181) the correct prolongation of the ordinary vector field. This was verified by taking the semicontinuous limit of a difference scheme in [57].

11.5.4 General algorithm for calculating Lie point symmetries of a differential-difference equation

Let us consider a DΔE involving $L + M + 1$ points and t derivatives up to order K as in (11.162). The Lie point symmetries of this equation can be obtained using the evolutionary formalism by imposing

$$\text{pr}\,\widehat{X}_E \mathcal{E}_n \big|_{\mathcal{E}_n=0, D_t^k \mathcal{E}_n=0} = 0, \qquad k = 1, \ldots, K \qquad (11.184)$$

(we drop the superscript SD). Thus the expression $\text{pr}\,\widehat{X}_E \mathcal{E}_n$ is annihilated on the solution set of the equation (11.162) and of all differential consequences of the equation.

The vector field $\widehat{X}_E$ has the form (11.177) with Q_n as in (11.180)). The prolongation of $\widehat{X}_E$ is

$$\text{pr}\,\widehat{X}_E = \sum_j Q_j \partial_{u_j} + \sum_{k=1}^K \sum_j (D_t^k Q_j) \partial_{u_j^{(k)}}, \qquad (11.185)$$

where the j summation is over all points figuring in (11.162) and $u_j^{(k)}$ denotes the kth t-derivative of u_j.

The generalization of (11.182), (11.183) for the prolongation of the standard vector field $\widehat{X}$ generating Lie point symmetries is given by

$$\text{pr}\,\widehat{X} = \phi_n \partial_{u_n} + \tau_n \partial_t + \sum_{j \neq n} \phi_j \partial_{u_j}$$

$$+ \sum_j \sum_{k=1}^K \phi_j^{[k]} \partial_{u_j^{[k]}} + \sum_j \sum_{k=1}^K (\tau_n - \tau_j)(D_t^{k+1} u_j) \partial_{u_j^{[k]}}, \qquad (11.186)$$

$$\phi_j^{[k]} = D_t \phi_j^{[k-1]} - (D_t \tau_j) u_j^{[k]}, \qquad D_t^k u_j \equiv u_j^{[k]}. \qquad (11.187)$$

Notice that $\phi_j^{[k]}$ is the same as for a differential equation but the last term in (11.186) has no analog in the continuous case. The coefficients ϕ_n and τ_n in the vector field $\widehat{X}$ itself are a priori functions of n, t and u_n. In the following section we will examine some cases when $\tau_n(t, u_n)$ simplifies.

Equation (11.186) is also obtained as the semicontinuous limit of the discrete prolongation (see [57]).

11.5.5 Theorems simplifying the calculation of symmetries of DΔE

Lie point symmetries of DΔE of the form (11.162) or (11.163) are generated by vector fields of the form (11.164) or (11.165) respectively. We shall now investigate 3 important cases when the coefficients $\tau_n(t, u_n)$ or $\xi(x, y, u_n), \eta(x, y, u_n)$ actually depend on t alone.

The 3 classes of DΔE are

$$\dot{u}_n = f_n(t, u_{n-1}, u_n, u_{n+1}), \tag{11.188}$$

$$\ddot{u}_n = f_n(t, \dot{u}_n, u_{n-1}, u_n, u_{n+1}), \tag{11.189}$$

$$u_{n,xy} = f_n(x, y, u_{n,x}, u_{n,y}, u_{n-1}, u_n, u_{n+1}). \tag{11.190}$$

Equation (11.188) contains integrable Volterra, modified Volterra and discrete Burgers type equations [85]. A list of integrable Toda type equations of the form (11.189) can be found in the reference [86]. The class (11.190) involves 2 continuous variables and contains the two dimensional Toda model [24, 63]. A list of integrable cases exists [75] and Lie point symmetries of this class have been studied [40, 62].

For equations (11.188) and (11.189) Lie point symmetries correspond to commuting flows of the form (11.174) with Q_n given by (11.180). For (11.190) the form is

$$u_{n,\lambda} = \psi_n(x, y, u_n, u_{n,x}, u_{n,y}), \tag{11.191a}$$

$$\psi_n(x, y, u_n, u_{n,x}, u_{n,y}) = \phi_n(x, y, u_n) - \xi_n(x, y, u_n)u_{n,x}$$
$$- \eta_n(x, y, u_n)u_{n,y}. \tag{11.191b}$$

For all equations (11.188), (11.189), (11.190), we assume everywhere below that at least one of the following two conditions is satisfied:

$$\frac{\partial f_n}{\partial u_{n+1}} \neq 0, \quad \text{for all } n, \quad \text{or} \quad \frac{\partial f_n}{\partial u_{n-1}} \neq 0, \quad \text{for all } n. \tag{11.192}$$

Let us consider each class separately.

11.5.6 Volterra type equations and their generalizations

Let us consider (11.188).

Theorem 11.2 *If* (11.188) *satisfies at least one of the conditions* (11.192) *and* (11.174) *represents a point symmetry of* (11.188) *then we have*

$$\tau_n(t, u_n) = \tau(t). \tag{11.193}$$

Proof The compatibility condition (11.175) of equations (11.188) and (11.174), (11.180) implies

$$\sum_{l=-1}^{1} f_{n,u_{n+l}}[\phi_{n+l} - \tau_{n+l}f_{n+l}] + (\tau_{n,t} + \tau_{n,u_n}f_n)f_n$$
$$+ \tau_n[f_{n,t} + \sum_{l=-1}^{1} f_{n,u_{n+l}}f_{n+l}] - \phi_{n,t} - \phi_{n,u_n}f_n = 0, \tag{11.194}$$

where indices t, u_{n+l} denote partial derivatives. Taking the derivative of (11.194) with respect to u_{n+2} and separately with respect to u_{n-2}, we obtain two relations:

$$f_{n+1,u_{n+2}} f_{n,u_{n+1}} (\tau_n - \tau_{n+1}) = 0, \tag{11.195}$$

$$f_{n-1,u_{n-2}} f_{n,u_{n-1}} (\tau_n - \tau_{n-1}) = 0.$$

In view of the conditions (11.192), equations (11.195) imply

$$\tau_{n+1}(t, u_{n+1}) = \tau_n(t, u_n) \quad \text{or} \quad \tau_{n-1}(t, u_{n-1}) = \tau_n(t, u_n). \tag{11.196}$$

Each of these conditions must be satisfied for any n and they are equivalent. Since $u_0, u_1, u_{-1}, \ldots$ are independent, we find that $\tau_n(t, u_n)$ depends on t alone and this proves Theorem 11.2. $\qquad\square$

A somewhat weaker theorem can be proved for a more general differential-difference equation, namely,

$$\dot{u}_n = f_n(t, u_{n+k}, u_{n+k+1}, \ldots, u_{n+m}), \qquad k \le m. \tag{11.197}$$

Theorem 11.3 *Let (11.174), (11.180) represent a symmetry of (11.197). If the function f_n in (11.197) satisfies*

$$m > 0, \qquad \frac{\partial f_n}{\partial u_{n+m}} \ne 0 \quad \text{for all } n, \tag{11.198}$$

then the function $\tau_n(t, u_n)$ is such that

$$\tau_n(t, u_n) = \tau_n(t), \qquad \tau_{n+m}(t) = \tau_n(t). \tag{11.199}$$

If the function f_n satisfies

$$k < 0, \qquad \frac{\partial f_n}{\partial u_{n+k}} \ne 0 \quad \text{for all } n, \tag{11.200}$$

then we have

$$\tau_n(t, u_n) = \tau_n(t), \qquad \tau_{n+k}(t) = \tau_n(t). \tag{11.201}$$

Proof The compatibility condition for equations (11.174), (11.180) and (11.197) will be the same as (11.194) but all sums will be from $l = k$ to $l = m$. If (11.198) is satisfied we can differentiate (11.194) with respect to u_{n+2m} and obtain $\tau_n(t, u_n) = \tau_{n+m}(t, u_{n+m})$ which implies (11.199). If (11.200) is satisfied we differentiate (11.194) with respect to u_{n+2k} and obtain $\tau_n(t, u_n) = \tau_{n+k}(t, u_{n+k})$ which implies (11.201). $\qquad\square$

This result is valid, in particular, for Burgers type equations for which $k = 0$, $m > 0$ or $k < 0$, $m = 0$ For all equations in the class (11.197), under the assumptions of this theorem, the function τ_n is independent of u_n and is periodic in n. In particular, if $k = -2$, $m = 2$ it is two-periodic and we can write

$$\tau_n(t) = \frac{1 + (-1)^n}{2} \tau_0(t) + \frac{1 - (-1)^n}{2} \tau_1(t). \tag{11.202}$$

11.5.7 Toda type equations

The compatibility condition for equations (11.189) and (11.174), (11.180) is $u_{n,tt\lambda} = u_{n,\lambda tt}$ and implies

$$f_{n,\dot{u}_n}[\phi_{n,t} + (\phi_{n,u_n} - \tau_{n,t})\dot{u}_n - \tau_{n,u_n}(\dot{u}_n)^2] + \sum_{k=-1}^{1} f_{n,u_{n+k}}[\phi_{n+k} + (\tau_n - \tau_{n+k})\dot{u}_{n+k}]$$

$$- \phi_{n,tt} + (-2\phi_{n,tu_n} + \tau_{n,tt})\dot{u}_n + (-\phi_{n,u_n u_n} + 2\tau_{n,tu_n})(\dot{u}_n)^2$$

$$+ \tau_{n,u_n u_n}(\dot{u}_n)^3 + \tau_n f_{n,t} + (2\tau_{n,t} - \phi_{n,u_n} + 3\tau_{n,u_n}\dot{u}_n)f_n = 0. \tag{11.203}$$

We use (11.203)) to prove the following theorem.

Theorem 11.4 *Let* (11.174), (11.180) *represent a point symmetry of* (11.189) *and let the function f_n in* (11.189) *satisfy at least one of the conditions* (11.192) *for all n. Then the function $\tau_n(t, u_n)$ in* (11.180) *satisfies* (11.193), *i.e., $\tau_n(t, u_n)$ depends on t alone.*

Proof None of the functions f_n, ϕ_n, τ_n figuring in (11.203) depends on $\dot{u}_{n+1}$ or $\dot{u}_{n-1}$. These two expressions do however figure explicitly in (11.203). Their coefficients must hence vanish and we obtain

$$f_{n,u_{n+1}}(\tau_n - \tau_{n+1}) = 0, \qquad f_{n,u_{n-1}}(\tau_n - \tau_{n-1}) = 0. \tag{11.204}$$

In view of the conditions (11.192), we can use one of equations (11.204), and both of them provide the same:

$$\tau_n(t, u_n) = \tau_{n+1}(t, u_{n+1}) \tag{11.205}$$

for any n. Hence we again obtain the result (11.193), as stated in Theorem 11.4.

$\square$

11.5.8 Toda field theory type equations

Let us now consider the equation (11.190) and assume that it has a Lie point symmetry represented by (11.191).

Theorem 11.5 *Let* (11.191) *represent a Lie point symmetry of the field equation* (11.190) *and let the function* $f_n(x, y, u_{n,x}, u_{n,y}, u_{n-1}, u_n, u_{n+1})$ *satisfy at least one of the conditions*

$$\frac{\partial f_n}{\partial u_{n-1}} \neq 0, \quad or \quad \frac{\partial f_n}{\partial u_{n+1}} \neq 0. \tag{11.206}$$

The functions ξ_n *and* η_n *in the symmetry* (11.191) *then are given by*

$$\xi_n(x, y, u_n) = \xi(x, y), \qquad \eta_n(x, y, u_n) = \eta(x, y). \tag{11.207}$$

Proof The compatibility condition $u_{n,xy\lambda} \doteq u_{n,\lambda xy}$ in this case can be written as

$$\sum_{k=-1}^{1} \frac{\partial f_n}{\partial u_{n+k}} \psi_{n+k} + \frac{\partial f_n}{\partial u_{n,x}} D_x \psi_n + \frac{\partial f_n}{\partial u_{n,y}} D_y \psi_n - D_x D_y \psi_n = 0 \tag{11.208}$$

with ψ_n as in (11.191b); D_x and D_y are the total derivative operators. The terms $u_{n\pm1,x}$, $u_{n\pm1,y}$ only figure in $\psi_{n\pm1}$ and in $D_x D_y \psi_n$ where we have

$$D_x D_y \psi_n = -\xi_n(D_x f_n) - \eta_n(D_y f_n) + \cdots$$

with

$$D_x f_n = \frac{\partial f_n}{\partial u_{n-1}} u_{n-1,x} + \frac{\partial f_n}{\partial u_{n+1}} u_{n+1,x} + \cdots$$

$$D_y f_n = \frac{\partial f_n}{\partial u_{n-1}} u_{n-1,y} + \frac{\partial f_n}{\partial u_{n+1}} u_{n+1,y} + \cdots .$$

Substituting into (11.208) and setting the coefficients of $u_{n\pm1,x}$ and $u_{n\pm1,y}$ equal to zero separately, we obtain

$$(\xi_{n-1} - \xi_n)\frac{\partial f_n}{\partial u_{n-1}} = 0, \qquad (\eta_{n-1} - \eta_n)\frac{\partial f_n}{\partial u_{n-1}} = 0,$$

$$(\xi_{n+1} - \xi_n)\frac{\partial f_n}{\partial u_{n+1}} = 0, \qquad (\eta_{n+1} - \eta_n)\frac{\partial f_n}{\partial u_{n+1}} = 0. \tag{11.209}$$

Thus, under the assumption (11.206) we obtain (11.207) and this completes the proof. $\qquad\square$

11.6 Examples of symmetries of DΔE

Let us now consider examples of each of the classes of differential-difference equations discussed in Section 11.5.5.

11.6.1 The YdKN equation

The Krichever–Novikov equation [38]

$$\dot{u} = \frac{1}{4}u_{xxx} - \frac{3}{8}\frac{u_{xx}^2}{u_x} + \frac{3}{2}\frac{P(u)}{u_x}, \tag{11.210}$$

where $P(u)$ is an arbitrary fourth degree polynomial with constant coefficients, is an integrable PDE with many interesting properties [1, 23, 38, 39, 42, 55, 64, 67, 68, 77, 78].

Yamilov and collaborators have proposed integrable discretizations of (11.210) [48, 84, 85]. The original form of the YdKN equation [84–86] is

$$u_{n,t} = \frac{P_n u_{n+1} u_{n-1} + Q_n(u_{n+1} + u_{n-1}) + R_n}{u_{n+1} - u_{n-1}}, \tag{11.211}$$

$$P_n = \alpha u_n^2 + 2\beta u_n + \gamma,$$
$$Q_n = \beta u_n^2 + \lambda u_n + \delta, \tag{11.212}$$
$$R_n = \gamma u_n^2 + 2\delta u_n + \omega,$$

where $\alpha, \ldots, \omega$ are pure constants.

A complete symmetry analysis of this equation and its generalizations is in preparation [58]. Here we will just consider one special case as an example of a Volterra type equation. Let us set $\alpha = 1, \beta = \cdots = \omega = 0$ in (11.212). The YdKN equation reduces to

$$u_{n,t} = \frac{u_n^2 u_{n+1} u_{n-1}}{u_{n+1} - u_{n-1}}. \tag{11.213}$$

According to Theorem 11.2 a compatible flow corresponding to a point symmetry will have the form

$$u_{n,\lambda} = \Phi_n(t, u_n) - \tau(t)u_{n,t}. \tag{11.214}$$

We replace $u_{n,t}$ in (11.214) using (11.213) and then impose the compatibility condition $u_{n,t\lambda} = u_{n,\lambda t}$. First of all, from terms containing u_{n+2} and u_{n-2} we find that Φ_n and τ must satisfy

$$\tau = \tau_0 + \tau_1 t, \qquad \Phi_n = a_n + b_n u_n + c_n u_n^2,$$
$$a_n = a + \hat{a}(-1)^n, \qquad b_n = b + \hat{b}(-1)^n, \qquad c_n = c + \hat{c}(-1)^n, \tag{11.215}$$

where $\tau_0, \tau_1, a, \hat{a}, b, \hat{b},$ and $c, \hat{c}$ are pure constants. This is actually the case for the general YdKN equation (11.211). Substituting (11.215) into the compatibility condition we obtain an equation that is polynomial in u_{n+k}. Setting coefficients of $u_{n-1}^a u_n^b u_{n+1}^c$ equal to zero for each independent term we obtain the following basis of the Lie point symmetry algebra of (11.213)

$$X_1 = \partial_t, \qquad X_2 = u_n^2 \partial_{u_n}, \qquad X_3 = (-1)^n u_n^2 \partial_{u_n},$$

$$X_4 = t\partial_t - \frac{1}{2} u_n \partial_{u_n}, \qquad X_5 = (-1)^n u_n \partial_{u_n}. \qquad (11.216)$$

This is a solvable Lie algebra with $\{X_1, X_2, X_3\}$ its Abelian nilradical. The two nonnilpotent elements satisfy $[X_4, X_5] = 0$ and their action on the nilradical is given by

$$\begin{pmatrix} [X_4, X_1] \\ [X_4, X_2] \\ [X_4, X_3] \end{pmatrix} = \begin{pmatrix} -1 & 0 & 0 \\ 0 & -\frac{1}{2} & 0 \\ 0 & 0 & -\frac{1}{2} \end{pmatrix} \begin{pmatrix} X_1 \\ X_2 \\ X_3 \end{pmatrix},$$

$$\begin{pmatrix} [X_5, X_1] \\ [X_5, X_2] \\ [X_5, X_3] \end{pmatrix} = \begin{pmatrix} 0 & 0 & 0 \\ 0 & 0 & 1 \\ 0 & 1 & 0 \end{pmatrix} \begin{pmatrix} X_1 \\ X_2 \\ X_3 \end{pmatrix}. \qquad (11.217)$$

11.6.2 The Toda lattice

The Toda lattice itself

$$u_{n,tt} = \exp(u_{n-1} - u_n) - \exp(u_n - u_{n+1}), \qquad (11.218)$$

is the best known example of an equation of the type (11.189). According to Theorem 11.4 the flow corresponding to its point symmetries will satisfy (11.214). From the compatibility condition $u_{n,tt\lambda} = u_{n,\lambda tt}$ we obtain the Lie point symmetry algebra

$$X_1 = \partial_t, \quad X_2 = t\partial_{u_n}, \quad X_3 = \partial_{u_n}, \quad X_4 = t\partial_t + 2n\partial_{u_n}. \qquad (11.219)$$

This Lie algebra is solvable, its nilradical $\{X_1, X_2, X_3\}$ is isomorphic to the Heisenberg algebra. We note that the Ansatz made in [44] was not correct and lead to $X_3 = q(n)\partial_{u_n}$ in (11.219) with $q(n)$ arbitrary. It was however noted there that a closed Lie algebra is obtained only for $q(n) = \text{const}$.

11.6.3 The two-dimensional Toda lattice equation

The equation to be considered [24, 63] is

$$u_{n,xy} = \exp(u_{n-1} - u_n) - \exp(u_n - u_{n+1}). \qquad (11.220)$$

According to Theorem 11.5 the flow corresponding to point symmetries will take the form

$$u_{n,\lambda} = \phi_n(x, y, u_n) - \xi(x, y)u_{n,x} - \eta(x, y)u_{n,y}. \qquad (11.221)$$

The Lie point symmetry algebra obtained from the compatibility condition $u_{n,xy\lambda} = u_{n,\lambda xy}$ is infinite-dimensional and depends on 4 arbitrary function of one variable each

$$X(f) = f(x)\partial_x + f'(x)n\partial_{u_n}, \qquad U(k) = k(x)\partial_{u_n},$$
$$X(g) = g(y)\partial_y + g'(y)n\partial_{u_n}, \qquad W(l) = l(y)\partial_{u_n}.$$

This is a Kac–Moody–Virasoro algebra as is typical for integrable equations with more than 2 independent variables (in this case x, y and n). It was first obtained, analyzed and applied to perform symmetry reduction in [46].

Acknowledgments

The author's research was supported by a research grant from NSERC of Canada.

References

[1] Adler, V. E., Shabat, A.B., and Yamilov, R.I. 2000. Symmetry approach to the integrability problem. *Theor. Math. Phys.*, **125**(3), 1603–1661.

[2] Anderson, R. L., Harnad, J., and Winternitz, P. 1981. Group theoretical approach to superposition rules for systems of Riccati equations. *Lett. Math. Phys.*, **5**(2), 143–148.

[3] Bakirova, M. I., Dorodnitsyn, V. A., and Kozlov, R. V. 1997. Symmetry-preserving difference schemes for some heat transfer equations. *J. Phys. A*, **30**(23), 8139–8155.

[4] Bessel-Hagen, E. 1921. Über die Erhaltungssatze der Electrodynamic. *Math. Ann.*, **84**, 258–276.

[5] Bluman, G. W., and Anco, S. C. 2002. *Symmetry and Integration Methods for Differential Equations*. Appl. Math. Sci., vol. 154. New York: Springer.

[6] Bluman, G. W., and Kumei, S. 1989. *Symmetries and Differential Equations*. Appl. Math. Sci., vol. 81. New York: Springer.

[7] Bourlioux, A., Cyr-Gagnon, C., and Winternitz, P. 2006. Difference schemes with point symmetries and their numerical tests. *J. Phys. A*, **39**(22), 6877–6896.

[8] Bourlioux, A., Rebelo, R., and Winternitz, P. 2008. Symmetry preserving discretization of $SL(2, \mathbb{R})$ invariant equations. *J. Nonlinear Math. Phys.*, **15**(suppl. 3), 362–372.

[9] Budd, C., and Dorodnitsyn, V. 2001. Symmetry-adapted moving mesh schemes for the nonlinear Schrödinger equation. *J. Phys. A*, **34**(48), 10387–10400.

[10] Budd, C. J., and Iserles, A. 1999. Geometric integration: numerical solution of differential equations on manifolds. *R. Soc. Lond. Philos. Trans. Ser. A Math. Phys. Eng. Sci.*, **357**(1754), 945–956.

[11] Budd, C. J., and Piggott, M. D. 2003. Geometric integration and its applications. Pages 35–139 of: Cucker, F. (ed), *Handbook of Numerical Analysis, Vol. XI*. Amsterdam: North-Holland.

[12] Clarkson, P. A., and Nijhoff, F. W. (eds). 1999. *Symmetries and Integrability of Difference Equations*. London Math. Soc. Lecture Note Ser., vol. 255. Cambridge: Cambridge Univ. Press.

[13] Doliwa, A., Korhonen, R., and Lafortune, S. (eds). 2007. *Symmetries and Integrability of Difference Equations*. J. Phys. A **40**(42).

[14] Dorodnitsyn, V. 1996. Continuous symmetries of finite-difference evolution equations and grids. Pages 103–112 of: *Symmetries and Integrability of Difference Equations*. CRM Proc. Lecture Notes, vol. 9. Providence, RI: Amer. Math. Soc.

[15] Dorodnitsyn, V. 2001a. Noether-type theorems for difference equations. *Appl. Numer. Math.*, **39**(3-4), 307–321.

[16] Dorodnitsyn, V., and Kozlov, R. 2003. A heat transfer with a source: the complete set of invariant difference schemes. *J. Nonlinear Math. Phys.*, **10**(1), 16–50.

[17] Dorodnitsyn, V., and Winternitz, P. 2000. Lie point symmetry preserving discretizations for variable coefficient Korteweg-de Vries equations. *Nonlinear Dynam.*, **22**(1), 49–59.

[18] Dorodnitsyn, V., Kozlov, R., and Winternitz, P. 2000. Lie group classification of second-order ordinary difference equations. *J. Math. Phys.*, **41**(1), 480–504.

[19] Dorodnitsyn, V., Kozlov, R., and Winternitz, P. 2004. Continuous symmetries of Lagrangians and exact solutions of discrete equations. *J. Math. Phys.*, **45**(1), 336–359.

[20] Dorodnitsyn, V. A. 1991. Transformation groups in net spaces. *J. Math. Sci. (N. Y.)*, **55**(1), 1490–1517.

[21] Dorodnitsyn, V. A. 1993. A finite-difference analogue of Noether's theorem. *Dokl. Akad. Nauk*, **328**(6), 678–682. English translation: *Phys. Dokl.*, 38:66–68, 1993.

[22] Dorodnitsyn, V. A. 2001b. *The Group Properties of Difference Equations*. Moscow: Fizmatlit. (in Russian).

[23] Dubrovin, B. A., Krichever, I. M., and Novikov, S. P. 1985. Integrable systems. I. Pages 179–285 of: Gamkrelidze, R. V. (ed), *Current problems in mathematics. Fundamental directions, Vol. 4*. Itogi Nauki i Tekhniki. Moscow: VINITI.

[24] Fordy, A. P., and Gibbons, J. 1980. Integrable nonlinear Klein–Gordon equations and Toda lattices. *Comm. Math. Phys.*, **71**, 21–30.

[25] Gaeta, G. 1994. *Nonlinear Symmetries and Nonlinear Equations*. Math. Appl., vol. 299. Dordrecht: Kluwer.

[26] Gómez-Ullate, D., Lafortune, S., and Winternitz, P. 1999. Symmetries of discrete dynamical systems involving two species. *J. Math. Phys.*, **40**(6), 2782–2804.

[27] Grammaticos, B., Ramani, A., and Winternitz, P. 1998. Discretizing families of linearizable equations. *Phys. Lett. A*, **245**(5), 382–388.

[28] Harnad, J., Winternitz, P., and Anderson, R. L. 1983. Superposition principles for matrix Riccati equations. *J. Math. Phys.*, **24**(5), 1062–1072.

[29] Hernández Heredero, R., Levi, D., and Winternitz, P. 1999. Symmetries of the discrete Burgers equation. *J. Phys. A*, **32**(14), 2685–2695.

[30] Hernández Heredero, R., Levi, D., Rodríguez, M. A., and Winternitz, P. 2000. Lie algebra contractions and symmetries of the Toda hierarchy. *J. Phys. A*, **33**(28), 5025–5040.

[31] Hernández Heredero, R., Levi, D., Rodríguez, M. A., and Winternitz, P. 2001a. Relation between Bäcklund transformations and higher continuous symmetries of the Toda equation. *J. Phys. A*, **34**(11), 2459–2465.

[32] Hernández Heredero, R., Levi, D., and Winternitz, P. 2001b. Symmetries of the discrete nonlinear Schrödinger equation. *Theoret. and Math. Phys.*, **127**(3), 729–737.

[33] Hietarinta, J., Nijhoff, F. W., and Satsuma, J. (eds). 2001. *Symmetries and Integrability of Difference Equations*. *J. Phys. A* **34**(48).

[34] Hoarau, E., and David, C. 2007. Lie group computation of finite difference schemes. *Dyn. Contin. Discrete Impuls. Syst. Ser. A Math. Anal.*, **14**(Advances in Dynamical Systems, suppl. S2), 180–184.

[35] Hydon, P. E. 2000. *Symmetry Methods for Differential Equations*. Cambridge Texts Appl. Math. Cambridge: Cambridge Univ. Press.

[36] Ibragimov, N. H. 1985. *Transformation Groups Applied to Mathematical Physics*. Math. Appl. (Soviet Ser.). Dordrecht: D. Reidel.

[37] Krantz, S. G., and Parks, H. R. 2008. *Geometric integration theory*. Cornerstones. Boston, MA: Birkhäuser.

[38] Krichever, I. M., and Novikov, S. P. 1979. Holomorphic bundles and nonlinear equations. Finite-gap solutions of rank 2. *Sov. Math. Dokl.*, **20**, 650–654.

[39] Krichever, I. M., and Novikov, S. P. 1980. Holomorphic bundles over algebraic curves and non-linear equations. *Russ. Math. Surv.*, **35**, 53–80.

[40] Lafortune, S., Winternitz, P., and Martina, L. 2000. Point symmetries of generalized Toda field theories. *J. Phys. A*, **33**(12), 2419–2435.

[41] Lafortune, S., Tremblay, S., and Winternitz, P. 2001. Symmetry classification of diatomic molecular chains. *J. Math. Phys.*, **42**(11), 5341–5357.

[42] Latham, G., and Previato, E. 1995. Darboux transformations for higher-rank Kadomtsev–Petviashvili and Krichever–Novikov equations. *Acta Appl. Math.*, **39**, 405–433.

[43] Levi, D., and Ragnisco, O. (eds). 2000. *SIDE III – Symmetries and Integrability of Difference Equations*. CRM Proc. Lecture Notes, vol. 25. Providence, RI: Amer. Math. Soc.

[44] Levi, D., and Winternitz, P. 1991. Continuous symmetries of discrete equations. *Phys. Lett. A*, **152**(7), 335–338.

[45] Levi, D., and Winternitz, P. 1993. Symmetries and conditional symmetries of differential-difference equations. *J. Math. Phys.*, **34**(8), 3713–3730.

[46] Levi, D., and Winternitz, P. 1996. Symmetries of discrete dynamical systems. *J. Math. Phys.*, **37**(11), 5551–5576.

[47] Levi, D., and Winternitz, P. 2006. Continuous symmetries of difference equations. *J. Phys. A*, **39**(2), R1–R63.

[48] Levi, D., and Yamilov, R. 1997. Conditions for the existence of higher symmetries of evolutionary equations on the lattice. *J. Math. Phys.*, **38**, 6648–6674.

[49] Levi, D., Vinet, L., and Winternitz, P. (eds). 1996. *Symmetries and Integrability of Difference Equations*. CRM Proc. Lecture Notes, vol. 9. Providence, RI: Amer. Math. Soc.

[50] Levi, D., Vinet, L., and Winternitz, P. 1997. Lie group formalism for difference equations. *J. Phys. A*, **30**(2), 633–649.

[51] Levi, D., Tremblay, S., and Winternitz, P. 2000. Lie point symmetries of difference equations and lattices. *J. Phys. A*, **33**(47), 8507–8523.

[52] Levi, D., Tremblay, S., and Winternitz, P. 2001. Lie symmetries of multidimensional difference equations. *J. Phys. A*, **34**(44), 9507–9524.

[53] Levi, D., Tempesta, P., and Winternitz, P. 2004a. Lorentz and Galilei invariance on lattices. *Phys. Rev. D*, **69**(10), 105011.

[54] Levi, D., Tempesta, P., and Winternitz, P. 2004b. Umbral calculus, difference equations and the discrete Schrödinger equation. *J. Math. Phys.*, **45**(11), 4077–4105.

[55] Levi, D., Petrera, M., Scimiterna, C., and Yamilov, R. 2008. On Miura transformations and Volterra-type equations associated with the Adler–Bobenko–Suris equations. *SIGMA*, **4**, 077.

[56] Levi, D., Olver, P. J., Thomova, Z., and Winternitz, P. (eds). 2009. *Symmetries and Integrability of Difference Equations. J. Phys. A* **42**(45).

[57] Levi, D., Winternitz, P., and Yamilov, R. I. 2010. Lie point symmetries of differential–difference equations. *J. Phys. A*, **43**(29), 292002.

[58] Levi, D., Winternitz, P., and Yamilov, R. *Symmetries of the continuous and discrete Krichever–Novikov equation.* To be published.

[59] Maeda, S. 1980. Canonical structure and symmetries for discrete systems. *Math. Japon.*, **25**(4), 405–420.

[60] Maeda, S. 1981. Extension of discrete Noether theorem. *Math. Japon.*, **26**(1), 85–90.

[61] Maeda, S. 1987. The similarity method for difference equations. *IMA J. Appl. Math.*, **38**(2), 129–134.

[62] Martina, L., Lafortune, S., and Winternitz, P. 2000. Point symmetries of generalized Toda field theories. II. Symmetry reduction. *J. Phys. A*, **33**(36), 6431–6446.

[63] Mikhailov, A. V. 1979. Integrability of two-dimensional Toda chain. *Sov. Phys. JETP Lett.*, **30**, 414–418.

[64] Mokhov, O. I. 1991. Canonical Hamiltonian representation of the Krichever–Novikov equation. *Math. Notes*, **50**, 939–945.

[65] Nijhoff, F. W., Suris, Yu. B., and Viallet, C.-M. (eds). 2003. *Symmetries and Integrability of Difference Equations (SIDE V). J. Nonlinear Math. Phys.* **10**(suppl 2).

[66] Noether, E. 1918. Invariante Variationsprobleme. *Nachr. v. d. Ges. d. Wiss. zu Göttingen*, 235–257.

[67] Novikov, D. P. 1999. Algebraic-geometric solutions of the Krichever–Novikov equation. *Theoretical and Mathematical Physics*, **121**, 1567–1573.

[68] Novikov, S. P., Manakov, S. V., Pitaevsky, L. P., and Zakharov, V. E. 1984. *Theory of Solitons: The Inverse Scattering Method.* New York: Plenum.

[69] Olver, P. J. 2000. *Applications of Lie Groups to Differential Equations.* Grad Texts in Math., vol. 107. New York: Springer.

[70] Penskoi, A. V., and Winternitz, P. 2004. Discrete matrix Riccati equations with superposition formulas. *J. Math. Anal. Appl.*, **294**(2), 533–547.

[71] Rand, D. W., and Winternitz, P. 1984. Nonlinear superposition principles: a new numerical method for solving matrix Riccati equations. *Comput. Phys. Comm.*, **33**(4), 305–328.

[72] Rebelo, R., and Winternitz, P. 2009. Invariant difference schemes and their application to sl(2, ℝ) invariant ordinary differential equations. *J. Phys. A*, **42**(45), 454016.

[73] Rodríguez, M. A., and Winternitz, P. 2004. Lie symmetries and exact solutions of first-order difference schemes. *J. Phys. A*, **37**(23), 6129–6142.

[74] Rogers, C., Schief, W. K., and Winternitz, P. 1997. Lie-theoretical generalization and discretization of the Pinney equation. *J. Math. Anal. Appl.*, **216**(1), 246–264.

[75] Shabat, A. B., and Yamilov, R. I. 1997. To a transformation theory of two-dimensional integrable systems. *Phys. Lett. A*, **227**, 15–23.

[76] Shokin, Yu. I. 1983. *The Method of Differential Approximation*. Springer Ser. Comput. Phys. New York: Springer.

[77] Sokolov, V. V. 1984. Hamiltonian property of the Krichever–Novikov equation. *Sov. Math. Dokl.*, **30**, 44–46.

[78] Svinolupov, S. I., Sokolov, V. V., and Yamilov, R. I. 1983. Bäcklund transformations for integrable evolution equations. *Sov. Math. Dokl.*, **28**, 165–168.

[79] Turbiner, A., and Winternitz, P. 1999. Solutions of nonlinear differential and difference equations with superposition formulas. *Lett. Math. Phys.*, **50**, 189–201.

[80] Valiquette, F., and Winternitz, P. 2005. Discretization of partial differential equations preserving their physical symmetries. *J. Phys. A*, **38**(45), 9765–9783.

[81] van Diejen, J. F., and Halburd, R. (eds). 2005. *Symmetries and Integrability of Difference Equations (SIDE VI)*. *J. Nonlinear Math. Phys.* **12**(suppl 2).

[82] Winternitz, P. 1993. Lie groups and solutions of nonlinear partial differential equations. Pages 429–495 of: Ibort, L. A., and Rodríguez, M. A. (eds), *Integrable Systems, Quantum Groups, and Quantum Field Theories (Salamanca, 1992)*. NATO Adv. Sci. Inst. Ser. C Math. Phys. Sci., vol. 409. Dordrecht: Kluwer.

[83] Winternitz, P. 2004. Symmetries of discrete systems. Pages 185–243 of: Grammaticos, B., Kosmann-Schwarzbach, Y., and Tamizhmani, T. (eds), *Discrete Integrable Systems*. Lecture Notes in Phys., vol. 644. Berlin: Springer.

[84] Yamilov, R. 2006. Symmetries as integrability criteria for differential-difference equations. *J. Phys. A*, **39**, R541–R623.

[85] Yamilov, R. I. 1983. Classification of discrete evolution equations. *Uspekhi Mat. Nauk*, **38**(6), 155–156.

[86] Yamilov, R. I. 1993. Classification of Toda type scalar lattices. In: Makhankov, V., Puzynin, I., and Pashaev, O. (eds), *Proceedings of Int. Workshop NEEDS'92*. World Scientific Publishing.

[87] Yanenko, N. N., and Shokin, Yu. I. 1976. The group classification of difference schemes for a system of one-dimensional equations of gas dynamics. Pages 259–265 of: *Some Problems of Mathematics and Mechanics*. Amer. Math. Soc. Transl. Ser. 2, vol. 104. Providence, RI: Amer. Math. Soc.

Printed in the United States
by Baker & Taylor Publisher Services